EMULSIONS,

FOAMS, AND

THIN FILMS

EMULSIONS,

FOAMS, AND

THIN FILMS

EDITED BY

K. L. MITTAL

Hopewell Junction, New York

PROMOD KUMAR

Gillette Research Institute
Gaithersburg, Maryland

CRC Press
Taylor & Francis Group
Boca Raton London New York

CRC Press is an imprint of the
Taylor & Francis Group, an **informa** business

CRC Press
Taylor & Francis Group
6000 Broken Sound Parkway NW, Suite 300
Boca Raton, FL 33487-2742

First issued in paperback 2019

ISBN-13: 978-0-8247-0366-0 (hbk)
ISBN-13: 978-0-367-39865-1 (pbk)

Visit the Taylor & Francis Web site at
http://www.taylorandfrancis.com

and the CRC Press Web site at
http://www.crcpress.com

Darsh T. Wasan

Preface

This book embodies the proceedings of the Symposium on Emulsions, Foams and Thin Films held in honor of Professor Darsh T. Wasan on his 60th birthday in conjunction with the 72nd Colloid & Surface Science Symposium at The Pennsylvania State University, June 21–24, 1998.

When the idea of such a symposium was conceived, we knew there would be a tremendous response from the colloid and surface science community, as Darsh was held in high esteem by his peers. The symposium program contained 58 invited papers by internationally renowned colloid and surface science researchers. The sheer magnitude of the symposium was a testimonial to the high regard for Darsh, and the program reflected a diversity of research topics.

The book opens with some personal vignettes about Darsh by Professor Stig E. Friberg. The rest of the book contains 27 chapters and is divided into six sections: Part I, General Overviews; Part II, Emulsions; Part III, Foams; Part IV, Thin Films; Part V, Adsorption and Monolayers; and Part VI, General Papers. Chapter 1 provides a comprehensive overview of the contributions of Professor Wasan and coworkers, at The Illinois Institute of Technology, Chicago, to enhance the understanding of interfacial phenomena. The development of novel experimental techniques by Wasan's group has provided a new dimension to the understanding of the complex processes at interfaces that govern the stability of colloidal dispersions such as emulsions and foams. A unity in the many diverse areas of interfacial science is elegantly described by Shah and coworkers in Chapter 2, where they have found a common denominator in many different interfacial phenomena.

The flocculation, coalescence, and floc fragmentation coupled together determine the rate of destabilization and phase separation in emulsion systems. A theory of coupling coalescence and flocculation in dilute oil-in-water emulsions is presented in Chapter 3 by Dukhin and colleagues. The effects of protein–surfactant complex and fat particle structure formation in oil-in-water food emulsion systems (whipped cream) are described in Chapter 4 by Kumar and coworkers. If supersaturation can be achieved through diffusion when the oil and aqueous phases are brought in contact, an emulsion can form spontaneously without mixing. This behavior, called self-emulsification, is quite important in drug delivery, in emulsifiable concentrates for agrochemicals, and in detergency. A concise account of spontaneous emulsification in oil–water–surfactant systems is presented in Chapter 5 by Miller and Rang. The potential application of multiple emulsions for drug delivery is presented in Chapter 6 by Zheng and coworkers.

The stability of foams is critically dependent on the properties of interfacial films in general and elastic and viscous properties of films in particular. A great deal of information on the interfacial rheology of aqueous surfactant systems is available, but such information for nonaqueous systems is scarce. The interfacial rheology for a nonaqueous system and its relation to foam stability is presented in Chapter 7 by Islam and Bailey. The stability of thin liquid films of aqueous polyelectrolyte solutions is discussed in Chapter 8 by Langevin. In Chapter 9, the development of silicone-based antifoams is summarized by Tamura and coworkers. In the following chapter a preliminary investigation of the potential application of perfluoroalkyl alkanes in hydrocarbon-based antifoams is presented by Curtis and coworkers.

The extension of the theory of liquids to thin liquid films is described in Chapter 11 by Henderson. The problem of the instability of thin liquid films is common in various systems and processes, such as emulsions, foams, flotation, and coatings. In Chapter 12, Sharma and coworkers describe theoretical and experimental studies on instability and pattern formation in thin liquid films. The behavior of thin liquid films at solid surfaces is critical to froth flotation of minerals. The thinning and rupture of aqueous surfactant films at the silica surface is presented by Somasundaran and coworkers in Chapter 13. The Newton black films are free-standing bilayers representing the final stages of thinning of soap films due to drainage. In Chapter 14 the possible role of hydration forces in black films is investigated by disjoining pressure measurement and x-ray reflectance technique by Benattar and coworkers. They have shown here, for the first time, the confinement of a single layer of protein in surfactant Newton black films.

The reorganization of self-assembled aggregates occurs in thin liquid films formed from bulk aqueous surfactant solutions containing self-assembled entities. The micellar structures form at a lower concentration in thin liquid films. A theoretical scheme is proposed by Mileva and Exerowa in Chapter 15, which connects the theory of bulk self-assembly to a particular characteristic of thin liquid films. The formation of two-dimensional structures of latex particles inside thinning foam films is presented in Chapter 16 by Velikov and Velev. The potential applications for the controlled growth of two-dimensional ordered arrays on solid substrates are suggested.

Many food products, such as salad dressing, mayonnaise, and ice cream, are basically foams and emulsions, which are thermodynamically unstable. Their stability can be enhanced by the adsorption of amphiphilic molecules at the interface. Despite considerable research efforts, little is known about the protein–surfactant interactions at interfaces. In Chapter 17 the adsorption and exchange of whey proteins onto spread lipid (monoglyceride) monolayers is presented by Narsimhan and coworkers. An understanding of the dynamic interfacial processes requires a knowledge of the equilibrium behavior of the surfactant layers at liquid interfaces. The diffusion-controlled adsorption kinetic model to describe the adsorption isotherm, which takes into account the reorientation of molecules at an interface, is presented by Miller and associates in Chapter 18.

The natural lung or pulmonary surfactant consists of various lipids and proteins. The adsorption of the lung surfactant at the interface of the alveolar lining layer is critical for the stabilization of the lung for proper breathing. The DPPC is the major component of the lung surfactant, and its presence is essential for proper biophysical function. The association behavior of the SP-C and its possible interaction with the DPPC is presented in Chapter 19 by Franses and associates. The adsorption of single surfactants at interfaces in terms of the equilibrium surfactant concentration is well understood. However, the description of the adsorption behavior of mixed surfactants at interfaces is in its infancy. In Chapter 20, Sundaram and Stebe discuss the penetration of a surfactant into a pre-existing monolayer as a means of studying mixed surfactant systems. The delivery of proteins is a current challenge for drug delivery due to the loss of biological activity of proteins in current delivery vehicles. Bartzoka and coworkers in Chapter 21 discuss the interaction of proteins with silicone at interfaces in order to develop silicone-based emulsions as a carrier for protein delivery.

The depletion interactions arise in the dispersion of colloidal particles in equilibrium with nonadsorbing macromolecular species. An overview of the depletion model is presented by Walz in Chapter 22. Zaman and Moudgil report in Chapter 23 on the rheological properties of concentrated bimodal colloidal dispersions and show that shear thickening can be delayed by the addition of polymers or by introducing polydispersity into the system.

The attachment of gas bubbles to hydrophobic solid particles or liquid droplets is critical in the flotation process. A comprehensive study by Dabros and coworkers on the attachment of gas bubbles to solid surfaces is reported in Chapter 24. In Chapter 25 Qutubuddin and coworkers report two types of rheological instabilities: irreversible particle flocculation and reversible shear thickening. The preparation and characterization of thin metal oxide layers to achieve angle-dependent optical effects is described by Stech and Reynders in Chapter 26. The polymerization of oleic acid and the physical properties of polymer soaps are presented in Chapter 27 by Friberg and coworkers. To conclude, this book, which comments on an array of research topics, reflects the rich diversity of the research interests of Professor Wasan.

We will not expatiate here upon the many and varied contributions of Professor Darsh T. Wasan, as these can be gathered from the tribute by Professor Friberg as well as from Chapter 1. But we would definitely like to record that Darsh is a multidimensional personality and a live wire. It is a pleasure for us to dedicate this book to him and we are sure that colleagues all over the world will join us in wishing him another 60 years of professionally stimulating and personally gratifying life.

Now we would like to acknowledge those who helped us in this endeavor. First, we are thankful to the appropriate officials, in particular Professor R. Nagarajan, of the Division of Colloid & Surface Chemistry of the American Chemical Society, who gave us the opportunity to organize this symposium. Our thanks are also extended to the session chairpersons. We sincerely appreciate the interest, enthusiasm, time, and efforts of the contributors to this volume. Finally, our sincere thanks go to Anita Lekhwani for her continued interest in this project and to Joseph Stubenrauch (both of Marcel Dekker, Inc.) for taking care of the details regarding the production of this book.

K. L. Mittal
Promod Kumar

Contents

Contributors

Patricia A. Aikens   Uniqema, Wilmington, Delaware

Eugene V. Aksenenko   Institute of Colloid Chemistry and Chemistry of Water, Kiev, Ukraine

John E. Baatz   Department of Pediatrics, Medical University of South Carolina, Charleston, South Carolina

Anita I. Bailey   Department of Chemical Engineering, Imperial College of Science, Technology, and Medicine, London, England

Vasiliki Bartzoka   Department of Chemical Engineering, McMaster University, Hamilton, Ontario, Canada

R. L. Beissinger   Department of Chemical and Environmental Engineering, Illinois Institute of Technology, Chicago, Illinois

J. J. Benattar   Service de Physique de l'Etat Condensé, Commissariat à l'Energie Atomique, Saclay, France

Michael A. Brook   Department of Chemistry, McMaster University, Hamilton, Ontario, Canada

Daechul Cho   Department of Agricultural and Biological Engineering, Purdue University, West Lafayette, Indiana

Michel Cornec   Department of Agricultural and Biological Engineering, Purdue University, West Lafayette, Indiana

Richard John Curtis   Port Sunlight Laboratory, Unilever Research, Wirral, Merseyside, England

Jan Czarnecki   Edmonton Research Center, Syncrude Canada Ltd., Edmonton, Alberta, Canada

Tadeusz Dabros   Energy Technology Branch, Advanced Separation Technologies, Natural Resources Canada, Devon, Alberta, Canada

Q. Dai   Energy Technology Branch, Advanced Separation Technologies, Natural Resources Canada, Devon, Alberta, Canada

Surekha Devi*   Department of Chemical Engineering, Center for Surface Science and Engineering, University of Florida, Gainesville, Florida

Stanislav Dukhin   Civil and Environmental Engineering, New Jersey Institute of Technology, Newark, New Jersey

Dotchi Russeva Exerowa   Department of Colloid and Interface Science, Institute of Physical Chemistry, Bulgarian Academy of Sciences, Sofia, Bulgaria

Valentin B. Fainerman   Institute of Technical Ecology, Donetsk, Ukraine

Elias I. Franses   School of Chemical Engineering, Purdue University, West Lafayette, Indiana

Stig E. Friberg   Department of Chemistry, Clarkson University, Potsdam, New York

P. R. Garrett   Port Sunlight Laboratory, Unilever Research, Wirral, Merseyside, England

---

* *Current affiliation*: Department of Chemistry, M. S. University, Baroda, India.

Hassan A. Hamza Energy Technology Branch, Advanced Separation Technologies, Natural Resources Canada, Devon, Alberta, Canada

Robert E. Hannemann School of Chemical Engineering, Purdue University, West Lafayette, Indiana

Douglas Henderson Department of Chemistry and Biochemistry, Brigham Young University, Provo, Utah

Mojahedul Islam Witco Corporation, Dublin, Ohio

R. K. Jain NSF IUCR Center for Advanced Studies in Novel Surfactants, Langmuir Center for Colloids and Interfaces, Columbia University, New York, New York

Alexander M. Jamieson Department of Macromolecular Science, Case Western Reserve University, Cleveland, Ohio

Motohiro Kageyama Beauty-Care Research Laboratories, Lion Corporation, Tokyo, Japan

Yukihiro Kaneko Material Science Research Center, Lion Corporation, Tokyo, Japan

Rajesh Khanna Institut de Chimie des Surfaces et Interfaces, Mulhouse, France

Rahul Konnur Department of Chemical Engineering, Indian Institute of Technology, Kanpur, India

Jürgen Krägel Max-Planck-Institut für Kolloid- und Grenzflächenforschung, Golm, Germany

K. Kumar Department of Chemical and Environmental Engineering, Illinois Institute of Technology, Chicago, Illinois

Dominique Langevin Laboratoire de Physique des Solides, Université Paris Sud, Orsay, France

Hyeon K. Lee*   Department of Chemical Engineering, Center for Surface Science and Engineering, University of Florida, Gainesville, Florida

Fang Li   Department of Chemistry, Clarkson University, Potsdam, New York

Libero Liggieri   Institute of Physical Chemistry of Materials, National Research Council, Genoa, Italy

Tien-Feng Ling†   Department of Chemical Engineering, Center for Surface Science and Engineering, University of Florida, Gainesville, Florida

Giuseppe Loglio   Institute of Organic Chemistry, University of Florence, Florence, Italy

Alexander V. Makievski   Max-Planck-Institut für Kolloid- und Grenzflächenforschung, Golm, Germany, and Institute of Technical Ecology, Donetsk, Ukraine

Mark R. McDermott   Department of Pathology and Molecular Medicine, McMaster University, Hamilton, Ontario, Canada

Elena Mileva   Department of Colloid and Interface Science, Institute of Physical Chemistry, Bulgarian Academy of Sciences, Sofia, Bulgaria

Clarence A. Miller   Department of Chemical Engineering, Rice University, Houston, Texas

Reinhard Miller   Max-Planck-Institut für Kolloid- und Grenzflächenforschung, Golm, Germany

F. Millet   Service de Physique de l'Etat Condensé, Commissariat à l'Energie Atomique, Saclay, France

B. M. Moudgil   Department of Materials Science and Engineering, Engineering Research Center for Particle Science and Technology, University of Florida, Gainesville, Florida

Current affiliations:
* Development Center, Daihan Swiss Chemical, Ulsan, Korea.
† Technology and Development/Industrial Water Treatment, BioLab, Inc., Decatur, Georgia.

Ganesan Narsimhan   Department of Agricultural and Biological Engineering, Purdue University, West Lafayette, Indiana

M. Nedyalkov*   Service de Physique de l'Etat Condensé, Commissariat à l'Energie Atomique, Saclay, France

M. Nicholls   Port Sunlight Laboratory, Unilever Research, Wirral, Merseyside, England

Masanori Nikaido   Chemical Research Laboratories, Lion Corporation, Tokyo, Japan

A. D. Nikolov   Department of Chemical and Environmental Engineering, Illinois Institute of Technology, Chicago, Illinois

Seong-Geun Oh†   Department of Chemical Engineering, Center for Surface Science and Engineering, University of Florida, Gainesville, Florida

Sun Young Park   International Paper Corporate Research Center, Tuxedo, New York

Alexander Patist   Department of Chemical Engineering, Center for Surface Science and Engineering, University of Florida, Gainesville, Florida

Syed Qutubuddin   Department of Chemical Engineering, Case Western Reserve University, Cleveland, Ohio

V. Raghuraman   NSF IUCR Center for Advanced Studies in Novel Surfactants, Langmuir Center for Colloids and Interfaces, Columbia University, New York, New York

Moon-Jeong Rang‡   Department of Chemical Engineering, Rice University, Houston, Texas

Francesca Ravera   Institute of Physical Chemistry of Materials, National Research Council, Genoa, Italy

---

Current affiliations:
* University of Sofia, Sofia, Bulgaria.
† Department of Chemical Engineering, Hanyang University, Seoul, Korea.
‡ Household and Personal Care R&D Institute, LG Chemical Ltd., Taejon, Korea.

Günter Reiter   Institut de Chimie des Surfaces et Interfaces, Mulhouse, France

Peter Reynders   Pigments Division, Merck KGaA, Darmstadt, Germany

Øystein Sæther   Department of Chemistry, University of Bergen, Bergen, Norway

Lakshman R. Sehgal   Department of Surgery, Evanston Northwestern Healthcare, Evanston, Illinois

D. Sentenac   Service de Physique de l'Etat Condensé, Commissariat à l'Energie Atomique, Saclay, France

D. O. Shah   Department of Chemical Engineering, Center for Surface Science and Engineering, University of Florida, Gainesville, Florida

Ashutosh Sharma   Department of Chemical Engineering, Indian Institute of Technology, Kanpur, India

M. K. Sharma*   Department of Chemical Engineering, Center for Surface Science and Engineering, University of Florida, Gainesville, Florida

S. Y. Shiao   Department of Chemical Engineering, Center for Surface Science and Engineering, University of Florida, Gainesville, Florida

S. Simpson   NSF IUCR Center for Advanced Studies in Novel Surfactants, Langmuir Center for Colloids and Interfaces, Columbia University, New York, New York

Johan Sjöblom   Department of Chemistry, University of Bergen, Bergen, Norway

P. Somasundaran   NSF IUCR Center for Advanced Studies in Novel Surfactants, Langmuir Center for Colloids and Interfaces, Columbia University, New York, New York

Kathleen J. Stebe   Department of Chemical Engineering and Department of Materials Science and Engineering, The Johns Hopkins University, Baltimore, Maryland

---

* *Current affiliation*: Research Laboratories, Eastman Chemical Company, Kingsport, Tennessee.

Martin Stech   Darmstadt University of Technology, Darmstadt, Germany

Sekhar Sundaram   Corporate Science and Technology Center, Air Products and Chemicals, Inc., Allentown, Pennsylvania

T. Tagawa   Mitsubishi Chemical Corporation, Yokohama, Japan

Takamitsu Tamura   Material Science Research Center, Lion Corporation, Tokyo, Japan

Orlin D. Velev   Department of Chemical Engineering, University of Delaware, Newark, Delaware

Krassimir P. Velikov   Van't Hoff Laboratory for Physical and Colloid Chemistry, Debye Institute, Utrecht University, Utrecht, The Netherlands

John Y. Walz   Department of Chemical Engineering, Yale University, New Haven, Connecticut

Shi-Qing Wang   Department of Macromolecular Science, Case Western Reserve University, Cleveland, Ohio

Darsh T. Wasan   Department of Chemical and Environmental Engineering, Illinois Institute of Technology, Chicago, Illinois

Rainer Wüstneck   Institute of Solid State Physics, University of Potsdam, Teltow, Germany

Jingrong Xu   Department of Chemical Engineering, Case Western Reserve University, Cleveland, Ohio

John W. H. Yorke   Port Sunlight Laboratory, Unilever Research, Wirral, Merseyside, England

Abbas A. Zaman   Department of Chemical Engineering, Engineering Research Center for Particle Science and Technology, University of Florida, Gainesville, Florida

Zhiqiang "George" Zhang* Department of Chemistry, Clarkson University, Potsdam, New York

Shuming Zheng† Department of Chemical Engineering, Hampton University, Hampton, Virginia

Current affiliations:
* The Valvoline Company, Lexington, Kentucky.
† Harvard University, Cambridge, Massachusetts.

About Darsh T. Wasan

Excellence in Science,
Excellence in Academic Leadership,
and
Excellence as a Human

"Life is short, its time span highly limited and, hence, efforts must be focused to reach far in any area of it." These are true words of wisdom and, with some exceptions, one finds that successful persons have, in their lives, concentrated on one area of excellence. For people in academia this means research, or teaching, or academic leadership. Infrequently, one finds a person with excellence in both of the two first areas. And sometimes someone appears who, because of his intelligence, drive, and enthusiasm, reaches excellence in all three areas.

Someone like Darsh T. Wasan.

The present book is intended to honor Darsh as a scientist, and this introduction will mainly discuss that part of his life, but it is certainly fitting and proper to say a few words about Darsh T. Wasan as a teacher and as an academic leader.

Anyone who has had the opportunity to listen to Darsh's lectures can understand his popularity as an instructor. As a somewhat envious colleague expressed it, "The guy could be on television!" Darsh Wasan is a lecturer of true excellence: an excellence founded on both knowledge of and enthusiasm for his subject, and he has been truly recognized as such. His award for Excellence in Teaching from his university, the Illinois Institute of Technology, has been followed by a series of honors and awards: Excellence in Instruction of Engineering Students (Western Electric Fund), 3M Lectureship Award for Outstanding Contributions to Fundamental Chemical Engineering Theory and Practice (American Society for Engineering Education), Distinguished Faculty Lecturer in a long series of Memorial Lectures at Case Western Reserve University (Jacob J. Bikerman), Clarkson University (Robert Gilqin), Rensselaer Polytechnic Institute (Sydney Ross), Yale University (Barrett F. Doge), University of Cincinnati (Procter & Gamble), and Syracuse University (Donald Gage Stevens).

His academic leadership is best illustrated by the fact that for almost 30 years, since the age of 33, he has constantly been in academic leadership positions at his university, accepting more and more responsibility over the years.

His contributions in the academic world are at a level that would be considered a life well spent. However, during all this activity, research has remained at the forefront in his mind, and his contributions in this area have resulted in a long series of awards, much too long to be listed in full on these few available pages. Suffice it to mention that Darsh T. Wasan is the only engineer and one of the few scientists to have twice received the National Science Foundation Special Creativity Award.

Darsh Wasan's scientific contributions include several scientific "firsts": His discovery of ordered colloidal microstructures in thinning liquid foams, emulsions, and polymeric latex films may be the most outstanding, because a completely new mechanism for stabilizing dispersed-phase systems is certainly a rare phenomenon. In addition, one finds, among his more technologically oriented discoveries, two novel methods to separate colloidal particles from synthetic fuels: cross-flow electrofiltration and lamella electrosettling methods, based on electrokinetic phenomena, and a new process for synthesizing hemoglobin multiple emulsions for use as a red blood substitute. The novel instrumentation he and his students developed for thin liquid film research and interfacial rheological measurements is widely used in industry.

The new colloidal stabilization mechanism in which the formation of "ordered" surfactant micelles inside the film over distances on the order of 100 nm resulting in enhanced stability was discovered by Wasan and his group using the confined geometry of free-standing thin liquid films. He showed that micelle microlayering was a universal phenomenon in a series of papers between 1984 and 1997. In fact, the particles may be any kind of isotropic structures in the 10–100 nm range, including micelles, fine solid particles, globular protein molecules, random coil-shaped polysaccharide molecules, and protein aggregates. It should be observed that this stabilization fundamentally differs from the classical film thinning mechanisms involving common black film/Newton black film transitions.

This discovery prompted Wasan to investigate theoretically the structure formation inside a liquid film by Monte Carlo numerical simulations (1994) and analytical methods using the Ornstein–Zernike statistical mechanics theory (1995). Results of this research not only confirmed the experimentally observed micellar/particle layering phenomenon and oscillatory disjoining pressure inside the thinning film but also revealed, for the first time, an ordered two-dimensional hexagonal structure within the particle layers parallel to the film surfaces. This phenomenon was later confirmed by his group by using a low-angle transmitted light diffraction technique.

This discovery of particle-induced oscillatory structural forces arising from the self-organization of nano-sized particles and their dynamic structure formation resulting from particle–particle interactions inside the confined geometry of thin films, especially with fluid surfaces, has opened up new vistas in dispersion science and technology.

These investigations, in turn, led Wasan to develop new techniques to measure the unique properties of thin liquid films, e.g., a surface force apparatus in conjunction with differential interference microscopy for measuring structural forces in curved films with fluid interfaces as encountered in polydispersed systems. The method subsequently permitted the development of improved oil recovery processes using foams as the mobility control agents in porous media, as well as advancing our knowledge of antifoaming action.

A nondestructive back-light-scattering technique was developed to quantitatively characterize long-range structural and non-DLVO forces in supramolecular fluids such as concentrated suspensions of nano-sized particles, surfactant micellar solutions, microemulsions, and systems of fat particles, emulsifiers, and gums (hydrocolloids).

Wasan's contributions to clarifying the role of surface shear and dilatational viscosities and elasticities in the stabilization of thin liquid films in coalescence phenomena and interfacial mass transfer processes necessitated the development of precise, reproducible, and meaningful methods for measuring interfacial viscosities and elasticities. His work has resulted in three commercial instruments now used worldwide: the deep-channel interfacial viscometer, the controlled-drop dynamic tensiometer, and the film rheometer.

Finally, his fundamental colloid stabilization research gave a physical explanation for the experimentally observed ''condensed'' phenomena, in which a condensed phase and an expanded phase (voids) co-existed. Wasan showed theoretically that an attractive interaction could exist between similarly charged colloidal particles in contrast with the purely repulsive force predicted by earlier theories.

Professor Wasan incorporated nearly three decades of his research efforts in his 1991 graduate-level textbook, *Interfacial Transport Processes and Rheology*, co-authored with his former doctoral student, David Edwards, and Professor Howard Brenner. The text, highly praised by reviewers, is a seminal contribution to the field of interfacial hydrodynamics and rheology.

His contributions to fundamental colloid science have been appreciated by his peers. In addition to his many awards, he was the first chemical engineer to edit the *Journal of Colloid and Interface Science* and has served as its editor-in-chief since January 1993.

Early on, Wasan showed a specific talent in applying fundamental knowledge to create new technologies. He and his colleagues were the first to consider simultaneously both hydrodynamic and molecular forces for the capture of small particles by fibrous and granular media, and the role of colloid chemistry in modeling deep-bed filtration, cross-flow electrofiltration, and lamella electrosettling for separating suspended particles from aqueous and nonaqueous media. In 1975, Wasan (with his colleagues Wnek and Gidaspow) presented a simple, yet complete, formulation describing the filtration of colloidal particles in granular beds taking into account the changes in surface interactions between the filter grains and the colloidal particles to be filtered down to the colloidal deposition. This remains the only practical model useful in colloidal filtration, when surface interaction change due to deposition is significant.

Wasan's enhanced filtration technology based on the application of electrostatic force fields covered in two U.S. patents has made it possible to separate colloidal particles from nonaqueous media such as synthetic liquid fuels derived from coal, shale, and tar sands. He and his co-authors also have extended the application of lamellar electrosettlers to clarify fine particles from nonaqueous suspensions.

To summarize a life of so many contributions is, of course, as difficult as to give a summary description of a scientist with such a wealth of achievements at such a high and sophisticated level.

In my opinion, Eli Ruckenstein said it best: Darsh Wasan is "an imaginative experimentalist, a versatile theoretician and a man with technological insight."

Let me add, he is also an enthusiastic and dynamic person with whom it has always been a pleasure to interact.

Stig E. Friberg
Department of Chemistry
Clarkson University
Potsdam, New York

EMULSIONS,

FOAMS, AND

THIN FILMS

1
Foam and Emulsion Stability: Interfacial Rheology and Thin Liquid Film Phenomena

Darsh T. Wasan
Illinois Institute of Technology, Chicago, Illinois

ABSTRACT

This chapter presents an overview of the research work performed in my laboratory over the past three decades in the subject area of interfacial rheology and thin liquid film phenomena affecting emulsion and foam stability. A critical thrust of our research program has been the development of instrumental techniques for measuring rheological or flow properties of fluid–fluid interfaces and dynamic behavior of thin liquid films containing surfactants, macromolecules, and solid particles. These techniques included both the biconical bob and modified deep-channel oscillatory interfacial rheometers, the combined capillary and longitudinal wave methods for measurements of dilatational rheological properties, controlled-drop dynamic tensiometer, and a film rheometer. The development of reliable measurement techniques has been followed up by us in a series of studies both theoretical and experimental which were aimed at understanding the role of these dynamic interfacial properties in the stability of thin liquid films associated with bubbles and drops as in foam and emulsion systems.

We have developed a capillary force balance to study the phenomenon of nanosized particle structuring inside the free-standing liquid films, the so-called foam and emulsions films, and discovered the presence of long-range (non-DLVO) oscillatory structural forces (oscillatory disjoining pressures) induced by the confined boundaries of the film with fluid surfaces. We carried out a theoretical analysis of these forces using the statistical mechanics approach and Monte Carlo simulations. At low particle concentrations, the long-range oscillatory structural force leads to an attractive depletion effect which gives rise to phase separation in colloidal dispersions. However, at high particle concentrations, the oscillatory structural force induces particle structural transitions inside the film and the formation of two-dimensional crystalline layers with hexagonal interplanar ordering which offers a new mechanism for stabilizing particle suspensions, foams, and emulsions.

I. INTRODUCTION

Emulsions and foams are complex fluid dispersions generally formed and stabilized by adsorbing surfactants, polymers, proteins, solid particles, or their mixtures onto fluid–fluid interfaces. These fluid dispersions containing small droplets or bubbles possess large specific surfaces. Surfactants stabilize such systems largely by imparting an intrinsic rheological behavior to the fluid interfaces, with respect to both its viscous and elastic natures. At surfactant concentrations below or around the critical micelle concentration (cmc), the adsorption of surface-active molecules on the film surfaces and the properties of the adsorbed layers control the drainage and stability of the film separating the droplets or bubbles. The lifetime (stability) of foam or emulsion film is determined primarily by its rate of thinning. The thinning rate and stability of the thin lamella are governed by the hydrodynamic and thermodynamic interactions between the two film surfaces. The first stage of film thinning, at a thickness greater than 100 nm, is determined by hydrodynamic interactions, which are greatly influenced by the deformation and mobility of the surfaces. Consequently, the interfacial and film rheological properties play a significant role. When the film has thinned to less than 100 nm, thermodynamic interactions begin to dominate.

At surfactant concentrations much above cmc, the surfactant micelles or other nanometer-sized colloidal particles self-organize in the confined geometry of the film, and oscillatory structural forces (oscillatory disjoining pressures) appear to rise by ordered micellar/particle layering within the thin liquid films.

In this chapter, we will briefly review the work performed in our laboratory on both the basic interfacial rheological properties that arise by virtue of the adsorption of surfactants onto fluid–fluid interfaces and the new mechanism of film stability induced by the formation of ordered surfactant micelles inside the film over distances of the order of 100 nm. We show that micelle microlayering is a universal phenomenon which fundamentally differs from the classical film-thinning mechanisms by common black film/Newton black film transitions. A major focus of our research work in the last decade or so has been to establish a theoretical link between oscillatory disjoining pressures in thin liquid films and oscillatory structural forces in fluid dispersions such as foams, emulsions, and solid-particle suspensions.

We will first discuss the interfacial and film rheological properties, their methods of measurements, and their importance in emulsion and foam stability, in macroscopic rheology of foams, and in the process of demulsification. The latter part of this chapter deals with our discovery of the ordered organization of micelles or solid particles in foam and emulsion films, and the stability they provide to such systems.

II. INTERFACIAL RHEOLOGY

Interfacial rheology deals with the shear and dilatational mechanical behavior of adsorbed and deposited layers of surfactants, proteins, polymers, and other mixtures at fluid–fluid interfaces and of monolayers at solid surfaces. The field of interfacial rheology has developed as a science largely in the past 30 years. Our research group at the Illinois Institute of Technology recognized early on its importance in dynamic processes such as emulsification (making emulsions) and demulsification (breaking emulsions), foam generation and its decay, interfacial mass transfer, enhanced oil recovery by surfactant/polymer processes, and dynamic wetting, spreading, and coating processes [1–49]. Incorporating over two decades of research effort, our textbook *Interfacial Transport Processes and Rheology* [1] addresses the theoretical framework, measurement techniques, and practical applications of dynamic interfacial properties such as dynamic interfacial tension, both shear and dilatational viscosities, and surface elasticities. First, a very brief review of novel instrumentation developed by us for precise, reproducible, and meaningful methods for measuring interfacial viscoelastic properties is given in the following subsections.

A. Interfacial Shear Measurement

We modified the instrument called the deep-channel surface shear viscometer developed by Mannheimer and Schechter [7,42] and extended its capabilities to measure interfacial shear viscoelasticity at liquid–liquid interfaces from data at liquid–gas interfaces alone in a liquid–liquid–gas experiment. Therefore, we made this technique particularly useful for opaque systems such as crude-oil–aqueous solutions [3]. We then used this oscillatory deep-channel interfacial viscometer to examine the shear-rate dependence of the surface shear viscosity to

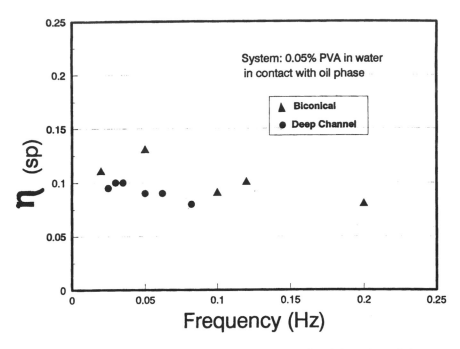

Figure 1 Comparison of interfacial viscosity for biconical and deep-channel rheometers.

demonstrate the nonlinear surface rheological behavior for systems which displayed linear rheological behavior in the bulk. A commercial version of our oscillatory interfacial shear viscometer which can operate at high temperatures and pressure is now available.

More recently, we have developed the bioconical bob oscillatory interfacial rheometer to measure the dynamic viscoelastic response of a liquid–liquid interface subjected to a small-amplitude oscillatory shear stress. Measurements of interfacial viscosity for the same liquid–liquid system with two geometrically different rheometers, the biconical bob and the oscillatory deep channel rheometers, are shown in Figure 1. It is noteworthy that these two instruments, which operate on different principles, yield comparable results. Thus, we demonstrated the intrinsic nature and, therefore, the instrument independence of these dynamic interfacial rheological properties. Furthermore, we found that the sensitivity of the biconical bob interfacial rheometer is better than that of the deep-channel rheometer for very elastic interfaces and higher upper-phase viscosity fluids [49]. Therefore, accurate measurements of interfacial shear viscoelasticity can be carried out for a wide range of systems by combining measurements with the biconical bob and deep-channel oscillatory interfacial rheometers.

B.   Interfacial Dilatational Measurement

The dilatational rheology is based on area changes due to an expansion or compression of a fluid surface and stress relaxation experiments. We developed a novel method utilizing a capillary wave propagating orthogonally to a longitudinal wave to probe the dilatational viscoelastic measurement at gas–liquid and liquid–liquid interfaces [23–25]. Figure 2 compares the surface dilatational and shear viscosities of an aqueous octanoic acid solution–air interface. The surface dilatational viscosities were determined by using the combined capillary and longitudinal wave techniques and the shear viscosities by using the deep-channel surface viscometer. These results are very significant in that, for the first time, we observed that the surface dilatational properties can be much higher than the surface shear properties for the same system.

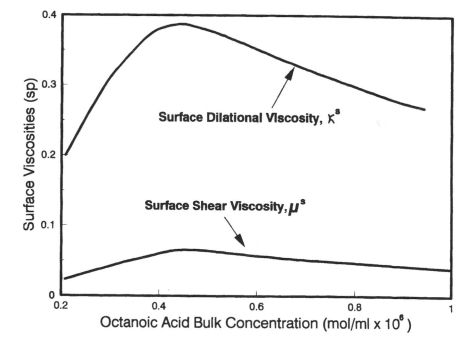

Figure 2   Comparison of surface shear and dilatational viscosities for an aqueous octanoic acid solution–air interface.

A major limitation of the longitudinal surface-wave technique discussed earlier, as well as several other experimental methods for measuring dilatational surface properties, is that these methods are effective only for fluid surfaces with a bulk concentration near or below the critical micelle concentration (cmc). To overcome this limitation, we developed a new methodology to determine surface dilatational viscosity using the maximum pressure method [38]. We compared these results with results obtained from longitudinal measurements and found similar behavior using two different techniques. We have also extended the applicability of the maximum bubble-pressure method for measurements of dilatational properties to liquid–liquid systems [39].

Another instrument (the controlled drop tensiometer) has been developed by us to determine the dynamic interfacial tension under both interface expansion and contraction conditions [41,46]. In this accurate and controlled method, the time-dependent pressure difference across the curved interface of a spherical drop, immersed in a second immiscible phase, is measured as a function of time. The capabilities of this versatile, automatic instrument for studying interfaces containing macromolecules has been demonstrated by us recently. We are now making this instrument commercially available.

III. FILM RHEOLOGY

When the droplets or bubbles interact in an emulsion or foam system, a film is formed from the continuous phase between the drops or bubbles. The stability of any emulsion or foam depends on the response of the thin liquid film and the plateau borders during shear and dilatation. In real polydispersed foam and emulsion systems, thin liquid films formed between drops or bubbles are not flat, but have a spherical, curved shape.

Recently, we have developed a versatile film rheometer [43] in which a curved, spherical-shaped liquid film is formed at the tip of the capillary with its meniscus adhering to the capillary tip (Fig. 3). The film tension is determined by measuring the capillary pressure of the film. Similar to our controlled drop technique, for relatively thick films (greater than 30 nm) the film tension is related to the capillary pressure and the film curvature by the Young–Laplace equation. This film rheometer can be purchased through our laboratory.

Figure 4 shows a film stress relaxation experiment with an aqueous emulsion film formed between dodecane drops. The film was suddenly expanded by 22% in area and then the film size was kept constant. The stress relaxation curve provides information about the kinetics of emulsifier adsorption on the film surfaces. Figure 4 also shows that the reproducibility of the film stress relaxation experiment was very good. The initial (maximum) film tension after the expansion in the film stress relaxation experiments can be used to determine the film elasticity.

Principle of Rheometer

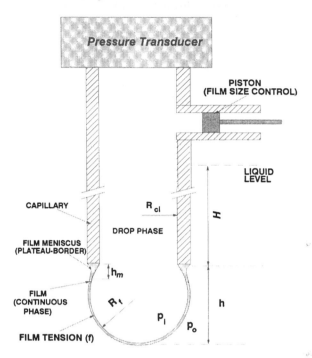

Figure 3 Principle of studying film rheology of a liquid film.

We used our new film rheometer to measure dynamic film tension and film elasticity in both the emulsion and foam systems [48]. In the dynamic film tension experiments, the film area was continuously increased by a constant rate and the dynamic film tension was monitored. The measured film tensions were compared with the interfacial tensions of the oil–water interfaces. It was found that under dynamic conditions, the film tension was higher than twice the single interfacial tension. These results have important implications for modeling the stability and rheology of emulsions with high dispersed-phase ratios.

IV. APPLICATIONS: ROLE OF INTERFACIAL AND FILM RHEOLOGY

The development of reliable measurement techniques has been followed up by us in a series of studies both theoretical and experimental which are aimed at

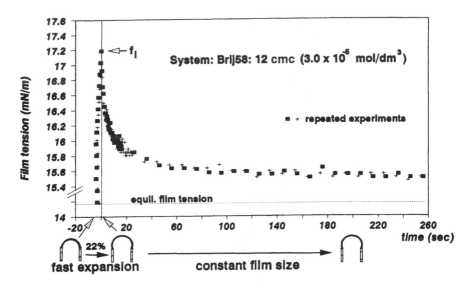

Figure 4   Film stress relaxation experiment.

understanding the role of interfacial viscosities and elasticities in the stabilizing of thin liquid films associated with bubbles and drops, as in foam and emulsion systems [14–18]. Our findings provided the theoretical foundation whereby the interfacial rheological considerations could be included in the droplet–bubble coalescence phenomenon and interfacial mass-transfer processes [1,29].

The stability of foams and emulsions strongly depends on the stability of the liquid films which form between approaching bubbles and/or drops. We have recently used our newly developed interfacial and film rheometer to model the continuous film expansion process accompanying film drainage and the relaxation process after film rupture in both foam and emulsions systems. We found a direct correlation between the film rheological properties and the stability of foam and emulsion systems stabilized by low-molecular-weight surfactant, proteins, and macromolecules [48].

In another application, we have demonstrated the important role film rheological properties play in the demulsification of water-in-crude-oil emulsions. Our data on dynamic film tension and film elasticity showed a direct correlation between film interfacial properties and emulsion stability. As the film elasticity or the film tension gradient is reduced, the emulsion stability is reduced.

Surface rheological properties such as surface viscosities and surface tension gradients have a direct significance to the macroscopic rheology of foams, owing both to the adsorption of surfactant to bubble surfaces and the large specific surface encountered with foams. We were the first [35,36] to establish the relationship between the macroscopic foam rheological behavior for both dilatational and shear deformation and surface dilatational viscosity and surface and film tension, as well as the foam film parameters such as disjoining pressure, contact angle, and film thickness. For wet foams (i.e., a relatively low dispersed-phase volume), the foam dilatational viscosity is found to be directly proportional to surface dilatational viscosity and inversely to foam film thickness. However, for dry foams (i.e., a dispersed phase volume fraction approaching unity), the foam dilatational viscosity is directly proportional to the interface's tension gradients (i.e., Gibbs' elasticity) acting along the foam bubble surfaces.

The flow resistance of a foam is much higher than that of the liquid or gas phases alone. This characteristic is utilized for foam-based mobility control in enhanced oil recovery. The oil displacement efficiency of the foam in the porous medium is higher if the blocking efficiency of the foam is higher. We have investigated the relationship between the gas mobility in porous media and the surface dilatational modulus of a surfactant system such as α-olefin sulfonates and found a direct correlation between the surface dilatational elasticity (i.e., dilatational modulus) and the gas-blocking performance of foam (i.e., higher dilatational modulus results in a low gas mobility) [63–64].

In summary, our research over the past three decades has clearly demonstrated the importance of interfacial and film rheology and has developed new powerful experimental tools to resolve complex interfacial flow behavior in dynamic stress conditions. For more detailed analyses and applications, the reader is referred to our text [1] and a series of our publications listed at the end of this chapter [1–157].

V. THIN LIQUID FILMS CONTAINING MICELLES OR PARTICLES

Thin liquid films are formed between liquid droplets or bubbles and solid or liquid surfaces, or by a liquid spread on a solid or liquid substance. When two emulsion drops or foam bubbles approach each other, they hydrodynamically interact, which generally results in the formation of a dimple. After the dimple moves out, a thick lamella with parallel interfaces forms. If the continuous phase (i.e., the film phase) contains only surface-active components at relatively low concentrations (not more than a few times their critical micellar concentration), the thick lamella thins continually (see Fig. 5, left side) until it becomes very thin, about 20–30 nm, causing it to appear dark gray in reflected light (common black film). The thickness of the common black film depends on the capillary pressure and electrolyte concentration. During continuous thinning, the film generally reaches a critical thickness where it either ruptures or black spots appear

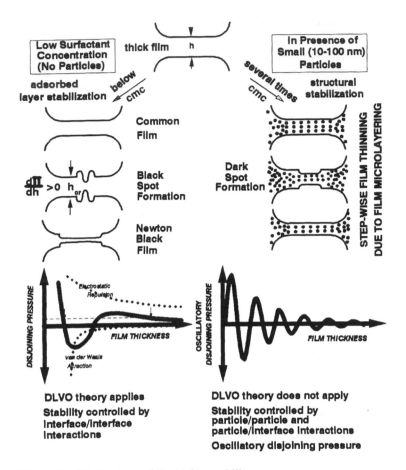

Figure 5 Mechanisms of liquid film stability.

in it and then, by the expansion of these black spots, it transforms into a very thin film, the so-called Newton black film (5–10 nm thick).

A thin-film drainage mechanism has been studied by several researchers and it has been found that the classical DLVO theory of dispersion stability can be qualitatively applied to it by taking into account the electrostatic and van der Waals interactions between the film interfaces. The hydrodynamic stability of such films is controlled by the capillary pressure, film area, the interfacial rheological properties (such as surface shear viscosity), and the surface tension gradients (Gibbs–Marangoni effect) in the surfactant adsorption layer at the film interfaces [96,108,110].

THINNING FILM SHOWS ORDERED STRUCTURE

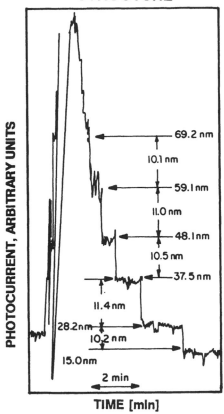

Figure 6 Interferogram of a thinning film.

Our recent film studies have revealed the existence of another film-stability mechanism: If the continuous phase contains not only a small amount of surface-active substances but also a ''sufficient amount'' of micelles or other colloidal particles, these particles can form layers inside the draining film (see Fig. 5, right side). As a result, such films thin stepwise by several step transitions (also called stratification) when at a step transition a layer of colloidal particles or micelles leaves the film. The process can be followed in Figure 6, which shows a photocurrent (film thickness) versus time interferogram of a horizontal flat film formed from the micellar solution of a nonionic surfactant (ethoxylated alcohol) [71,87].

We developed a novel experimental technique called capillary force balance to study the thinning behavior of both flat and curved liquid films with fluid–fluid interfaces [136]. The method uses a specially designed glass cell for forming flat or curved films having different sizes and, thereby, different capillary pressures (Fig. 7). Our capillary force balance equipped with this new cell in conjunction with differential interference microscopy was used to measure film thickness and disjoining pressure. This method is applicable to both symmetric foam and emulsions films and asymmetric films such as those between approaching small oil drops and the oil–water surface, or water droplets and an air–oil surface, or a wedge formed by a small bubble or drop approaching a solid surface immersed in a liquid solution.

As soon as the film is formed, it starts to thin. After it is thinner than 104 nm (the highest-order interferential maximum corresponding to the applied monochromatic 546-nm light reflected from the film), the film thickness is observed to change. The stratification process is temporarily resolved, in that the film resides for a time in each uniformly thick state, prior to thinning to the next level. Dark spots, thinner than the rest of the film, appear and gradually grow in size as shown in Figure 8. Eventually, the spots cover the entire film and the film "rests" for a time in a new state. Then, even darker spots appear, and after their expansion, a subsequent metastable state ensues. This process continues until the film finally reaches a stable state with no more stepwise changes. Each thickness state of the film appears in the interferogram (Fig. 6) as an extended plateau which, after some time, drops steplike to the next plateau. The width of the plateau is indicative of the lifetime of the respective state. The height of the steps is nearly equal (in this case about 10.6 nm) and it approximately corresponds to the bulk intermicellar distance or the effective micellar diameter. However, the time for occurrence of the thickness transition increases with the thinning of the film.

The particle self-layering phenomenon inside the film was observed by us not only in horizontal, but in large, vertical films, and foams and emulsions as well [71,98,102,108,157]. Figure 9 is a photograph of a vertical film formed from a 44 vol% latex suspension with 156-nm particles with continued stratification. A series of uniform stripes of different colors were observed at the upper part of the film. The boundaries between the stripes were very short, a consequence of the stepwise profile of the film thickness in this region, and the liquid meniscus below the film appears as a region with gradually changing color when observed in reflected (polychromatic) light; the top stripes have the following, sharply distinguished colors: white, yellow, blue, red, and green-yellow. Following these stripes, a sequence of diffuse, alternating green and red bands indicates the gradual change in film thickness (the film has a wedge profile) where the particles' layered structure gradually becomes disordered. The thickness of the stripes as determined by the difference in film reflectivity is marked in Figure 9.

Stratification was photographed by us [82,102,104,157] in a real foam gen-

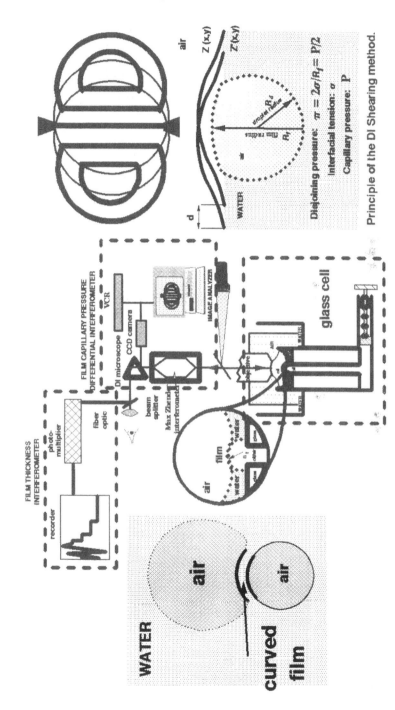

Figure 7 Capillary force balance apparatus to study thinning behavior of curved and flat films.

Figure 8 Stratification of films: 0.1 mol/L sodium dodecyl sulfate solution.

erated from a solution of 20 vol% silica particles with diameter of 19 nm containing 0.1 vol/L sodium dodecyl sulfate as a surfactant. The photograph (Fig. 10) clearly captures the multibanded pattern associated with stratification inside the curved lamella.

The step-transition phenomenon discussed does not resemble the common black film/Newton black film transition (Fig. 5) and there are basic differences between the two processes. The film stratification is due to the particle layering transition inside the film. However, the film-thickness transition from common to Newton black film is due to the film–surface interactions. The step transitions,

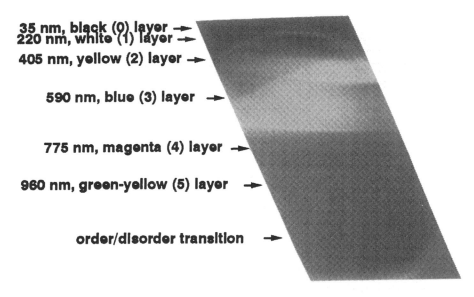

Figure 9 Stripes in a vertical stratifying film formed from a 44 vol% latex suspension with a particle diameter of 156 nm. Each stripe represents a different number of particle layers inside the thinning film.

Figure 10 Aqueous foam stabilized due to the stratification in the foam bubble lamellae generated from a solution of 20 vol% silica particles with diameter of 19 nm containing 10^{-2} mol/L sodium dodecyl sulfate.

due to microlayering, can occur at a very high thickness (depending on the size and concentration of the small particles), and the number of the stepwise transitions can be much greater than unity. The investigations in our laboratory showed that the film microlayering was a universal phenomenon which fundamentally differed from the classical film-thinning mechanism by the common black film/ Newton film transition as summarized in Figure 5. We have found that the "small particles" can be virtually any kind of particles about 5–1000 nm in size, including spherical micelles of ionic or nonionic surfactants [71,82,87,108,123], fine solid particles, such as silica or polymeric latexes [98,104], macromolecules such as globular protein molecules or random coil-shaped polysaccharide molecules, and protein aggregates, such as caseinate submicelles [123] for the occurrence of layering transitions inside thin liquid films.

We also studied [63,75,81] the drainage and stability of pseudoemulsion films (i.e., asymmetrical aqueous micellar films between the oil and gas phases) at concentrations much above the critical micellar concentration by using reflected light interferometry. Lobo and co-workers [84,106] observed that for a 4 wt% ethoxylated alcohol, the film thinned stepwise by stratification in a fashion similar to the symmetrical foam or emulsion films formed from micellar solu-

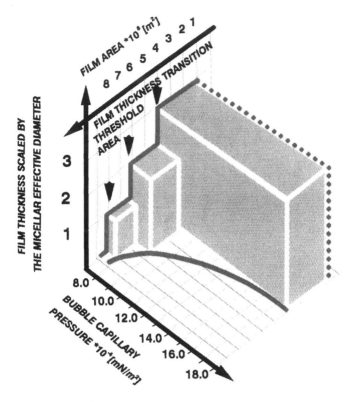

Figure 11   The effects of film area and capillary pressure on film thickness stability of a curved film formed from a CTAB solution.

tions. Three thickness transitions were observed at a 4 wt% concentration with *n*-octane as an oil phase, which was the same number of steps observed by us in foam films at the same concentration. These observations on the micellar layering in the pseudoemulsion films confirm, again, the universality of the stratification phenomenon.

Figure 11 shows the film thickness stability for the curved foam film formed from a hexadecyltrimethylammonium bromide (CTAB) solution as a function of the film area and the capillary pressure. It is noted that the smaller the film area, the higher the capillary pressure, but the film remains at equilibrium with a higher thickness. These experimental observations showed, for the first time, that the film thickness stability in the presence of particles was controlled by the film size.

Liquid film stability in the presence of particle layers cannot be explained by the classical DLVO approach. The oscillatory structural force (structural dis-joining pressure) arises due to the particle-layering phenomenon. In this case, the height of the step transition depends less on the nature of the film surfaces; the same steps (by number and height) can be observed in a foam as well as in an oil-in-water or water-in-oil emulsion films if the two types of films contain the same amount of particles. However, the critical film size for the occurrence of the thickness transition will not be the same.

VI. ROLE OF LONG-RANGE OSCILLATORY STRUCTURAL FORCES

We were the first to recognize that the stratification in thin liquid films was a manifestation of the long-range oscillatory structural forces [63], whose periodic distance was about the effective diameter of the particle [67]. We carried out a theoretical analysis of these forces using the statistical mechanics approach (i.e., the Ornstein–Zernike theory) and Monte Carlo simulation. Results of these stud-ies are summarized below.

A. Statistical Mechanics Approach

It is known that in a many-body system, the pair potential of the mean force, or the effective pair interaction $u(r)$, is related to the packing structure (radial distribution function) of particles. First, we calculated the effective pair interac-tion between hard spheres from the Ornstein–Zernike equation under the Percus–Yevick (P–Y) closure. We found that not only was there an attractive well in the effective interaction due to the depletion (i.e., particle-excluded volume) ef-fect, but the depth of this attractive well could be of the order of $1 \, \kappa T$ or even deeper, depending on the particle concentration.

Next, following the work by Henderson, we calculated [114] by P–Y theory the structural disjoining pressure (i.e., the pressure exerted by the particles on the surfaces of the film) in the film for the hard sphere–hard wall model at differ-ent bulk concentrations, as shown in Figure 12. It can be seen that the structural disjoining pressure exerted by particles in a thin film oscillates around zero and that it can be positive or negative, depending on the thickness of the film. The period of the oscillation is the effective diameter of the particle. A very short peak near the surface located one particle diameter away is apparent. When the film thickness is about an integer multiple of the particle diameter, the disjoining pressure is positive, and when the film thickness is around 1.5, 2.5, 3.5 times the effective particle diameter, the disjoining pressure is negative. The states with a negative particle disjoining pressure are intrinsically unstable from a thermody-namic point of view; thus, the film changes its thickness during a thinning process

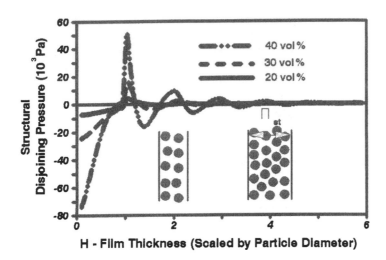

Figure 12   Structural disjoining pressure exerted on a film of particle concentrations of 20, 30, and 40 vol%.

by stepwise transitions and stays at the thickness with a positive particle disjoining pressure. As seen in Figure 12, the amplitude of the oscillation of the particle disjoining pressure becomes weaker as the film thickness increases, and beyond a certain film thickness, oscillation ceases. From a thermodynamic point of view, the oscillatory disjoining pressure gives the necessary condition for the stepwise thickness transition during the film-thinning process. Therefore, a system of given particle concentrations has only a finite number of maxima in the particle disjoining pressure isotherm and one should expect to observe the same corresponding number of stepwise transitions during the actual film-draining experiments. Indeed, experimentally we observed that the number of film-thickness transitions depended on the effective concentration of the colloidal particles and that there was a good agreement between experimental observations and predicted results [114].

In order to study the effect of polydispersity on the structural disjoining pressure, a model film system was studied, in which the particles were polydisperse and the size distribution followed the Gaussian distribution. Figure 13 shows the calculated structural disjoining pressure of two different cases. In both cases, the average particle size was the same and the concentration in volume fraction was kept constant. It is clearly shown in the figure that the polydispersity has significant effects on the structural disjoining pressure. The higher the polydispersity is, the weaker the oscillation of the disjoining pressure will be.

We compared our theoretical results with the experimentally measured

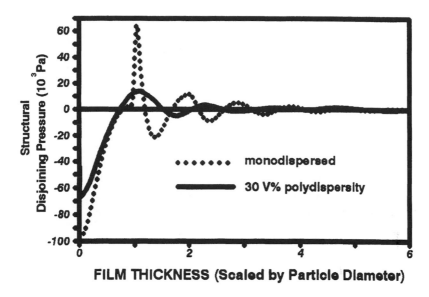

Figure 13   Effect of particle polydispersity on the disjoining pressure isotherm obtained by a Monte Carlo simulation (particle effective volume 42%, particle diameter 10 nm). The dashed curve is the result of the monodispersed system. The continuous curve is the result for the 30% polydispersed system.

force–distance curves by Parker et al. [158] for the colloidal system consisting of reversed micelle (water-in-oil) prepared from the AOT/water/heptane micro-emulsion system [Aerosol "OT" = AOT]. Due to the electroneutrality which holds for each individual micelle, the effect of electrostatic interaction can be included by using the effective particle volume. Hence, this system is very close to an ideal "hard sphere." Micelles are found to be spherical with a polydispersity of about 25%. Parker et al. could measure the force–distance relation at all distances, including both the repulsive and the attractive parts. The result was found to be oscillatory with the period of oscillation to be about the diameter of the micelles. We compared the measured force with the calculation for a polydisperse micellar system and found the theoretical prediction to be in good agreement with the experimental measurement [118].

B.   Monte Carlo Simulations

In order to reveal the in-layer structure formation and structural transitions inside a thin film containing particles or micelles, we performed the Grand Canonical

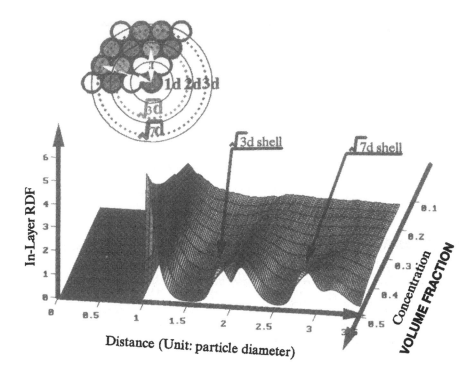

Figure 14 Surface plot of in-layer RDF versus in-layer particle distance and concentration. At low concentrations, the particles inside a layer pack randomly and form a liquidlike 2D structure. At the high concentration limit ($>$ 43 vol%), two new peaks near $\sqrt{3}$ and $\sqrt{7}$ can be seen, indicating the hexagonal packing. The film thickness is two particle diameters.

Ensemble Monte Carlo (GCEMC) simulation in a model film, which consisted of two plane surfaces with micelles/particles sandwiched between them.

In order to reveal the particle structure inside the layer parallel to the film surface, the radial distribution function (RDF) of the particles in a layer has been examined. Figure 14 shows a surface plot of in-layer RDF versus in-layer distance and concentration. When the particle concentration is low, the particles inside a layer pack randomly and form a liquidlike two-dimensional (2D) structure without order [i.e., damped peaks near the integers 1σ, 2σ, and 3σ (in the unit of particle diameter)].

When the average effective concentration increases to somewhere between 40 and 45 vol%, new peaks begin to appear in the in-layer RDF, especially near

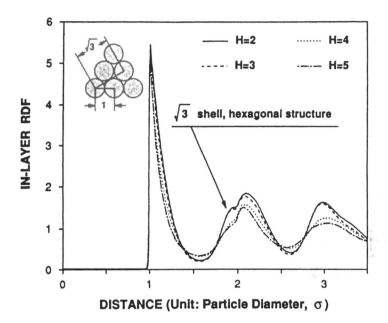

Figure 15 In-layer radial distribution function for the surface layer of a hard sphere–hard wall film.

$r = \sqrt{3}$ and $r = \sqrt{7}$, which indicates the formation of 2D hexagonal structures inside the layer. These peaks grow as the concentration increases, indicating a structural transition from a liquidlike in-layer structure to the colloid crystallike in-layer structure. Both the hard sphere–hard wall and the Leonard–Jones models predict that the disorder–order transition for the surface layers for a film thickness of two particle diameters starts at about 40–45 vol% particle concentration.

The in-layer structure depends not only on the position of the layer in the film but also on the film thickness. Figure 15 is a plot of the in-layer RDF of surface layers for different film thicknesses, 2σ, 3σ, 4σ, and 5σ, at the same concentration of 46 vol%. The degree of ordered 2D hexagonal structure can be determined by the height of the peak near $\sqrt{3}$. It is clearly seen that the in-layer particles in thinner films of thickness 2σ and 3σ are better organized than those in films of thickness 4σ and 5σ. The numerical prediction of in-layer 2D hexagonal parallel structuring is consistent with our transmitted light-diffraction experiments using monodispersed, highly charged latex particles of 150 nm (Fig. 16).

Figure 17 shows the in-layer RDFs for different layers in a film of particle concentration 46 vol%. One can see a peak near $\sqrt{3}$ for the surface layer, which

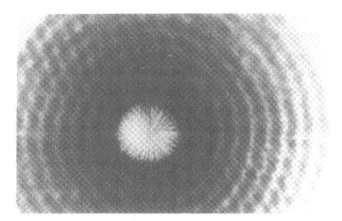

Figure 16 Diffraction pattern from a macroscopic vertical foam film at a film thickness of two particle layers formed from a 44 vol% latex suspension with a particle diameter of 156 nm.

indicates the formation of 2D hexagonal packing. This peak becomes weaker in the next layer and disappears in the middle layer. The difference in the in-layer RDFs of different layers illustrates that there exists, in a single film, more ordered structure in the surface layer and disordered structure in the middle layer.

Figure 18 is the in-layer RDF for a film of thickness of 1.7σ and a particle concentration of 46 vol%. Contrary to the in-layer RDF for free liquid films (Figs. 14 and 15), there is no $\sqrt{3}$ peak in this case, but there is a peak at $\sqrt{2}$ instead, indicating the existence of a 2D square structure.

Pieranski [159] observed a square-type in-layer packing structure for the case of a thin film confined by two solid walls where he could control the film thickness at an arbitrary value. Such a square-type in-layer structure is predicted by our Monte Carlo simulations (Fig. 18). However, it should be noted that for a liquid film with fluid surfaces, such a thickness is not favorable and, hence, there is no peak at $\sqrt{2}$.

VII. CONCLUDING REMARKS

The phenomenon of nanosized particle structuring and structural and phase transitions due to confinement in thin liquid films is of considerable interest in both science and technology. Our experimental observations on thinning of single liquid films containing nanosized colloidal particles with diameters varying from 5 nm to 1 μm clearly established the presence of long-range (non-DLVO) oscilla-

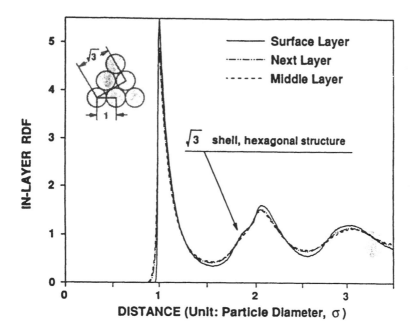

Figure 17 In-layer RDF for different layers in hard sphere–hard wall film of the thickness of five particle diameters. The solid curve is the RDF for the surface layer, the broken curve for the layer next to it, and the dotted curve for the layer in the middle. The average concentration is 46 vol%.

tory structural forces induced by the confined boundaries of the film. The effective particle interaction potential is oscillatory, with the period of oscillation equal to about an effective particle diameter, and the amplitude decays exponentially with the increase of the film thickness.

At low particle concentrations, the long-range oscillatory depletion force leads to an attraction (i.e., the excluded volume effect) which is found to destabilize various colloidal dispersions [137]. However, at high particle concentrations, the colloidal particles can form structured layers between the film surfaces, giving rise to a significant oscillatory interparticle effective pair potential that can be large enough to actually stabilize a dispersion [142].

The oscillatory force induces particle structural transitions inside the film and the formation of two-dimensional crystalline layers with hexagonal interplanar ordering which offers a new mechanism for stabilizing particle dispersions, foams, and emulsions. Our experimental, theoretical, and computer modeling results have shown that the phenomenon of particle structuring and phase transitions in confined films depends not only on the effective micelle/particle concen-

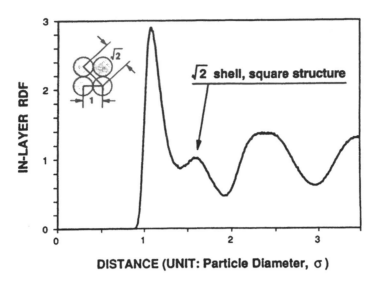

DISTANCE (UNIT: Particle Diameter, σ)

Figure 18 In-layer RDF for a film of thickness of 1.7 particle diameters at a concentration of 46 vol%. A film of such thickness can only be realized in a system with two solid walls; a peak near $\sqrt{2}$ is an indication of the square-type in-layer structure.

tration, added electrolyte, temperature, film thickness, and the position of the layer inside the film but also on the film area.

From a practical point of view, the foam and emulsion stability can be improved by increasing the micellar/particle concentration and decreasing polydispersity in the micellar size, decreasing the individual film area (e.g., by decreasing the bubble or droplet size), or decreasing the electrolyte concentration.

For nearly 20 years, the oscillatory forces have been measured by means of the surface-force apparatus developed by Israelachivili et al. This method has been successfully applied for measuring solvation oscillatory forces, depletion forces, and oscillatory forces for the colloidal systems consisting of micelles and microemulsions. The surface-force apparatus, which uses two solid mica sheets or cross cylinders, is not particularly useful for investigating oscillatory structural forces in thin liquid films with fluid surfaces because our recent research work has clearly shown that the ''vacancies condensation'' mechanism for micelle/particle layering is greatly influenced by the capillarity phenomenon normally associated with fluid surfaces.

As discussed earlier in this chapter, we have developed a novel capillary force balance apparatus and use it to study the dynamic process of stratification or multilayer microstructuring in submicrometer thin liquid films especially with curved fluid surfaces such as those associated with foam and emulsions systems. This technique can serve as an important tool for probing the long-range oscillatory structural/depletion forces in confined particle suspensions and colloidal dispersions.

Work is in progress in our laboratory to investigate the particle structural and phase-transition phenomena in the liquid film–meniscus region (i.e., in a wedge film). We have found the coexistence of hexagonal and cubic packing structures inside the wedge. The wedge was formed by a small bubble or drop approaching a solid surface immersed in a liquid solution containing charged latex particles of the order of 1 μm. The studies using a wedge geometry provides information on structural ordering and phase transitions in lower dimensions (1D and 2D) as well as three dimensions (3D). The results obtained from these studies are helping us to better understand the mechanisms of oily soil removal, soil remediation, enhanced oil recovery by surfactant/polymer processes, and in designing new formulations containing colloidal particles for dynamic wetting and spreading processes.

ACKNOWLEDGMENTS

The author gratefully acknowledges the support provided by Dr. Alex Nikolov in the preparation of this chapter. Much of this chapter is the product of research done by the many students and postdoctoral associates who worked with me over the years. This work was supported by the National Science Foundation and the U.S. Department of Energy.

REFERENCES

1. D Edwards, H Brenner, DT Wasan. Interfacial Transport Processes and Rheology. Boston: Butterworth–Heinemann, 1991.
2. AJ Pintar, AB Israel, DT Wasan. J Colloid Interf Sci 37:52, 1971.
3. DT Wasan, L Gupta, MK Vora. AIChE J 17:1287, 1971.
4. L Gupta, DT Wasan. Ind Eng Chem Fundam 13:26, 1974.
5. DT Wasan. Proceedings of the Symposium on Advances in Petroleum Recovery. New York: Division of Petroleum Chemistry, American Chemistry Society, 1976, p. 326.
6. V Mohan, L Gupta, DT Wasan. J Colloid Interf Sci 57:496, 1976.
7. V Mohan, DT Wasan. Colloid Interf Sci 4:439, 1976.
8. V Mohan, BK Malviya, DT Wasan. Can J Chem Eng 54:515, 1976.
9. DT Wasan, V Mohan. In: DO Shah, RS Schechter, eds. Improved Oil Recovery by Surfactant and Polymer Flooding. New York: Academic Press, 1977, pp. 161–203.

10. DO Shah, NF Djabbarah, DT Wasan. Colloid Polym Sci 256:1002, 1978.

11. DT Wasan, SM Shah, M Chan, K Sampath, R Shah. In: RT Johnson, RL Berg, eds. Chemistry of Oil Recovery, ACS Symposium Series, 91. Washington, DC: American Chemical Society, 1979, p. 115.

12. DT Wasan. J Rheol 23:181, 1979.

13. DT Wasan, NF Djabbarah, MK Vora, ST Shah. In: T Sorenson, ed. Dynamics and Instability of Fluid Interfaces. Berlin: Springer-Verlag, 1979, p. 205.

14. SS Agarwal, DT Wasan. Chem Eng J 18:215, 1979.

15. DT Wasan. State-of-the-Art in Interfacial Phenomena Preprints, AIChE Meeting, San Francisco, 1979.

16. HC Maru, V Mohan, DT Wasan. Chem Eng Sci 34:1283, 1979.

17. DT Wasan, K Sampath, N Aderangi. Recent Advances in Separation Techniques— II, AIChE Symposium Series 192, 1980, p. 93.

18. DT Wasan, M Chan, SM Shah, K Sampath, R Shah. In: Physical Separations. New York: Engineering Foundation, 1981, p. 117.

19. CH Pasquarelli, DT Wasan. In: DO Shah, ed. Surface Phenomena in Enhanced Oil Recovery. New York: Plenum Press, 1981, p. 237.

20. NF Djabbarah, DT Wasan. Ind Eng Chem Fundam 21:27, 1982.

21. NF Djabbarah, DT Wasan. Chem Eng Sci 37:175, 1982.

22. CI Christov, L Ting, DT Wasan. J Colloid Interf Sci 85:363, 1982.

23. K Miyano, B Abrahams, L Ting, DT Wasan. J Colloid Interf Sci 92:297, 1983.

24. L Ting, DT Wasan, K Miyano, S-Q Xu. J Colloid Interf Sci 102:248, 1984.

25. L Ting, DT Wasan, K Miyano. J Colloid Interf Sci 107:345, 1985.

26. NF Djabbarah, DT Wasan. AIChE J 81:1041, 1985.

27. AK Malhotra, DT Wasan. Chem Eng Commun 55:95, 1987.

28. DO Shah, M Ginn, DT Wasan, eds. Surfactants in Chemical and Process Engineering, Surfactant Science Series Vol. 28. New York: Marcel Dekker, 1988.

29. D Edwards, DT Wasan. In: DT Wasan, M Ginn, DO Shah, eds. Surfactants in Chemical and Process Engineering. New York: Marcel Dekker, 1988, p. 1.

30. AK Malhotra, DT Wasan. In: IB Ivanov, ed. Thin Liquid Films. New York: Marcel Dekker, 1988, p. 829.

31. DA Edwards, DT Wasan. J Rheol 32:429, 1988.

32. DA Edwards, DT Wasan. J Rheol 32:447, 1988.

33. DA Edwards, DT Wasan. J Rheol 32:473, 1988.

34. P Berger, C Hsu, A Jimenez, D Wasan, S Chung. In: B Cross, HB Scher, eds. Pesticide Formulations: Innovations and Developments, ACS Symposium Series 371, Washington, DC: American Chemical Society, 1988, p. 142.

35. DA Edwards, H Brenner, DT Wasan. J Colloid Interf Sci 130:266, 1989.

36. DA Edwards, DT Wasan. J Colloid Interf Sci 139:479, 1990.

37. DA Edwards, DT Wasan. Chem Eng Sci 46:1247, 1991.

38. RL Kao, DA Edwards, DT Wasan, E Chen. J Colloid Interf Sci 48:247, 1992.

39. RL Kao, DA Edwards, DT Wasan, E Chen. J Colloid Interf Sci 48:257, 1992.

40. L Lobo, I Ivanov, DT Wasan. AIChE J 39:353, 1993.

41. R Nagarajan, DT Wasan. J Colloid Interf Sci 159:164, 1993.

42. R Nagarajan, DT Wasan. Rev Sci Instrum 65:2675, 1994.

43. JM Soos, K Koczo, E Erdos, DT Wasan. Rev Sci Instrum 65:3555, 1994.

44. DT Wasan, K Koczo, AD Nikolov. In: A Gaonkar, ed. Characterization of Food: Emerging Methods. Amsterdam: Elsevier, 1995, p. 1.
45. N Aderangi, DT Wasan. Chem Eng Commun 132:207, 1995.
46. R Nagarajan, K Koczo, E Erdos, DT Wasan. AIChE J 41:915, 1995.
47. JD Michaels, JE Nowak, AK Mallik, K Koczo, DT Wasan, ET Papoutsakis. Biotechnol Bioeng 47:420, 1995.
48. YH Kim, K Koczo, DT Wasan. J Colloid Interf Sci 187:29, 1997.
49. R Nagarajan, SL Chung, DT Wasan. J Colloid Interf Sci 204:53, 1998.
50. JI Rosenfeld, DT Wasan. J Colloid Interf Sci 47:27, 1974.
51. ED Manev, SV Sazdanova, AA Rao, DT Wasan. J Dispers Sci Technol 3:435, 1982.
52. AA Rao, DT Wasan, ED Manev. Chem Eng Commun 15:63, 1982.
53. Z Zapryanov, AK Malhotra, N Aderangi, DT Wasan. Int J Multiphase Flow 9:105, 1983.
54. VB Menon, DT Wasan. J Separ Sci Technol 19:555, 1984.
55. ED Manev, SV Sazdanova, DT Wasan. J Colloid Interf Sci 97:591, 1984.
56. ED Manev, SV Sazdanova, DT Wasan. J Dispers Sci Technol 5:111, 1984.
57. ZM Gu, DT Wasan, NN Li. Separ Sci Technol 20:599, 1985.
58. JP Perl, HE Bussy, DT Wasan. J Colloid Interf Sci 108:528, 1985.
59. WB Krantz, RK Jain, DT Wasan. AIChE Symp Series 82(252), 1986.
60. AK Malhotra, DT Wasan. Chem Eng Commun 48:35, 1986.
61. VB Menon, DT Wasan. Colloids Surf 19:89, 1986.
62. VB Menon, DT Wasan. Colloids Surf 19:107, 1986.
63. AD Nikolov, DT Wasan, DD Huang, D Edwards. The effect of oil on foam stability: Mechanisms and implications for oil, displacement by foam in porous media. Proc. 61st Annual Technical Conf. of the SPE of AIME, Preprint SPE 15443, 1986.
64. DD Huang, AD Nikolov, DT Wasan. Langmuir 2:672, 1986.
65. DT Wasan, AK Malhotra. AIChE Symposium Series on Thin Film Phenomena No. 82, 1986, p. 5.
66. RP Borwankar, SI Chung, DT Wasan. J Appl Polym Sci 32:5749, 1986.
67. DT Wasan, AD Nikolov. Sixth International Symposium on Surfactants in Solution, New Delhi, 1986.
68. RP Borwankar, DT Wasan, NN Li. Proceedings of the International Symposium on Metal Speciation, Separation and Recovery, Chicago, 1986.
69. VB Menon, DT Wasan. Colloids Surf 23:353, 1987.
70. AK Malhotra, DT Wasan. AIChE J 33(9):1533, 1987.
71. AD Nikolov, DT Wasan, PA Kralchevsky, IB Ivanov. In: N Ise, I Sogami, eds. Yamada Conference Proceedings on Ordering and Organization in Ionic Solutions. Singapore: World Scientific, 1987, pp. 302–314.
72. VB Menon, DT Wasan. Colloids Surf 29:7, 1988.
73. VB Menon, AD Nikolov, DT Wasan. J Colloid Interf Sci 124:317, 1988.
74. VB Menon, DT Wasan. Separ Sci Technol 23:2131, 1988.
75. DT Wasan, AD Nikolov, DD Huang, DA Edwards. In: DH Smith, ed. Use of Surfactants in Mobility Control. ACS Symposium Series 373. Washington, DC: American Chemical Society, 1988, pp. 136–162.

76. CM Borwankar, SB Pfeiffer, S Zheng, RL Beissinger, DT Wasan. Biotechnol Prog 4:210, 1988.
77. RP Borwankar, CC Chan, DT Wasan, RM Kurzeja, ZM Gu, NN Li. AIChE J 34: 753, 1988.
78. SI Chung, DT Wasan. Colloids Surf 29:323, 1988.
79. AK Malhotra, DT Wasan. AIChE J 34:1407, 1989.
80. VB Menon, AD Nikolov, DT Wasan. J Dispers Sci Technol 9:575, 1989.
81. L Lobo, AD Nikolov, DT Wasan. J Dispers Sci Technol 10:143, 1989.
82. AD Nikolov, DT Wasan. J Colloid Interf Sci 133:1, 1989.
83. AD Nikolov, PA Kralchevsky, IB Ivanov, DT Wasan. J Colloid Interf Sci 133:13, 1989.
84. L Lobo, AD Nikolov, DT Wasan. Langmuir 6:995, 1990.
85. PA Kralchevsky, AD Nikolov, DT Wasan, IB Ivanov. Langmuir 6:1180, 1990.
86. AS Dimitrov, PA Kralchevsky, AD Nikolov, DT Wasan. Colloids Surf 47:299, 1990.
87. AD Nikolov, DT Wasan, ND Denkov, PA Kralchevsky, IB Ivanov. Prog Colloid Polym Sci 87:87, 1990.
88. LA Lobo, DT Wasan. In: PJ Wan, ed. Food Emulsions and Foams—Theory & Practice, AIChE Symposium Series 277. American Institute of Chemical Engineers, 1990.
89. JP Perl, C Thomas, DT Wasan. J Colloid Interf Sci 137:425, 1990.
90. C Thomas, JP Perl, DT Wasan. J Colloid Interf Sci 139:1, 1990.
91. MA Krawczyk, DT Wasan, CS Shetty. Ind Eng Chem Res 30:367, 1991.
92. S Zheng, RL Beissinger, DT Wasan. J Colloid Interf Sci 144:72, 1991.
93. K Koczo, L Lobo, DT Wasan. J Colloid Interf Sci 150:492, 1992.
94. CS Shetty, AD Nikolov, DT Wasan, BR Bhattacharyya. J Dispers Sci Technol 13: 121, 1992.
95. S Zheng, RL Beissinger, DT Wasan. J Dispers Sci Technol 13:133, 1992.
96. DT Wasan, AD Nikolov, L Lobo, K Koczo, DA Edwards. Prog Surf Sci 39:119, 1992.
97. L Lobo, DT Wasan, M Ivanova. In: KL Mittal, DO Shah, eds. Surfactants in Solution, Vol. 11. New York: Plenum Press, 1992, p. 395.
98. ES Basheva, AD Nikolov, PA Kralchevsky, IB Ivanov, DT Wasan. In: KL Mittal, DO Shah, eds. Surfactants in Solution, Vol. 11. New York: Plenum Press, 1992, p. 467.
99. DT Wasan. In: J Sjöblom, ed. Emulsions—A Fundamental and Practical Approach, NATO ASI Series C363. Dordrecht: Kluwer, 1992, p. 283.
100. DT Wasan. Chem Eng Ed 104, Spring 1992, pp. 104–112.
101. ND Denkov, IB Ivanov, PA Kralchevsky, DT Wasan. J Colloid Interf Sci 150:589, 1992.
102. DT Wasan, AD Nikolov, P Kralchevsky, IB Ivanov. Colloids Surf 67:139, 1992.
103. RP Borwankar, L Lobo, DT Wasan. Colloids Surf 69:135, 1992.
104. AD Nikolov, DT Wasan. Langmuir 8:2985, 1992.
105. DT Wasan, AD Nikolov. In: MC Roco, ed. Particulate Two-Phase Flow. Boston: Butterworth-Heinemann, 1993, pp. 325–354.
106. L Lobo, DT Wasan. Langmuir 9:1668, 1993.

107. S Zheng, Y Zheng, RL Beissinger, DT Wasan, DL McCormick. Biochim Biophys Acta 1158:65, 1993.
108. DT Wasan, AD Nikolov. Proceedings of the First World Congress on Emulsions, Paris, Vol. 4, pp. 93–112, 1993.
109. DT Wasan, K Koczo, AD Nikolov. In: LL Schramm, ed. Foams, Fundamentals and Applications in the Petroleum Industry. Advances in Chemistry Series 242. Washington, DC: American Chemical Society, 1994, pp. 47–114.
110. AD Nikolov, PA Kralchevsky, IB Ivanov, DT Wasan. In: MG Velarde, CI Christov, eds. Fluid Physics, Lecture Notes of Summer Schools. World Scientific Series on Nonlinear Science Series B, Vol. 5. Singapore: World Scientific, 1994, pp. 209–228.
111. J Rudin, DT Wasan. J Colloid Interf Sci 162:252, 1994.
112. J Rudin, C Bernard, DT Wasan. Ind Eng Chem Res 33:1150, 1994.
113. K Koczo, JK Koczone, DT Wasan. J Colloid Interf Sci 166:225, 1994.
114. XL Chu, AD Nikolov, DT Wasan. Langmuir 10:4403, 1994.
115. AD Nikolov, DT Wasan. In: AT Hubbard, ed. Handbook of Surface Imaging and Visualization. Boca Raton, FL: CRC Press, 1995, pp. 209–214.
116. D Edwards, DT Wasan. In R Prud'homme, S Khan, eds. Foams: Theory, Experiments and Applications. New York: Marcel Dekker, 1995, pp. 189–215.
117. YH Kim, DT Wasan, PJ Breen. Colloids Surf 95:235, 1995.
118. XL Chu, AD Nikolov, DT Wasan. J Chem Phys 103:6653, 1995.
119. AD Nikolov, DT Wasan. Ind Eng Chem Res 34:3653, 1995.
120. R Beissinger, SM Zheng, DT Wasan. U.S. Patent 5,438,041 (August 1, 1995).
121. PJ Breen, DT Wasan, YH Kim, AD Nikolov, CS Shetty. In: J Sjöblom, ed. Emulsions and Emulsion Stability. New York: Marcel Dekker, 1996, pp. 237–286.
122. YH Kim, AD Nikolov, DT Wasan, H Diaz-Arauzo, CS Shetty. J Dispers Sci Technol 17:33, 1996.
123. K Koczo, AD Nikolov, DT Wasan, RP Borwankar, A Gonsalves. J Colloid Interf Sci 178:694, 1996.
124. YH Kim, DT Wasan. Ind Eng Chem Res 35:1141, 1996.
125. GK Rácz, K Koczo, DT Wasan. J Colloid Interf Sci 181:124, 1996.
126. J Chatterjee, AD Nikolov, DT Wasan. Ind Eng Chem Res 35:2933, 1996.
127. XL Chu, AD Nikolov, DT Wasan. Chem Eng Commun 148–150:123, 1996.
128. AD Nikolov, DT Wasan. Powder Technol 88:299, 1996.
129. AD Nikolov, M Randle, CS Shetty, DT Wasan. Chem Eng Commun 153:337, 1996.
130. XL Chu, DT Wasan. J Colloid Interf Sci 184:268, 1996.
131. XL Chu, AD Nikolov, DT Wasan. Langmuir 12:5004, 1996.
132. AD Nikolov, DT Wasan, SE Friberg. Colloids Surf 118:221, 1996.
133. DT Wasan, SP Christiano. In: KS Bridi, ed. Handbook of Surface and Colloid Chemistry. Boca Raton, FL: CRC Press, 1997, pp. 179–215.
134. D Henderson, XL Chu, DT Wasan. J Colloid Interf Sci 185:265, 1997.
135. JJ Kilbane II, P Chowdiah, KJ Kayser, B Misra, KA Jackowski, VJ Srivastava, GN Sethu, AD Nikolov, DT Wasan. Land Contamin Reclam 5:41, 1997.
136. AD Nikolov, DT Wasan. Colloids Surf 123–124:375, 1997.
137. W Xu, AD Nikolov, DT Wasan. AIChE J 43:3215, 1997.
138. D Henderson, A Kovalenko, O Pizio, DT Wasan. Physica A 245:276, 1997.

139. D Henderson, S Sokolowski, DT Wasan. J Statist Phys 89:233, 1997.
140. W Xu, AD Nikolov, DT Wasan. J Colloid Interf Sci 191:471, 1997.
141. AD Nikolov, DT Wasan. Colloids Surf 128:243, 1997.
142. DT Wasan, AD Nikolov, XL Chu. In: DO Shah, ed. Micelles, Microemulsions, and Monolayers, Science and Technology. New York: Marcel Dekker, 1998, pp. 124–144.
143. D Henderson, DT Wasan. In: JP Hsu, ed. Theory of Interfaces: Forces and Fields. New York: Marcel Dekker, 1999.
144. D Henderson, S Sokolowski, DT Wasan. J Mol Phys 93:295, 1998.
145. W Xu, AD Nikolov, DT Wasan. J Colloid Interf Sci 197:160, 1998.
146. D Henderson, S Sokolowski, DT Wasan. J Phys Chem B 102:3009, 1998.
147. D Henderson, S Sokolowski, DT Wasan. Phys Rev E 57:5539, 1998.
148. K Koczo, DT Wasan, RP Borwankar, A Gonvalves. Food Hydrocolloids 12:43, 1998.
149. W Xu, AD Nikolov, DT Wasan, A Gonvalves, R Borwankar. J Food Sci 63:183, 1998.
150. Yu Duda, D Henderson, O Pizio, DT Wasan. Mol Phys 94:341, 1998.
151. D Henderson, DL Boda, K Chan, DT Wasan. Mol Phys 95:131, 1998.
152. C Thomas, J Rudin, DT Wasan. SPE Reservoir Eng J (in press).
153. A Huerta, Yu Duda, O Pizio, DT Wasan. J Phys Chem (in press).
154. XD Zhang, CW Macosko, HT Davis, AD Nikolov, DT Wasan. J Colloid Interf Sci 215:270–279, 1999.
155. D Boda, K-Y Chan, D Henderson, DT Wasan, AD Nikolov. Langmuir 15(13): 4311–4313, 1999.
156. A Trokhymchuk, D Henderson, DT Wasan. J Colloid Interf Sci 210:320, 1999.
157. DT Wasan, AD Nikolov. In: G Warr, ed. Supramolecular Structure in Confined Geometries. ACS Symposium Series. Washington, DC: American Chemical Society 736:40–53, 1999.
158. JL Parker, P Richetti, P Kekicheff, S Sarman. Phys Rev Lett 68:1955, 1992.
159. P Pieranski. Contemp Phys 24:25, 1983.

2

Unity in Diversity in Interfacial Phenomena

**Alexander Patist, Seong-Geun Oh,* S. Y. Shiao, Tien-Feng Ling,†
Hyeon K. Lee,‡ M. K. Sharma,§ Surekha Devi,¶ and D. O. Shah**
*Center for Surface Science and Engineering, University of Florida,
Gainesville, Florida*

ABSTRACT

This chapter summarizes three decades of studies performed by our research group on monolayers, micelles, foams, emulsions and wetting phenomena, in which a commonality in the performance of surfactant systems was observed. Parameters such as chain-length compatibility, surfactant concentration, salinity, and molecular association were considered.

It was found that the effect of chain-length compatibility was particularly important to interfacial properties and technological applications. When mixed-surfactant systems have equal chain length, the average area per molecule is minimum, resulting in minimum interfacial tension, maximum surface viscosity, and maximum micellar stability. These results, in turn, influence technological processes,

---

Current affiliations:
* Hanyang University, Seoul, Korea.
† BioLab, Inc., Decatur, Georgia.
‡ Daihan Swiss Chemical, Ulsan, Korea.
§ Eastman Chemical Company, Kingsport, Tennessee.
¶ M. S. University, Baroda, India.

such as foaming, wetting, lubrication, enhanced oil recovery, and microemulsion stability.

It was observed that the concentration of sodium dodecyl sulfate (SDS) in aqueous solution strikingly influenced the dynamic interfacial properties of the system. The slow micellar relaxation time τ_2, which is directly related to the micellar stability, was found to be maximum at 200 mM SDS concentration (i.e., 25 times cmc), corresponding to the least foaming, largest bubble size, longest wetting time of textile, largest emulsion droplet size, and the most rapid solubilization of oil.

The importance of the molecular ratio in interfacial properties was investigated for a variety of mixed-surfactant systems. At the 1:3 and 3:1 molar ratios, minima and maxima in various interfacial as well as bulk properties were observed. A possible hexagonal arrangement of molecules allows one type of molecules to occupy the corners of the hexagon and the second type of the molecules to occupy the centers. This arrangement results in the tightest possible packing at the surface and, hence, minimum area per molecule, maximum surface viscosity, lowest rate of drainage, and minimum surface tension. These surface properties, in turn, influence the performance of foams, emulsions, and monolayers.

One parameter that has been discovered to be crucially important in the successful implementation of the surfactant–polymer flooding process for enhanced oil recovery is the salinity of the aqueous phase. For a given oil–water–surfactant system at a given temperature and pressure, a specific salinity is required to produce an ultralow interfacial tension. Parameters, such as interfacial tension, viscosity of emulsions or pressure difference in porous media, surfactant retention, and overall oil-recovery efficiency exhibit maxima or minima at the optimal salinity, indicating that these processes are all interrelated for the oil displacement in porous media by the surfactant–polymer flooding process.

The last study showing a commonality in interfacial properties is the effect of tetraalkyl ammonium salt on foaming properties of SDS solutions. At low concentrations, the molecular interaction between SDS and tetraalkylammonium ions increases the molecular packing in the adsorbed film, resulting in lower surface tensions and higher surface viscosity. However, above a critical concentration (depending on the alkyl chain length of the tetraalkylammonium ion), the interfacial properties go in the reverse direction, indicating a disruption of the molecular packing at the interface and resulting in a higher surface tension and a lower surface viscosity.

The parameters investigated over the years clearly show the existence of the unity in diversity as illustrated by the interfacial phenomena studied for several surfactant systems. This commonality determines various surface properties, microstructures, and macroscopic properties of surfactant solutions, monolayers, foams, and emulsions, yielding "master diagrams" in support of this statement.

I. INTRODUCTION

During the past three decades, the Center for Surface Science and Engineering at the University of Florida has carried out considerable studies on monolayers, micelles, foams, emulsions, and wetting phenomena [1]. From various theoretical considerations, as well as experimental results, we have observed a commonality in the performance of surfactant systems considering parameters such as chain-length compatibility, surfactant concentration, salinity, or molecular association. In this chapter, we will discuss the commonality observed for several mixed-surfactant systems studied over the years to show the existence of a unity in diversity.

Figure 1 illustrates how molecular properties of surfactants are related to the performance in technological processes. The structure of surfactant molecules influences the properties of adsorbed films as well as micellar characteristics. Both of these, in turn, influence the performance of technological processes such as foaming, emulsification, solubilization, wetting, and lubrication.

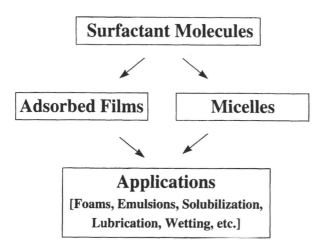

Figure 1   Correlation between molecular properties and macroscopic phenomena.

II. CHAIN-LENGTH COMPATIBILITY EFFECTS IN MIXED-SURFACTANT SYSTEMS

The chain-length compatibility is an important factor in systems involving interfacial films. As surface-active as well as other hydrocarbon molecules are aligned at interfaces, the properties of the interface are impacted, to a large extent, by the matching or mismatching of the alkyl chain lengths. The effect of chain-length compatibility is particularly important to interfacial properties and technologies such as surface tension, surface viscosity, foamability, lubrication, contact angle, bubble size, environmental remediation, enhanced oil recovery, water solubilization in microemulsions, and microemulsion stability [2]. Figure 2 shows schematically mixed-surfactant systems having equal or different chain lengths. It is evident that when the two chains are unequal, the excess hydrocarbon tails have more freedom to disrupt the molecular packing through conformational disorder and thermal motion. This causes a slight increase in the intermolecular distance and, hence, the average area per molecule. The greater the length of the excess segment, the greater the area per molecule.

An example of the influence of chain-length compatibility on interfacial properties for the mixed-surfactant system sodium dodecyl sulfate (SDS)/long-chain alcohols (C_nOH for $n = 8$, 10, 12, 14, and 16) is shown in Figure 3 [2,3]. It shows that the surface tension is minimum and the surface viscosity (in surface poise, SP) is maximum when both molecules have equal chain length (i.e., for SDS/$C_{12}OH$ mixtures). The stability of micelles, as reflected by the slow micellar relaxation time, τ_2, is maximum when both surfactants have the same chain length. It was shown recently [4] that any long-chain alcohol (C_nOH for $n = 8$, 10, 12, 14, and 16) will stabilize SDS micelles because of the introduction of ion–dipole interactions between the alcohol and SDS molecule, causing a tighter packing of the micelle. However, beyond a certain concentration of SDS (de-

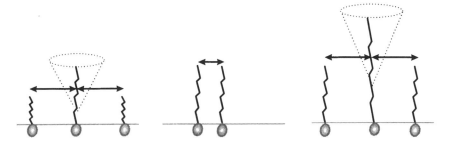

Figure 2   Effect of chain-length compatibility on molecular packing in mixed-surfactant systems.

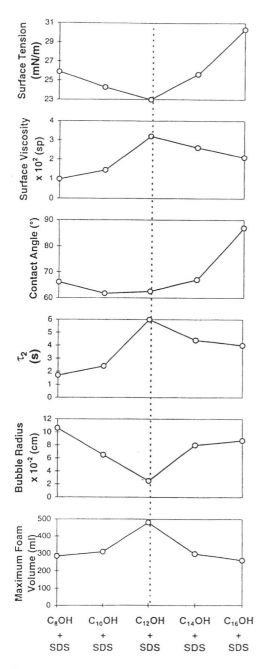

Figure 3   Effect of chain-length compatibility on interfacial properties of sodium dodecyl sulfate/long-chain alcohol (SDS/C$_n$OH) solutions.

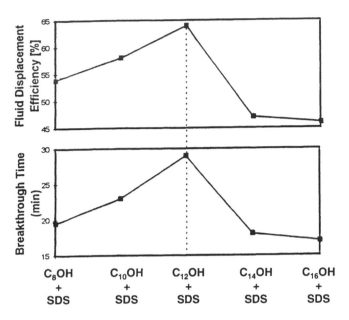

Figure 4 Chain-length-compatibility effect on fluid-displacement efficiency and break-through time for removal of residual trapped oil in porous media.

pending on the alcohol chain length), long-chain alcohols other than $C_{12}OH$ start destabilizing micelles, due to mismatching of the chains resulting in a disruption of the molecular packing causing the micelles to destabilize and, hence, shorter micellar relaxation times are obtained. Figure 3 also shows that a minimum bubble radius was found for the SDS/$C_{12}OH$ system, indicating that the lowest interfacial tension occurs when the chain lengths of the surfactant and the cosurfactant are equal. Because of the lowest interfacial tension, the foaming ability was maximum for the equal-chain-lengths system. Thus, these results show that the chain-length equality strikingly influences surface, micellar, and foam properties.

Previous research [3] has shown that the chain-length compatibility of mixed-surfactant systems also plays an important role in foaming systems, in particular for applications in enhanced oil recovery and environmental remediation of contaminated aquifer sites. Figure 4 shows the fluid displacement and breakthrough time in porous media (e.g., sand-pack). The time required for the gas to break through at the other end of the porous medium and the fluid-displacement efficiency show maxima when the chain lengths of the two components are equal. In the case of SDS/C_nOH systems, molecules interact with the ion–dipole association. A much stronger interaction is expected for ion–ion interactions.

Patist et al. [5] studied the effect of chain-length compatibility on interfacial properties for mixtures of sodium dodecyl sulfate and alkyltrimethylammonium bromides (C_nTAB for n = 8, 10, 12, 14, and 16). Figure 5 shows various surface properties as well as foam characteristics for the SDS/C_nTAB system at a 20:1 molar ratio. When the chain lengths of two surfactants are equal, the surface tension and foam ability show minima, whereas the surface viscosity, the micelle stability, and the foam stability show maxima. Thus, for mixed anionic/cationic surfactant mixtures, the Coulombic interaction as well as the chain-length compatibility determine the molecular packing in micelles and at the air–water interface. In conclusion, the equality of chain length offers a common denominator for adsorbed films, micelles, and foams.

III. EFFECT OF SDS CONCENTRATION ON VARIOUS INTERFACIAL PHENOMENA

It is well known that the critical micelle concentration of SDS is approximately 8.3 mM [6]. Gradually increasing the SDS concentration up to 300 mM leads to some interesting changes in interfacial properties of SDS solutions, showing minima and maxima at a concentration of 200 mM. Oh and Shah [7] determined the slow micellar relaxation time, τ_2 (which is directly related to the lifetime or stability of the micelle), by the pressure-jump technique [8]. Figure 6 shows the slow micellar relaxation time τ_2 as a function of SDS concentration. A maximum micellar stability was found at 200 mM (5 s). The micellar stability influences many processes of interest, such as foaming, wetting, emulsification, and solubilization [9]. Figure 7 presents the various phenomena exhibiting minima and maxima at the liquid–gas interface. At 200 mM SDS, minimum foamability, maximum single-film stability, maximum single-bubble volume, and a minimum frequency of bubble generation were found. These phenomena were explained based on the monomer flux to the newly created interface. If the micelles in solution are very stable, they cannot provide a monomer fast enough to the interface and thus the interfacial tension remains higher. Therefore, lower foamability, larger single-bubble foam volumes, and a minimum frequency of bubble generation were found [10,11]. A maximum single-film stability was found at 200 mM (i.e., when the micelles are most stable) [12]. An important factor influencing single-film stability is the micellar structure inside the thin liquid film, which has been investigated by Wasan and co-workers [13,14]. The stratification of thin liquid films can be explained as a layer-by-layer thinning of ordered structures of micelles inside the film. It is proposed that this structural effect is influenced by the micellar effective volume fraction, their stability, interaction, and polydispersity. Therefore, very stable micelles contribute to the stability of the thin liquid film.

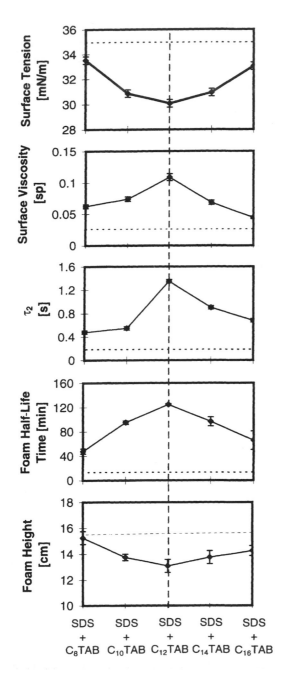

Figure 5 Chain-length-compatibility effect on interfacial properties of sodium dodecyl sulfate/alkyltrimethylammonium bromide (SDS/C_nTAB, 20:1) solutions. The horizontal dashed lines represent 100 mM pure SDS solutions.

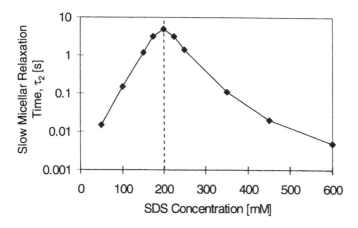

Figure 6 Slow micellar relaxation time τ_2 versus concentration of SDS at 25°C.

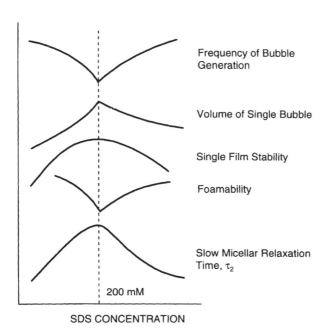

Figure 7 Liquid–gas phenomena exhibiting minima and maxima at 200 mM SDS concentration.

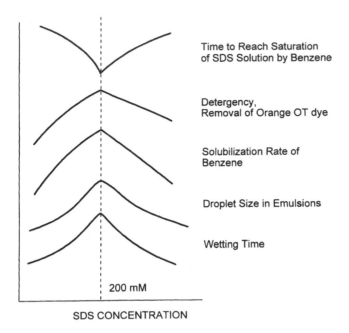

Time to Reach Saturation
of SDS Solution by Benzene

Detergency,
Removal of Orange OT dye

Solubilization Rate of
Benzene

Droplet Size in Emulsions

Wetting Time

200 mM

SDS CONCENTRATION

Figure 8 Liquid–liquid and solid–liquid phenomena exhibiting minima and maxima at 200 mM SDS concentration.

Very unstable micelles, however, provide monomers fast enough to the surface, resulting in lower interfacial tensions.

Interfacial phenomena occurring at the liquid–liquid and solid–liquid interfaces in SDS solutions at various concentrations are shown in Figure 8. The wetting time and droplet size in emulsions exhibit maxima at 200 mM. These phenomena can also be explained based on the monomer flux necessary to stabilize the newly created interface. Very stable micelles at 200 mM result in a high dynamic surface tension; hence, a larger droplet size and longer wetting time are obtained [15,16]. The solubilization rate of benzene in surfactant solutions as well as the detergency or removal of Orange OT dye from the fabric's surface also show maxima at 200 mM SDS concentration. The time required to reach saturation of the SDS solution upon the addition of benzene is minimum at 200 mM SDS concentration. This suggests that very stable micelles (i.e., tightly packed micelles) are more effective in solubilizing oil. This can be explained based on the interior of the micelles. The interior of rigid (i.e., tightly packed) micelles is more hydrophobic as compared to that of loosely packed micelles;

hence, the stronger hydrophobic core causes a more rapid solubilization of benzene and Orange OT dye into the micelles at 200 mM SDS concentration.

IV. INTERMICELLAR COULOMBIC REPULSION MODEL

In order to explain the maxima and minima in various phenomena occurring at 200 mM SDS solutions, the following Intermicellar Coulombic Repulsion Model (ICRM) was proposed. Knowing the aggregation number of SDS micelles and the total SDS concentration, one can calculate the number of micelles at a specific SDS concentration in the solution. By dividing the solution into identical cubes, equal to the number of micelles, one can equate the distance between the centers of the adjacent cubes as the average intermicellar distance. By this approach, we found that the intermicellar distance was 13, 10, and 7.86 nm respectively at 50 mM, 100 mM, and 200 mM SDS concentrations. This suggests that the adjacent micelles are one diameter apart at 200 mM concentration. The small gap of about 4 nm causes a Coulombic repulsion between SDS micelles and hence induces a rapid uptake of counterions to minimize the charge repulsion between adjacent micelles. This provides considerable stability to the micellar structure, resulting in a long relaxation time. Above 250 mM SDS concentration, a structural transition from spherical to cylindrical SDS micelles occurs. However, this structural transition is gradual, and in this concentration range (250–400 mM), the solution consists of a mixture of spherical and cylindrical micelles. Because the number of spherical micelles is less as compared to that at 200 mM concentration, as some of them have become cylindrical micelles, the distance between spherical micelles increases, which leads to shorter relaxation times. In a binary mixture, it is the most labile structure (i.e., spherical micelles), which responds more quickly to the change in pressure in the pressure-jump studies compared to cylindrical micelles [8]. The intermicellar distances obtained from this procedure at various SDS concentrations are shown in Figure 9.

In summary, SDS solutions exhibit maxima and minima for various properties at 200 mM concentration due to maximum stability of SDS micelles at this concentration. Most ionic surfactants may exhibit such a characteristic concentration at which the micellar stability is maximum due to a reduction in intermicellar distance.

V. SIGNIFICANCE OF THE 1:3 MOLECULAR RATIO IN MIXED-SURFACTANT SYSTEMS

Many mixed-surfactant systems exhibit unusual properties at the 1:3 or 3:1 molecular ratios. Shah [17] investigated interfacial properties, such as foamability,

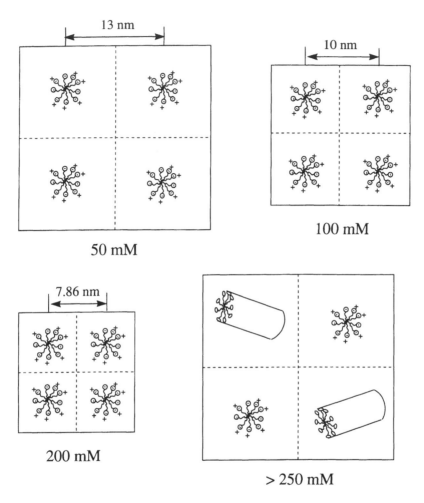

Figure 9 Schematic diagrams for micellar packing at 50, 100, 200, and 250 m*M* SDS concentration.

foam stability, and rate of evaporation of water for the system decanoic acid/ decanol. Figure 10 shows the foam volume in various cylinders containing mixtures of decanoic acid/decanol after 16 h of foam generation by vigorous shaking. It is evident that cylinder 7, which contains a 1:3 molecular ratio of the decanoic acid/decanol mixture, exhibits maximum foam stability. From Figure 11, it appears that the maximum foam stability coincides with a minimum rate of drainage and maximum surface viscosity. Apparently, decanoic acid and decanol associate

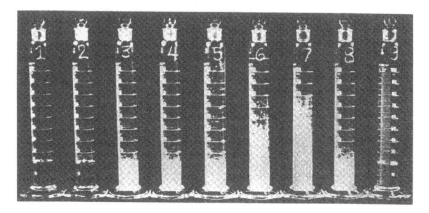

Figure 10 Foam volumes of decanoic acid/decanol mixtures, 16 h after foam generation. The mixture in cylinder 7 (1:3 ratio) shows the maximum foam stability.

at the 1:3 ratio, causing a tighter packing of molecules, resulting in an increase in surface viscosity, a decrease in drainage rate, and, therefore, an increase in foam stability, in agreement with the results reported by others [18–20].

Figure 12 shows the average area per molecule in mixed monolayers of stearic acid and stearyl alcohol [17]. The data indicate that the minimum area

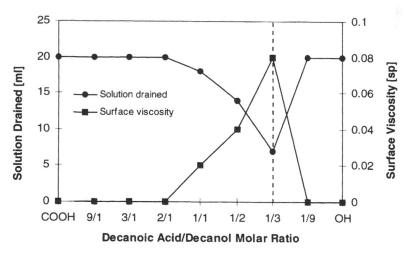

Figure 11 Relation between the amount of solution remaining under decanoic acid/decanol foams, 10 min after foam generation, and the surface viscosity.

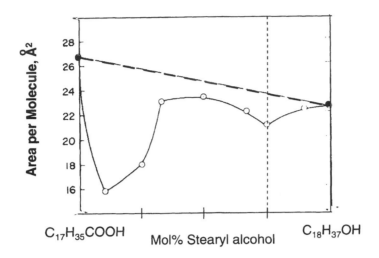

Figure 12 Average area per molecule in mixed monolayers of stearic acid and stearyl alcohol at a surface pressure of 10 mN/m. The vertical dashed line represents the 1:3 molar ratio.

per molecule occurs at the 1:3 and 9:1 molar ratios. In order to show that the minimum area per molecule at the 1:3 molar ratio is due to the tight packing of molecules, studies were carried out on the evaporation of water through these mixed monolayers. Figure 13 shows the evaporation rate of water through mixed monolayers of stearic acid and stearyl alcohol at various ratios [17]. It is evident that the minimum rate of evaporation occurs at the 1:3 molar ratio. This also suggests that the minimum area per molecule at the 9:1 molar ratio must be due to structural alterations within the mixed monolayers and not due to tight packing of the molecules.

Figure 14 shows the area per molecule in mixed monolayers of egg lecithin and cholesterol at various surface pressures [17,21]. It is evident that at all surface pressures, the optimum condensation occurs at the 1:3 and 3:1 molar ratios between egg lecithin and cholesterol. These results are in agreement with those reported by Dervichian [22]. Mixed monolayers of oleic acid and cholesterol [23] also show a minimum rate of evaporation at various surface pressures at the 1:3 molar ratio (Fig. 15). Recently [24], the heat transfer through the same monolayer system of cholesterol and oleic acid was studied using an infrared imaging camera. From Figure 16, it is evident that the heat-transfer process was most efficient through the mixed monolayer at the 1:3 molar ratio, due to the tight packing of molecules.

Booij [25] showed the importance of molecular association on mineral oil

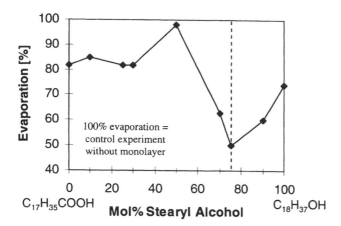

Figure 13 Rate of evaporation through mixed monolayers of stearic acid and stearyl alcohol. The vertical dashed line represents the 1:3 molar ratio.

extraction from emulsions stabilized by a mixture of sodium dodecyl sulfate and cetyl pyridinium chloride (Fig. 17). It is evident that at the 1:3 and 3:1 molar ratios, the maximum amount of oil is extracted from the emulsions, indicating that the 1:3 association also occurs at the oil–water interface. The same system was recently investigated for surface properties by Patist et al. [26] at low concentrations (total concentration: 1 mM). Figure 18 shows that again at the 1:3 and 3:1 molar ratios, minima and maxima are found due to tight packing of anionic and cationic surfactant molecules.

The observed minima and maxima in various properties of mixed-surfactant systems at the 1:3 or 3:1 ratio cannot be considered a coincidence. This must be due to a very fundamental mechanism, operating in mixed-surfactant systems. Figure 19 presents the proposed explanation for the molecular packing at the 1:3 and 3:1 molar ratios at the interface. This type of hexagonal arrangement allows one type of molecule to occupy the corners of the hexagon and the second type of molecule to occupy the centers. This arrangement results in the tightest possible packing at the surface and, hence, the minimum area per molecule, maximum surface viscosity, lowest rate of drainage, and minimum surface tension. These surface properties then, in turn, influence the performance of foams, emulsions, and monolayers.

VI. OPTIMAL SALINITY AND ENHANCED OIL RECOVERY

The efficient production of oil has been an important objective by the petroleum industry. Geological evolution has been kind to the industry insofar as producing

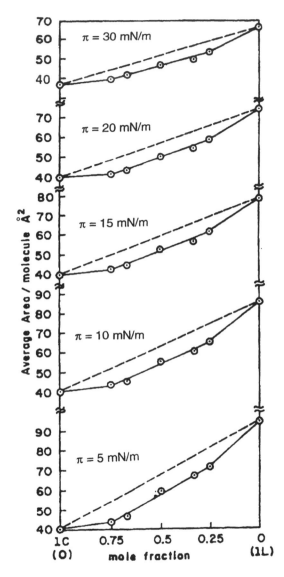

Figure 14 Average area per molecule in mixed monolayers of egg lecithin and cholesterol at various surface pressures.

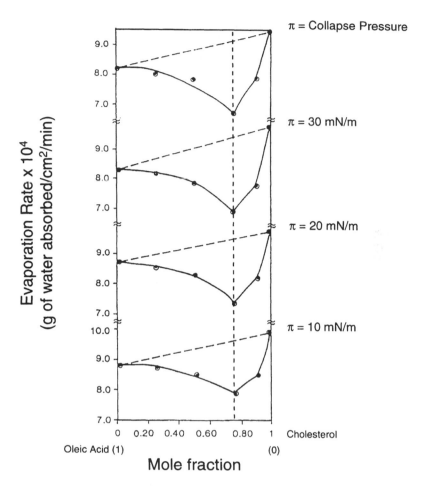

Figure 15 Rate of evaporation through mixed monolayers of oleic acid and cholesterol at various surface pressures. The vertical dashed line represents the 1:3 molar ratio.

high-pressure oil reserves in which primary and secondary oil-recovery methods can achieve close to 35% production efficiency. The remaining 65%, which is in the reservoir, has provided an engineering challenge. This area has been known as tertiary or enhanced oil recovery (EOR). Initial areas of research, brought about by the oil shortages of the 1970s, focused on the importance of reducing the interfacial tension (IFT) of the oil–water interface with the use of surfactants [6]. The petroleum industry seized this concept as a method to increase oil production through modified water-flushing strategies. Figure 20 shows a three-

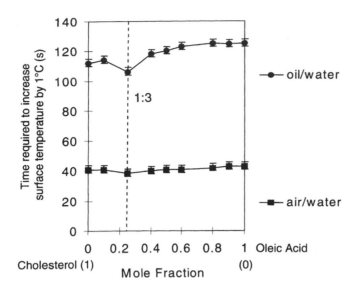

Figure 16 Time required for the surface temperature of water or oil (hexadecane) with mixed monolayers of oleic acid and cholesterol to increase by 1°C.

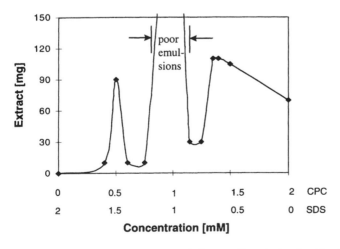

Figure 17 Extraction of mineral oil from sodium dodecyl sulfate/cetyl pyridinium chloride (SDS/CPC) mixtures.

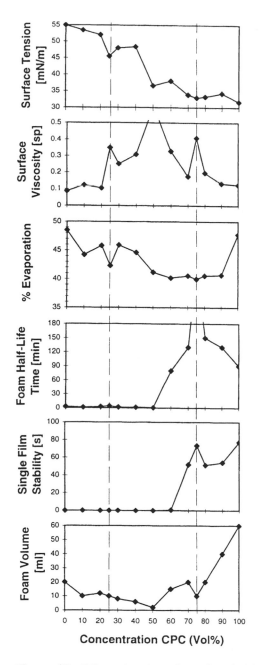

Figure 18 Effect of molar ratio on interfacial properties of sodium dodecyl sulfate/ cetyl pyridinium chloride (SDS/CPC) mixtures. Total concentration: 1 m*M*. The vertical dashed lines represent the 1:3 and 3:1 molecular ratios.

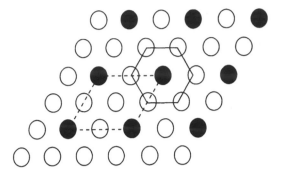

Figure 19 Proposed two-dimensional hexagonal arrangement of molecules at the 1:3 molar ratio in mixed-surfactant systems.

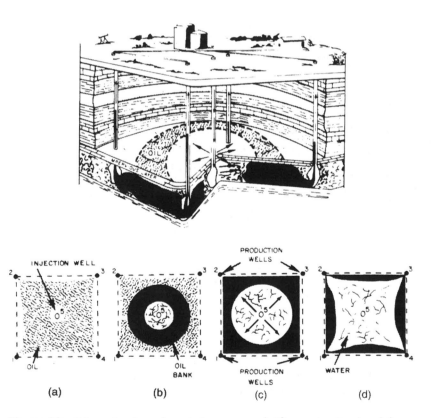

Figure 20 Schematic view of a petroleum reservoir (five-spot pattern) and the process of water or chemical flooding.

Figure 21 For the movement of oil through narrow necks of pores, an ultralow interfacial tension is required (≈ 0.001 mN/m).

dimensional view of a petroleum reservoir and the process of water or chemical flooding [27]. The objective of tertiary oil recovery is to reduce the interfacial tension at the oil–water interface (20–25 mN/m) to 10^{-3}–10^{-5} mN/m and hence promote the mobilization of oil in porous media. Figure 21 shows that an ultralow interfacial tension is required for the mobilization of oil ganglia in porous media [28–30]. The attainment of such an ultralow interfacial tension requires the tailoring of a surfactant to produce ultralow interfacial tensions under the reservoir conditions of temperature, pressure, and salinity. Conceptually, a surfactant formulation is injected into the porous media in the petroleum reservoir, so that upon mixing with the reservoir brine and oil, the surfactant produces a middle-phase microemulsion *in situ*. This middle-phase microemulsion, in equilibrium with excess oil and excess brine, propagates through the petroleum reservoir [31,32]. The design of the process is such that the oil bank maintains an ultralow interfacial tension with the reservoir brine until it arrives at the production wells. One parameter that has been discovered to be crucially important in the successful implementation of the surfactant–polymer flooding process is the salinity of the aqueous phase [27]. Figure 22 shows the effect of optimal salinity on various interfacial properties and phenomena relevant to enhanced oil recovery. For a given oil/water/surfactant system at a given temperature and pressure, a specific salinity is required to produce an ultralow interfacial tension. It is evident that all these parameters exhibit a maximum or minimum at the optimal salinity, indicating that these processes are all interrelated for the oil displacement in porous media by the surfactant–polymer flooding process. It also appears that the optimal salinity value is a critical parameter for oil recovery in this process. The field of enhanced oil recovery has received a great deal of attention over the last three decades. The insight and information obtained during these years has not only increased oil production but has led to other applications of microemulsions as well [33].

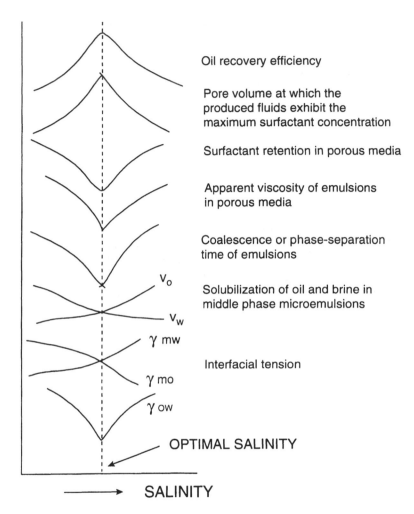

Figure 22 Effect of optimal salinity on various interfacial properties and phenomena relevant to EOR.

VII. EFFECT OF ORGANIC CATIONS ON FOAMING PROPERTIES OF SDS SOLUTIONS

It is well known that foaming properties can be greatly modified by the addition of organic materials [6,34,35]. It was shown recently that the influence of tetraalkylammonium salts on the SDS foam stability and foamability was a result of the effect of these compounds on the molecular packing at the air–water interface and in micelles [35,36]. Figure 23 shows the effect of concentration of tetraethylammonium chloride (TC$_2$AC) on surface tension, surface viscosity, foam stability, foaming ability, and slow micellar relaxation time, τ_2, in 150 mM SDS solutions. It is evident that the maxima and minima in various properties occur at a critical concentration of 5 mM of organic cations. Thus, it can be interpreted that at low concentrations of TC$_2$AC, the interaction between surfactant and TC$_2$A$^+$ ions increases and, hence, the surface tension decreases and surface viscosity increases, causing an increase in the foam stability as well as an increase in the slow relaxation time of micelles. However, above 5 mM of TC$_2$AC, all these properties go in the reverse direction, indicating that upon further penetration of TC$_2$A$^+$ ions, the SDS molecules increases the surface tension and foaming ability and decreases the surface viscosity, foam stability, and relaxation time τ_2. The results were explained in terms of the molecular packing at interfaces and in micelles. At concentrations lower than 5 mM, the ion–ion interaction between the SDS headgroups and the tetraalkylammonium salt causes the surfactant molecules to pack closer, thereby lowering the surface tension. However, beyond 5 mM of TC$_2$AC, the antifoamer salt starts penetrating between the SDS molecular packing at the air–water interface, resulting in a larger area per molecule and, hence, a higher surface tension is obtained. The effect of tetraalkylammonium salts on the molecular packing at interfaces and in micelles is schematically shown in Figure 24.

VIII. SUMMARY

This chapter illustrates that for various interfacial phenomena, a commonality exists which determines various surface properties, microstructures, and macroscopic properties of surfactant solutions, monolayers, foams, and emulsions.

It was shown that chain-length compatibility was an important factor in systems involving interfacial films. For the mixed-surfactant system sodium dodecyl sulfate (SDS)/long-chain alcohols, it was shown that the surface tension was minimum and the surface viscosity was maximum when both molecules had the same chain length (i.e., for SDS/C$_{12}$OH mixtures). The slow micellar relaxation time, τ_2, of the mixed micelles exhibits a maximum value, due to the ion–dipole interactions between the SDS and the long-chain alcohol, resulting in very stable micelles. Because of the lowest interfacial tension, the foaming ability was maximum for the equal-chain-length system. Thus, these results show that the chain-length equality strikingly influences surface, micellar, and foam properties.

The second parameter discussed in this chapter was the effect of concentration of SDS on various interfacial phenomena. The slow micellar relaxation time, τ_2, increased as a function of SDS concentration, showing a maximum at 200

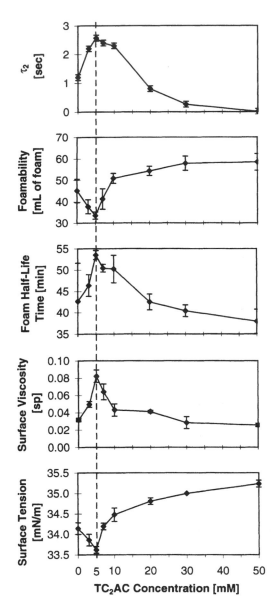

Figure 23 Effect of TC$_2$AC on foaming properties of 150 m*M* SDS solutions.

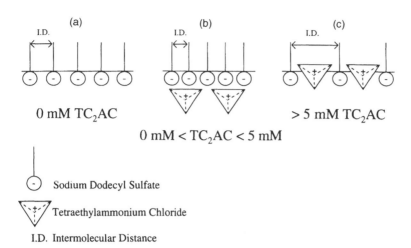

(a)

I.D.

0 mM TC$_2$AC

(b)

I.D.

0 mM < TC$_2$AC < 5 mM

(c)

I.D.

> 5 mM TC$_2$AC

⊖ Sodium Dodecyl Sulfate

▽ Tetraethylammonium Chloride

I.D. Intermolecular Distance

Figure 24 Schematic diagram showing the effect of tetraalkylammonium chlorides on the SDS molecular packing at interfaces [(a) pure SDS; (b) decrease in electrical repulsion due to adsorption of TC$_2$AC below the SDS monolayer results in a smaller area per molecule; (c) penetration of TC$_2$AC into the SDS monolayer increases the area per molecule].

mM (5 s). Beyond 200 mM of SDS, the stability decreased again. The micellar stability influenced many dynamic interfacial processes, such as foaming, wetting, emulsification, and solubilization. Very stable micelles were not able to break up fast enough to augment the flux of monomers necessary to stabilize the newly created interface. Hence, a higher interfacial tension was obtained. Therefore, at 200 mM, minimum foamability, maximum single-film stability, maximum single-bubble volume, and a minimum frequency of bubble generation were found. A similar process occurred at both the solid–liquid and liquid–liquid interface. Both the wetting time of fabrics and the droplet size in emulsions were maximum at 200 mM SDS solution.

The molecular ratio in mixed-surfactant system was an important factor that influenced many interfacial properties. Various binary systems, such as decanoic acid/decanol, stearic acid/stearyl alcohol, and SDS/cetyl pyridinium chloride (CPC), exhibited minima and maxima in interfacial properties at the 1:3 and 3:1 molecular ratios. The maximum foam stability, maximum surface viscosity, minimum surface tension, minimum rate of drainage, and minimum area per molecule were found. The observed minima and maxima were attributed to a hexagonal packing of molecules at the 1:3 and 3:1 molecular ratios at the interface. This type of hexagonal arrangement allowed one type of molecule to occupy the corners of the hexagon and the second type of molecule to occupy the centers.

This resulted in the tightest possible packing at the surface and, hence, the observed minima and maxima mentioned above.

One parameter that had been discovered to be crucially important in the successful implementation of the surfactant–polymer flooding process was the salinity of the aqueous phase. For a given oil/water/surfactant system at a given temperature and pressure, a specific salinity was required to produce an ultralow interfacial tension. At the optimal salinity, minimum interfacial tension, maximum oil recovery efficiency, minimum surfactant retention, and minimum viscosity of emulsions in porous media were observed. It was evident that all these processes are interrelated for the oil displacement in porous media by the surfactant–polymer flooding process.

The last parameter investigated in this chapter was the effect of tetraalkylammonium chloride on interfacial properties of SDS solutions. At low concentrations of tetraalkylammonium chloride, the micellar stability increased, due to shielding of the negative charges of the SDS headgroup. However, beyond a critical concentration (depending on the alkyl chain length of the tetraalkylammonium ion), the micellar stability decreased again, due to disruption of the molecular packing in the micelle. Similar results were found in interfacial properties. At the critical concentration (5 mM for tetraethylammonium chloride), minimum foamability, maximum foam stability, maximum surface viscosity, and minimum surface tension were observed. Thus, all the interfacial phenomena described above indicated a "common parameter" which influenced surface, micellar, and macroscopic properties of various surfactant systems.

ACKNOWLEDGMENTS

The authors wish to convey their sincere thanks to the National Science Foundation (Grant NSF-CPE 8005851), to the NFS–ERC Research Center for Particle Science and Technology (Grant EEC 94-02989), and to the ICI Corporation, Alcoa Foundation, and Dow Corning Company for their partial support of this research. D.O. Shah also expresses his indebtedness to his students and postdoctoral associates who contributed to this research.

REFERENCES

1. DO Shah, In: DO Shah, ed. Micelles, Microemulsions and Monolayers. New York: Marcel Dekker, 1998, Chap. 1.
2. SY Shiao, V Chhabra, A Patist, ML Free, PDT Huibers, A Gregory, S Patel, DO Shah. Adv. Colloid Interf Sci 74:1, 1998.
3. MK Sharma, DO Shah, WE Brigham. SPE Reservoir Eng. 253, May 1986.

4. A Patist, T Axelberd, DO Shah. J. Colloid Interf Sci 208:259, 1998.
5. A Patist, V Chhabra, R Pagidipati, R Shah, DO Shah. Langmuir 13:432, 1997.
6. MJ Rosen. Surfactants and Interfacial Phenomena. 2nd ed. New York: John Wiley, 1989.
7. SG Oh, DO Shah. J Am Oil Chem Soc 70:673, 1993.
8. PDT Huibers, SG Oh, DO Shah. In: AK Chattopadhyay, KL Mittal, eds. Surfactants in Solution. Marcel Dekker, New York, 1996, p. 105.
9. SG Oh, DO Shah. J Dispers Sci Technol 15:297, 1994.
10. SG Oh, DO Shah. Langmuir 7:1316, 1991.
11. SG Oh, SP Klein, DO Shah. AIChE J 38:149, 1992.
12. SS Patel, K Kumar, DO Shah, JJ Delfino. J Colloid Interf Sci 183:603, 1996.
13. AD Nikolov, DT Wasan. J Colloid Interf Sci 133:1, 1989.
14. AD Nikolov, PA Kralchevsky, IB Ivanov, DT Wasan. J Colloid Interf Sci 133:13, 1989.
15. SG Oh, M Jobalia, DO Shah. J Colloid Interf Sci 155:511, 1993.
16. SG Oh, DO Shah. Langmuir 8:1232, 1992.
17. DO Shah. J Colloid Interf Sci 37:744, 1971.
18. JT Davies, EK Rideal. Interfacial Phenomena. 2nd ed. New York: Academic Press, 1961.
19. AG Brown, WC Thuman, JW McBain. J Colloid Sci. 8:491, 1953.
20. DO Shah, NF Djabbarab, DT Wasan, Colloid Polym Sci 276:1002, 1978.
21. DO Shah, JH Schulman. J Lipid Res 8:215, 1967.
22. DG Dervichian. In: JF Danielli, KGA Pankhurst, AG Riddiford, eds. Surface Phenomena in Chemistry and Biology. New York: Pergamon Press, 1958, p. 70.
23. YK Rao, DO Shah. J Colloid Interf Sci 137:25, 1990.
24. H Fang, DO Shah. J Colloid Interf Sci 205:531, 1998.
25. HL Booij. J Colloid Interf Sci 29:365, 1969.
26. A Patist, S Devi, DO Shah. Langmuir 15:7403, 1999.
27. D.O. Shah. In: Proceedings of the European Symposium on Enhanced Oil Recovery, Bournemouth, England, Lausanne: Elsevier Sequoia S.A., 1981, p. 1.
28. WR Foster. J Petrol Technol 25:205, 1973.
29. VK Bansal, DO Shah. In: KL Mittal, ed. Micellization, Solubilization and Microemulsions. Vol. I. New York: Plenum Press, 1977, p. 87.
30. CH Pasquarelli, DT Wasan. In: DO Shah, ed. Surface Phenomena in Enhanced Oil Recovery. New York: Plenum Press, 1979, p. 237.
31. RL Reed, RN Healy. In: DO Shah, RS Schechter, eds. Improved Oil Recovery by Surfactant and Polymer Flooding. New York: Academic Press, 1977, p. 383.
32. WC Hsieh, DO Shah. International Symposium on Oilfield and Geothermal Chemistry, La Jolla, CA, 1977, Paper SPE 6594.
33. V Chhabra, ML Free, PK Kang, SE Truesdail, DO Shah. Tenside 34:156, 1997.
34. S Ross, RM Hauk. J Phys Chem 62:1260, 1958.
35. I Blute, M Jansson, SG Oh, DO Shah. J Am Oil Chem Soc 71:41, 1994.
36. A Patist, PDT Huibers, B Deneka, DO Shah. Langmuir 14:4471, 1998.

3
Coupling of Coalescence and Flocculation in Dilute O/W Emulsions

Stanislav Dukhin
New Jersey Institute of Technology, Newark, New Jersey

Øystein Sæther and Johan Sjöblom
University of Bergen, Bergen, Norway

ABSTRACT

A theory of the coupling of aggregation, coalescence, and floc fragmentation in dilute oil-in-water emulsions with narrow droplet-size distributions is proposed. The simplest singlet–doublet emulsion system exhibits singlet–doublet quasiequilibrium (SDE) and slow coalescence within doublets. At SDE the rate of decline in the total number of droplets (TND) is described by second-order kinetics in distinction to the exponential time dependence valid for coalescence at irreversible aggregation. The TND is measured by combining video-enhanced microscopy with the microslide preparative technique for the study of low-density contrast emulsions of dichlorodecane (DCD)-in-water, enabling the elimination of orthokinetic coagulation and droplets sticking to the microslide glass surface. Such determination of TND in a dilute DCD-in-water emulsion at SDE can be recommended as a standard method for measurement of the time of the elementary act of coalescence and the doublet fragmentation time. The perspective for emulsion dynamics modeling with the use of the measured coalescence time dependence on droplet dimension is discussed.

I. INTRODUCTION

In general, there are three coupled processes that will influence the rate of destabilization and phase separation in emulsions [1–8]. These are flocculation, coalescence, and floc fragmentation. Often, irreversible aggregation is called coagulation and the term "flocculation" is used for reversible aggregation [9,10]. Ostwald ripening [11,12] coupled with aggregation and fragmentation [13] is a separate topic which will not be considered here.

There have been successful attempts to elaborate kinetic models for a simultaneous coagulation and coalescence in emulsified systems [14,15]. However, recently there has been an increased interest in the flocculation in other colloidal systems [16–25].

Even a qualitative analysis to couple the processes of flocculation, floc fragmentation, and coalescence in emulsified systems has been lacking. In several articles [27–29], we attempted to take a first step toward a more comprehensive understanding of the coupling of the destabilization mechanisms in emulsions. The immense difficulties in the theory of flocculation [30–32] and in the theories of coupled coagulation and coalescence [15] are well known. Therefore, in several articles [27–29], we did not consider a coupling of all three processes in general, but we restricted ourselves to emulsions consisting mainly of singlets and doublets and an almost negligible concentration of multiplets [29].

With respect to aggregation and coalescence, we have followed established theories [14,15]. Similarly, the aggregation subprocess was described according to the Smoluchowski treatment [33]. Van den Tempel [34] extended Smoluchowski's theory to include the kinetics of coalescence.

The rate of coalescence was assumed [14,15,34] to be proportional to the number of interdroplet films and to the coalescence rate constant K_c. The rate constant for coalescence of a doublet has the meaning of inverse lifetime τ_c of the liquid film between the droplets.

Two interconnected topics will be considered, namely the coupling of coalescence and flocculation and the set of new applications arising from this coupling.

II. THEORY

A. Singlet–Doublet Quasiequilibrium

Each process among the three processes under consideration is characterized by a characteristic time. For the aggregation, it is the Smoluchowski time, τ_{Sm}, which gives the average time between droplet collisions. For coalescence, the time τ_c gives an average lifetime of the liquid film between the droplets. The analogous characteristic time for the fragmentation of flocs is the average doublet fragmentation time, τ_d [16–24,35]. This time characterizes the lifetime of a doublet, restricted by its spontaneous disruption. If the time between two collisions is less than τ_d, a doublet can transform into a triplet before it spontaneously disrupts. In the opposite case, that is, at

$$\tau_{Sm} \gg \tau_d \tag{1}$$

the probability for a doublet to transform into a triplet is very low because the disruption of the doublet occurs much earlier than its collision with a singlet. The rate of multiplet formation is very low for

$$Rev = \frac{\tau_d}{\tau_{Sm}} \ll 1 \tag{2}$$

where we introduce the notation Rev for small values of the ratio corresponding to the reversibility of aggregation and a singlet–doublet quasiequilibrium.

The kinetic equation for reversible flocculation in a dilute monodisperse oil-in-water (o/w) emulsion when neglecting coalescence is [13,29]

$$\frac{dn_2}{dt} = \frac{n_1^2}{\tau_{Sm}} - \frac{n_2}{\tau_d} \tag{3}$$

where n_1 and n_2 are the dimensionless concentrations of doublets and singlets, respectively, $n_1 = N_1/N_{10}$, $n_2 = N_2/N_{10}$, N_1 and N_2 are the concentrations of singlets and doublets, respectively, and N_{10} is the initial concentration, and

$$\tau_{Sm} = \left(\frac{4kT}{3\eta} N_{10}\right)^{-1} = (K_f N_{10})^{-1} \tag{4}$$

where k is the Boltzmann constant, T is the absolute temperature, and η is the viscosity of water. For aqueous dispersions at 25°C, $K_f = 4kT/3\eta = 6.13 \times 10^{-18}$ m³/s. The singlet concentration decreases with time due to doublet formation, whereas the doublet concentration increases. As a result, the total rate of aggregation and floc fragmentation will approach each other. Correspondingly, the change in the number of doublets $dn_2/dt = 0$. Thus, a dynamic singlet–doublet equilibrium is established:

$$n_{2eq} = \frac{\tau_d}{\tau_{Sm}} n_{1eq}^2 \tag{5}$$

For condition (2), it follows from Eq. (5) that

$$n_{1eq} \cong 1, \qquad N_{1eq} = N_{10} \tag{6}$$

$$N_2 = (Rev)N_{10} \tag{7}$$

Thus, at small values of Rev, the singlet–doublet equilibrium is established with only small deviations in the singlet equilibrium concentration from the initial concentration [Eq. (6)]. The doublet concentration is very low compared to the singlet concentration, and the multiplet concentration is very low compared to the doublet concentration. The last statement follows from a comparison of the production rates of doublets and triplets. The doublets appear as a result of singlet–singlet collisions, whereas the triplets appear as a result of singlet–doublet collisions. The latter rate is lower because of the low doublet concentration. The ratio of the number of singlet–doublet collisions to the number of singlet–singlet collisions is proportional to Rev.

B. Kinetic Equation for Coupling of Flocculation and Intradoublet Coalescence in Monodisperse Emulsions

Both the rate of doublet disaggregation and the rate of intradoublet coalescence are proportional to the momentary doublet concentration. This leads [13,29] to a generalization of Eq. (3):

$$\frac{dn_2}{dt} = \frac{n_1^2}{\tau_{Sm}} - n_2\left(\frac{1}{\tau_d} + \frac{1}{\tau_c}\right) \tag{8}$$

There are two unknown functions in Eq. (8), so an additional equation is needed. This equation describes the decrease in the droplet concentration caused by coalescence:

$$\frac{d}{dt}(n_1 + 2n_2) = -\frac{n_2}{\tau_c} \tag{9}$$

The system of Eqs. (8) and (9) can be transformed [13,29] into the second-order equation

$$\frac{d^2 n_2}{dt^2} + pD\frac{dn_2}{dt} + qn_2 = 0 \tag{10}$$

where

$$p = \frac{4}{\tau_{Sm}} + \frac{1}{\tau_d} + \frac{1}{\tau_c} \tag{11}$$

$$q = \frac{2}{\tau_{Sm}\tau_c} \tag{12}$$

The initial conditions are

$$n_2|_{t=0} = 0 \tag{13}$$

$$\left.\frac{dn_2}{dt}\right|_{t=0} = \frac{n_{10}}{\tau_{Sm}} \tag{14}$$

Condition (14) follows from Eqs. (8) and (9). The solution of Eq. (10), which account for boundary conditions (13) and (14), is a superposition of two exponents [13,29]. In the most interesting case,

$$\tau_c \gg \tau_d \tag{15}$$

the solution simplifies [13,29] to

$$n_2(t) = \frac{\tau_d}{\tau_{Sm}}\left[\exp\left(-\frac{2\tau_d t}{\tau_{Sm}\tau_c}\right) - \exp\left(-\frac{t}{\tau_d}\right)\right] \tag{16}$$

Equation (16), as compared to Eqs. (5) and (6), equals the single–doublet equilibrium if the expression in the second set of parentheses equals unity. In the time interval

$$\tau_d < t < \tau \tag{17}$$

where

$$\tau = \tau_c \frac{\tau_{Sm}}{2\tau_d} \tag{18}$$

the first term in the set of brackets approximately equals 1, whereas the second one decreases from 1 to a very small value. Thus, the singlet–doublet equilibrium is established during the time τ_d and is preserved during the longer time interval (17).

For times longer than τ, there is no reason to apply Eq. (16) because the condition to linearize Eq. (8) is no longer valid with the concentration decrease. At the beginning of the process, the doublet concentration increases; later, the coalescence predominates and the doublet concentration decreases. Thus, function (16) has a maximum [13,29] which corresponds to the condition (dn_2/dt) $(t_{max}) = 0$; that is,

$$\tau_{max} = \tau_d \frac{\ln(\tau/\tau_d)}{1 - \tau_d/\tau} \tag{19}$$

C. Coalescence in a Singlet–Doublet System at Quasiequilibrium

For condition (15), the following important property of the coalescence in the singlet–doublet system must be emphasized:

$$\tau \gg \tau_d \tag{20}$$

Thus, after a time t_{max}, a slow decrease in the doublet concentration takes place simultaneously with the more rapid processes of aggregation and disaggregation. Naturally, an exact singlet–doublet equilibrium is not valid due to the continuous decrease in the doublet concentration. However, the slower the coalescence, the smaller is the deviation from the momentary dynamic equilibrium with respect to the aggregation–disaggregation processes.

It is reasonable to neglect the deviation from the momentary doublet–singlet equilibrium with the condition

$$\frac{dn_2}{dt} \ll \frac{n_2}{\tau_d} \tag{21}$$

Indeed, for this condition, the derivative in Eq. (3) can be omitted, which corresponds to the singlet–doublet quasiequilibrium (SDE) characterized by Eq. (5). The left-hand side of condition (21) can be specified by the substitution of n_2 according to Eq. (5) and then using Eq. (9):

$$\frac{dn_2}{dt} = \frac{\tau_d}{\tau_{Sm}} 2n_1 \quad \frac{dn_1}{dt} \cong \frac{\tau_d}{\tau_{Sm}} 2n_1 \quad \frac{n_2}{\tau_c} < \frac{2\tau_d}{\tau_{Sm}} \frac{n_2}{\tau_c} \ll \frac{n_2}{\tau_d} \tag{22}$$

The use of Eq. (9) was simplified by means of condition (2), which enabled the omission of $2n_2$ in comparison with n_1:

$$\frac{dn_1}{dt} = -\frac{n_2}{\tau_c} \tag{23}$$

After modifying (22), it turns out that the deviation from singlet–doublet equilibrium is negligible when

$$\frac{\tau_d}{\tau} \ll 1 \tag{24}$$

[i.e., for conditions (2) and (15)].

The fragmentation of flocs influences the coalescence kinetics, which can be represented as a three-stage process for condition (15), as illustrated in Figure 1. During a rather short time τ_d, the approach to a singlet–doublet equilibrium takes place [i.e., a rather rapid increase in the doublet concentration (stage 1)]. During the next time interval $\tau_d < t < t_{max}$, the same process continues. However, the rate of doublet formation declines due to coalescence (stage 2). The exact equilibrium between the doublet formation and their disappearance due to coalescence takes place at the time t_{max} [Eq. (19)], when the doublet concentration reaches its maximum value $n_2(t_{max})$. During the third stage, when $t > t_{max}$, the rate of doublet fragmentation is lower than the rate of formation, because of the coalescence within doublets. This causes a slow monotonous decrease in the

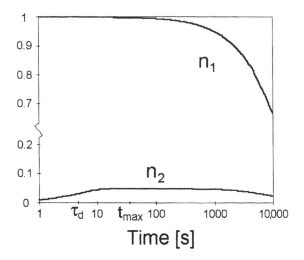

Figure 1 Three stages in the coupling of aggregation, fragmentation, and coalescence at the condition $\tau_d \ll \tau_{Sm} \ll \tau_c$. Initially, the doublet concentration n_2 is very low and the rates of doublet fragmentation and of coalescence are correspondingly low compared to the rate of aggregation (first stage, no coupling). Due to increasing n_2, the fragmentation rate increases and equals the aggregation rate at t_{max} (exact singlet–doublet equilibrium). The growth in n_2 stops at t_{max} (second stage, coupling of aggregation and fragmentation). Intradoublet coalescence causes a slight deviation from exact SDE to arise at $t > t_{max}$, and the singlet concentration n_1 and the doublet concentration decrease due to intradoublet coalescence (third stage, coupling of aggregation, fragmentation, and coalescence). n_1 and n_2 are dimensionless; $n_1 = N_1/N_{10}$, $n_2 = N_2/N_{10}$, where N_{10} is the initial singlet concentration.

concentration. Taking into account the singlet–doublet quasiequilibrium [Eq. (5)], the right-hand side of Eq. (23) can be expressed as

$$\frac{dn_1}{dt} = -\frac{\tau_d}{\tau_{Sm}\tau_c} n_1^2 \qquad (25)$$

The result of the integration of Eq. (25) can be simplified to

$$n_1(t) = \frac{n_1(t_{max})}{1 + n_1(t_{max})(t/2\tau)} \cong \left(1 + \frac{t}{2\tau}\right)^{-1} \qquad (26)$$

with a small deviation in $n_1(t_{max})$ from 1. Differing from the preceding stages when the decrease in the droplet concentration caused by coalescence is small, a large decrease is now possible during the third stage. Thus, this is the most important stage of the coalescence kinetics.

D. Reduced Role of Fragmentation with Decreasing τ_c

With decreasing τ_c, condition (15) is violated and new qualitative features of the destabilization process not discussed in Refs. 27–29 arise. As the ratio τ_c/τ_d diminishes and

$$\tau_c < \tau_d \tag{27}$$

the SDE is violated because a larger part of the doublets disappear due to coalescence. Correspondingly, the smaller the ratio τ_c/τ_d, the smaller is the fragmentation rate in comparison with the aggregation rate (i.e., the larger the deviation from SDE). In the extreme case,

$$\tau_c \ll \tau_d \tag{28}$$

the fragmentation role in a singlet–doublet emulsion can be neglected. This means that almost any act of aggregation is accompanied by coalescence after the short doublet lifetime. Neglecting this time in comparison with τ_{Sm} in agreement with condition (1), one concludes that any act of aggregation is accompanied by the disappearance of one singlet:

$$\frac{dn_1}{dt} = -\frac{n_1^2}{\tau_{Sm}} \tag{29}$$

This leads to a decrease in the singlet concentration described by an equation similar to the Smoluchovski equation for rapid coagulation:

$$n_1(t) = \left(1 + \frac{t}{\tau_{Sm}}\right)^{-1} \tag{30}$$

There are some differences. First, the convenient Smoluchovski equation describes the evolution in time for the total droplet concentration (TDC); in our case, $n = n_1 + 2n_2$. With Eq. (7) and for a small Rev [Eq. (2)], the difference between n and n_1 may be neglected. Second, in the Smoluchowski theory, a particle is an aggregate, and as an extreme case, it can be a singlet. The Smoluchowski equation describing the singlet time evolution does not coincide with Eq. (30). The peculiarity of Eq. (30) is that it describes the kinetics of coupled aggregation and coalescence with a negligible fragmentation rate.

Although the fragmentation was neglected in the derivation of Eq. (30), its role must not be disregarded. Due to fragmentation, doublet transformation into multiplets is almost impossible at condition (1). An additional peculiarity of the derived Eq. (30) compared with Smoluchowski theory is that it describes the irreversible growth of aggregates.

Keeping in mind that condition (28) is necessary for the application of Eq. (30), let us consider now the coupling of aggregation, fragmentation, and coalescence in the more general case described by condition (27). Exact SDE means the exact equality of the aggregation rate and the fragmentation rate. Due to coalescence, the fragmentation rate is lower than the aggregation rate. If this disbalance is weak, one may speak about SDE. The shorter the coalescence time, the larger is the deviation from SDE. Correspondingly, at the transition to condition (27), one lacks certainty about the SDE. This does not exclude the possibility of a quasi-steady-state characterized by the approximate equality of the doublet and singlet formation rates, because singlets are produced both by fragmentation and by coalescence. The condition of this quasi-steady-state is an analog of Eq. (21), following from Eq. (8):

$$\frac{dn_2}{dt} \ll n_2 \left(\frac{1}{\tau_d} + \frac{1}{\tau_c} \right) \tag{31}$$

It turns out that this main peculiarity of the coupling of aggregation, fragmentation, and coalescence is preserved at the transition from condition (15) to condition (27) because condition (1) is valid.

A characteristic of the steady state follows from Eq. (8) at condition (31):

$$n_2(t) = \tau_{Sm}^{-1} \left(\frac{1}{\tau_d} + \frac{1}{\tau_c} \right)^{-1} n_1^2(t) \quad (t > \tau_d + \tau_c) \tag{32}$$

Thus,

$$\frac{dn_2}{dt} = \tau_{Sm}^{-1} \left(\frac{1}{\tau_d} + \frac{1}{\tau_c} \right)^{-1} 2n_1 \quad \frac{dn_1}{dt} < 2\tau_{Sm}^{-1} \left(\frac{1}{\tau_d} + \frac{1}{\tau_c} \right)^{-1} \frac{n_2}{\tau_c} \ll n_2 \left(\frac{1}{\tau_d} + \frac{1}{\tau_c} \right) \tag{33}$$

taking into account Eq. (23) and conditions (1) and (27). The evaluation of Eq. (33) and its comparison with Eq. (8) justifies the quasisteady interconnection between the doublet and singlet concentrations (32). The substitution of the doublet concentration according to Eq. (32) into Eq. (23) and further integration leads to the generalization

$$n_1(t) = \frac{n_1(t_{max})}{1 + n_1(t_{max})t/2\tau_g} \cong \left(1 + \frac{t}{2\tau_g} \right)^{-1} \quad (t > \tau_d + \tau_c) \tag{34}$$

with a small deviation of $n_1(t_{max})$ from 1 and

$$\tau_g = \frac{\tau_{Sm}(\tau_d + \tau_c)}{2\tau_d} \tag{35}$$

At conditions (1) and (15), $\tau_g \sim \tau$ and Eq. (34) transforms into Eq. (26). At conditions (1) and (28), $\tau_g \sim \tau_{Sm}$ and Eq. (34) transforms into Eq. (30). Equation

(35) demonstrates the reduction of the role of fragmentation with decreasing τ_c. It is seen that at the transition from condition (27) to condition (28), τ_d cancels in Eq. (35) (i.e., the fragmentation role diminishes).

It is noteworthy that we succeeded in obtaining the approximate analytical solution for the system of Eqs. (8) and (9) even though Eq. (8) is nonlinear. This was possible because the condition of linearization of Eq. (8) and the condition for the quasi-steady-state (32) overlaps in the time interval (17).

E. Fragmentation of Primary Flocs in Emulsions and the Subsequent Reduction of Coalescence

For a comprehensive comparison, we need to use some results from the existing theory by Borwankar et al. [14]. In this theory, the difficult problem of the coupling of the coagulation and coalescence subprocesses is simplified by use of the Smoluchowski theory on Brownian coagulation. Smoluchowski has shown that at a first approximation, the complicated structure of aggregates can be neglected. This means that a coalescence process within a floc does not influence the coagulation kinetics. On the other hand, the coagulation influences the entire coalescence process. Correspondingly, in the coupling theory, the coagulation is described without accounting for coalescence. Afterward, it is sufficient to consider coalescence within aggregates which have a distribution obeying the Smoluchowski theory. Even with this simplification, large difficulties arise when the theory is extended to involve arbitrary aggregates, as shown in a recent theory [15]. This theory is reduced to the theory of Borwankar et al. [14] in the particular case of linear aggregates. Doublets are the simplest linearly built aggregates.

Two limiting conditions are defined in Ref. 14. First, the rate of coalescence is much higher than the rate of coagulation, that is,

$$\tau_{Sm} \gg \tau_c \tag{36}$$

and second, the rate of coagulation is much higher than that of coalescence, that is,

$$\tau_c \gg \tau_{Sm} \tag{37}$$

According to general rules of physicochemical kinetics, the slowest process is rate controlling. If the coagulation step is rate controlling, then the coalescence is rapid and the general equation of the theory in Ref. 14 is reduced to second-order kinetics; that is, to Smoluchowski's equation

$$n(t) = \left(1 + \frac{t}{\tau_{Sm}}\right)^{-1} \tag{38}$$

where $n(t)$ is the total number of particles (aggregates and primary droplets) at time t.

Flocs composed of three, four, and more droplets cannot be formed because of rapid coalescence within the floc. In this case, the structure of the flocs becomes irrelevant and the prediction of all theories ever for the difficult case of arbitrary flocs will match the numerical solution of Ref. 15. Thus, case (36) is the simplest one in the theory of coalescence kinetics. In this connection, let us note that the nontrivial case of slow coalescence is considered in Ref. 29.

The results of the theory in Ref. 14 concerning the case of slow coalescence are illustrated by curve 1 in Figure 3c in Ref. 15, which is redrawn as Figure 2a. It can be seen that for a low value of the coalescence rate constant, the semilogarithmic plot is linear, indicating that the process follows a coalescence-rate-controlled mechanism according to Eq. (14):

$$n(t) = \exp\left(-\frac{t}{\tau_c}\right) \tag{39}$$

A value of 1×10^{-11} cm^{-3}/s was used for the flocculation rate constant and the initial number of droplets was assumed to be 1×10^{10} cm^{-3}. This corresponds to $\tau_{Sm} = 10$ s, which is very small compared to $\tau_c = 10^3$ s. The slope of the line gives the K_c value. As τ_c decreases, the coalescence step becomes faster and because of the influence of the flocculation rate, the kinetic curve deviates from linearity. This is shown in Figures 2b ($\tau_c = 10^2$ s) and 2c ($\tau_c = 10$ s).

Let us pay attention to qualitative deviations in the reduction of the coalescence kinetics due to doublet fragmentation. Differing from a simple exponential time dependence (39), the second-order kinetics dominates at rapid doublet fragmentation even if the coalescence kinetics is very slow. The physical reason becomes clear when considering how Eq. (26) is derived. As seen from Eq. (25), the rate of decline of the droplet concentration is proportional to the doublet concentration. The latter is proportional to the square of the singlet concentration at singlet–doublet equilibrium, which causes the second-order kinetics. Thus, at slow coalescence, the disaggregation drastically changes the kinetic law of the coalescence (i.e., from the exponential law to second-order kinetics).

This statement has to be specified with respect to the two-stage kinetics in the coagulation. At the second stage, coagulation becomes the rate-controlling process because of the decrease in the collision rate accompanying the decrease in the droplet concentration. Thus, at sufficiently long times, second-order kinetics takes place for both reversible and irreversible aggregations. Nevertheless, a large difference exists even when identical functions describe the time dependence because the characteristic times are expressed through different equations for irreversible and reversible aggregations. In the first case, it is the Smoluchowski time; in the second case, it is the combination of three characteristic times [i.e., Eq. (18)].

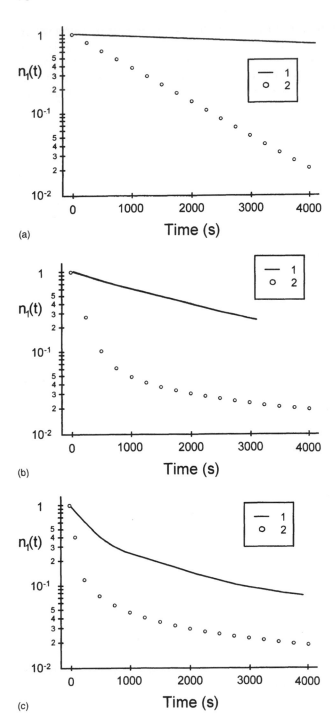

(a)

(b)

(c)

Let us now try to quantitatively characterize the reduction in coalescence caused by doublet disintegration. For this purpose, the calculations are performed according to Eq. (26) for conditions specified in Figures 2a–2c [i.e., at $\tau_{Sm} = 10$ s and $\tau_c = 10^3$ s (Fig. 2a), 10^2 s (Fig. 2b), and 10 s (Fig. 2c)]. For all these figures, the same value of the ratio $2\tau_d/\tau_{Sm} = 0.1$ is used, satisfying condition (2). In all these figures, the calculations according to Eq. (26) are illustrated by curve 1.

A comparison of curves 1 and 2 shows the reduction of coalescence caused by doublet disintegration. Even when not very small Rev values are used, the reduction is very strong. The lower the Rev values, the stronger the reduction. This is not shown in Figure 2 for small Rev values because in this case the small deviation of curve 1 from the horizontal level is not easily drawn. The simple curve 1 in Figure 2a can be used also for larger τ_c values because the condition of Eq. (39) is even better satisfied for larger τ_c values. Thus, if τ_{c1} and t_1 correspond to the data of Figure 2a and $\tau_{c2} = m\tau_{c1}$ with $m \gg 1$, the identity

$$\tau_{c2} t_2 \equiv \tau_{c1} m \left(\frac{t_1}{m} \right) \tag{40}$$

is useful. This means that

$$\frac{n_T}{n_0} \left(\tau_{c1} m, \frac{t_1}{m} \right) = \frac{n_T}{n_0} (\tau_{c1}, t_1) \tag{41}$$

[i.e., $t_2 = t_1/m$ where the right-hand side of Eq. (41) is drawn in Fig. 2a]. For example, Figure 3 is similar to Figure 2a and can be used 100 times longer, shown on the abscissa axis. The increase in τ_c enables us to increase τ_{Sm} without violating condition (37) and with Eq. (39) valid. Thus, $\tau_{Sm} = 1000$ s or smaller can be chosen as a condition for Figure 3. Curve 1, characterizing the rate of doublet disintegration, preserves as well if the value of $2\tau_d/\tau_{Sm} = 0.1$ preserves. Now, it corresponds to a larger τ_d value of 5 s.

Figure 2 Relative change in the total number of droplets versus time; initial number of droplets $N_{10} = 1 \times 10^{10}$ cm^{-3}; flocculation rate constant $K_f = 1 \times 10^{-11}$ cm^3 s^{-1}; curve 1: calculations according to Eq. (26); curve 2: the model of Borwankar et al. [14] for dilute emulsions, coalescence rate constants: (a) $K_c^{2,1} = 1 \times 10^{-1}$ s^{-1}, (b) $K_c^{2,1} = 1 \times 10^{-2}$ s^{-1}, (c) $K_c^{2,1} = 1 \times 10^{-3}$ s^{-1}. Coalescence time: (a) $\tau_c - 10^3$ s, (b) $\tau_c = 10^2$ s, (c) $\tau_c = 10$ s. Smoluchowski time: $\tau_{Sm} = 10$ s; doublet lifetime: $\tau_d = 0.5$ s. n is the dimensionless total droplet concentration, $n = N/N_{10}$.

Dukhin et al.

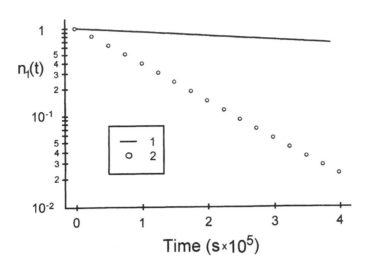

Figure 3 Similar to Figure 2, with other values for the characteristic times. Coalescence time: $\tau_c = 10^5$ s; Smoluchowski time: $\tau_{Sm} = 10^3$ s; doublet lifetime: $\tau_d = 50$ s. (O) data from Ref. 14; (——) data from Ref. 29.

F. Doublet Fragmentation Time

A doublet fragmentation was considered by Chandrasekhar [36] as the droplet diffusion from the potential pit, characterizing their attraction. The time scale for this process takes the form [37]

$$\tau_d = \frac{6\pi\eta a^3}{kT} \exp\left(-\frac{U_{min}}{kT}\right) \tag{42}$$

where U_{min} is the depth of the potential pit and a is the droplet radius.

    To derive the formula for the average lifetime of doublets, Muller [18,26] considers the equilibrium in a system of doublets and singlets (i.e., the numbers of doublets decomposing and appearing are equal). Both processes are described by the standard diffusion flux J of particles in the force field of the central particle.

    Each doublet is represented as an immovable particle with the second singlet "spread" around the central particle over a spherical layer, which corresponds to the region of the potential well. The diffusion flux J of "escaping" particles is described by the equation used in Fick's theory of slow coagulation. The first boundary condition corresponds to the assumption that the escaping particles do not interact with other singlets. The second condition reflects the fact that the potential well contains exactly one particle.

    At the small separation between the droplets in a doublet, the droplet diffusitivity reduces because of the increasing hydrodynamic resistance during droplet approach. A convenient interpolation formula was used [18] for the description of the influence of hydrodynamic interaction on the mutual diffusivity. The difference between the more exact Muller equation and Eq. (42) is caused mainly by taking into account hydrodynamic interaction. Retardation [38–40] and screening [41–43] of the molecular forces of attraction were treated [35] as a factor which tends to deteriorate secondary minimum aggregation in emulsions. Doublet fragmentation time is calculated in Refs. 35 and 44 as a function of droplet dimensions, electrolyte concentration, and the Stern potential with and without accounting for the retardation of molecular forces.

III. EXPERIMENTAL

A. Application of Video-Enhanced Microscopy Combined with the Microslide Technique for Investigation of Singlet–Doublet Equilibrium and Intradoublet Coalescence

Direct observation of doublets in the emulsion bulk is difficult because the doublets tend to move away from the focal plane. However, the microslide preparative technique can be successfully applied, providing pseudobulk conditions. A microslide is a plane-parallel glass capillary of rectangular cross section. The bottom and top sides of the capillary are horizontal, and the gravity-induced formation of a sediment or cream on one of the inner normal surfaces is rapidly completed due to the modest inner diameter of the slide. If both the volume fraction of droplets in an emulsion and the capillary height are small, the droplet coverage on the inside surface amounts to a few percent, and the analysis of results is rather simple. It can be seen through the microscope that the droplets which have sedimented onto the capillary surface participate in chaotic motion along the surface. This indicates that a thin layer of water separating the surface of the microslide from the droplets is preventing the main portion of droplets from adhering to the microslide surface, an action which would stop their Brownian motion.

During diffusion along the microslide ceiling, the droplets collide. Some collisions lead to the formation of doublets. Direct visual observation enables evaluation of the doublet fragmentation time which varies in a broad range [45]. Another approach to doublet fragmentation time determination is based on the evaluation of the average concentration of singlets and doublets and using the theory (Section II).

To obtain a representative number of droplets large enough for each sample to give a representative size distribution (at least 500 droplets), 10 images were made for each time [13,27]. The images were obtained arbitrarily over the whole length of the capillary and by the same procedure for all measurements.

The observations were restricted to droplets in the 2–6-μm size range. In this case, the droplets move preferentially owing to surface diffusion. The diffusion normal to the surface results in the orientation of the doublet axis being deflected from the horizontal by some small angle.

The application of the microslide preparative technique combined with video microscopy is promising and has enabled the measurement of the coupling of reversible flocculation and coalescence [13,27]. However, some experimental difficulties arose. Droplets could sometimes be seen sticking to the glass surface of the microslide. There were also examples of nonhorizontal slide surfaces, caused either by manufacture or positioning, causing droplets to roll and accumulate at one edge. These effects could result in droplet concentration inhomogeneities, prohibiting quantitative investigation. These effects can, however, be eliminated by the use of a low-density-contrast emulsion.

B. Improving the Experimental Technique with the Use of Low-Density-Contrast Emulsions

The lower the density contrast, the smaller is the gravitationally induced velocity of the droplets when moving along the near-to-horizontal glass wall. Sticking of droplets indicates a droplet–wall attraction and the existence of a secondary potential pit as for the droplet–droplet attraction in a doublet. The droplet concentration within the pit is proportional to the concentration on its boundary. The latter decreases with a decrease in the density contrast.

The electrostatic barrier between the potential pit and the wall retards the rate of sticking. The lower the droplet flux through this barrier, the lower is the potential pit occupancy by droplets. Thus, an essential decrease in the rate of sticking is possible with decreasing density contrast.

Oil-in-water emulsions were prepared [28] by mixing dichlorodecane (DCD, volume fraction 1%) into a 5×10^{-5} M sodium dodecyl sulfate (SDS) solution with a Silverson homogenizer. The oil phase was a 70:1 mixture of DCD, which is characterized by an extremely low-density contrast to water, and decane. This was done because the sedimentation equilibrium is obtained too slowly in the case of pure DCD droplets. On the other hand, using pure DCD can provide even better protection against the rolling/sticking of droplets and the resulting decrease in their surface diffusivity.

The droplet distribution along and across the slide was uniform [28]. This indicates that there was no gravity-induced rolling either. The microslides exhibit, to some extent, individual properties regarding the probability for droplets sticking. One slide among four was examined for 2 weeks without any sticking being observed [28]. The absence of the rolling and sticking phenomena allowed aquisition of rather accurate data concerning the time dependence of the droplet size distribution.

C. The Measurement of Coalescence Time and Doublet Fragmentation Time

The doublet fragmentation time was measured by direct real-time observation of the doublets on the screen and by analysis of series of images acquired with 1–3-min time intervals [45]. The formation and disruption/coalescence of a doublet can thus be determined.

The general form of the concentration dependence agrees with the theory. At $C \sim 3 \times 10^{-3} M$, both theory and experiment yield times of about 1 min; at $C = 9 \times 10^{-3} M$, these times exceed 10 min. For calculation of the doublet fragmentation time, the electrokinetic potential was measured [13,46].

In experiments with different droplet concentrations, it was established that the higher the initial droplet concentration, the higher the doublet concentration. This corresponds to the notion of singlet–doublet equilibrium. However, if the initial droplet concentration exceeds 200–300 per observed section of the microslide, multiplets predominate. Both the initial droplet concentration and size affect the rate of decrease in the droplet concentration. The larger the droplets, the smaller the concentration sufficient for the measurement of the rate of decrease in the droplet concentration. This agrees with the theory of doublet fragmentation time which increases with droplet dimension. Correspondingly, the probability for coalescence increases. These first series of experiments [13,27] were accomplished using toluene-in-water emulsions without the addition of a surfactant and decane-in-water emulsions stabilized by SDS. The obtained data concerning the influence of the electrolyte concentration and surface-charge density were in agreement with the existing notions about the mechanism of coalescence. With increasing SDS concentration, and correspondingly increasing surface potential, the rate of decrease in the droplet concentration diminishes.

Two methods were used for the measurement of the coalescence time [13,28]. Measurement of the time dependence for the concentrations of singlets and doublets and a comparison with Eq. (9) enables an evaluation of the coalescence time. Further, information about the time dependence for singlets and the doublet fragmentation time may be used as well. These results in combination with Eqs. (26) and (18) determine the coalescence time. The good agreement between results obtained by these very different methods indicates that the exactness of the theory and experiments is not low.

IV. APPLICATIONS

The restrictions in Eqs. (1) and (15) corresponding to strong retardation of the rate of multiplet formation and slow intradoublet coalescence are not frequently satisfied. Nevertheless, these conditions are important because they correspond to the case of very stable emulsions. As the kinetics of the retarded destabilization of rather stable emulsions is of interest, attention has to be paid to provide these conditions and, thus, the problem of coupled coalescence and flocculation arises.

There are large qualitative distinctions in the destabilization processes for the coupling of coalescence and coagulation, and coalescence and flocculation. In the first case, rapid aggregation causes rapid creaming and further coalescence within aggregates. In the second case, the creaming is hampered due to the low concentration of multiplets, and coalescence takes place both before and after creaming. Before creaming, singlets predominate for a rather long period of gradual growth of droplet dimensions due to coalescence within doublets. The discrimination of conditions for coupling of coalescence with either flocculation or coagulation is accomplished in Ref. 44.

Electrostatic repulsion decreases the depth of the potential pit and correspondingly decreases the doublet fragmentation time. As a result, flocculation becomes possible for submicrometer droplets as well as for micrometer-sized droplets, if the electrolyte concentration is not too high, the surface potential is rather high, and the droplet volume fraction is not too high.

Condition (1) can lose its physical sense due to creaming accompanied by a large increase in the droplet volume fraction, because the Smoluchowski theory assumes free Brownian diffusion. Meanwhile, there are different restrictions complicating Brownian diffusion of aggregates inside a cream and decreasing the exactness of the Smoluchowski theory in this case.

Thus, the theory of coupling of coalescence and flocculation needs further development regarding times exceeding the creaming time. However, similar restrictions exist in the theory of coupled coalescence and coagulation [14,15], which uses Smoluchowski theory as well. For the case of flocculation, this restriction is not as strict as for coagulation because of a much longer creaming time. Indeed, in the case of coagulation, the size of a creaming aggregate is larger because of the higher number of aggregated droplets. The creaming rate is, therefore, higher than in the case of flocculation, where the doublet fragmentation time is shorter and the resulting aggregate size is smaller. In addition, in the case of coagulation also, the droplet dimension is larger, which contributes to a higher creaming rate as well. This is seen from comparison of Eqs. (42) and (1).

Thus, the creaming time is much shorter in the coagulation case and, correspondingly, the equation describing the coupling of coalescence and flocculation preserves its physical sense for a longer time than is the case for coagulation. One concludes that the theory of the coupling of coalescence and flocculation provides a new opportunity for long-term prediction of emulsion stability, although creaming restricts the application of this theory as well. Note that this restriction weakens in the emulsions of low-density contrast and in water-in-oil emulsions with a high-viscosity continuum.

The long-term prediction is a two-step procedure. The first step is the determination of whether an emulsion exhibits coagulation or flocculation. It means that the characteristic time τ_d must be measured and compared with τ_{Sm}, the value of which is easily evaluated taking into account the measured concentration using Eq. (4). A comparison of these times enables the choice between condition (1) and the opposite condition ($\tau_{Sm} \ll \tau_d$). The second step is the prediction of the evolution in time for the total droplet concentration (TDC). If condition (1) is valid, Eq. (26) has to be used for the prediction. In the opposite case, Eq. (39) must be used. τ in Eq. (26) has to be specified in accordance with Eq. (18).

A. Long-Term Prediction of Emulsion Stability

It is possible, in principle, to give a long-term prediction of emulsion stability based on the first indications of aggregation and coalescence. The next example clarifies the principal difficulty in a reliable long-term prediction if a dynamic model of the emulsion is not available.

The first signs of aggregation and coalescence can always be characterized by a linear dependence, if the investigation time t is small in comparison with a characteristic time τ for the evolution of the total droplet concentration $n(t)$,

$$n(t) = n_0 \left(1 - \frac{t}{\tau} \right) \tag{43}$$

These short-time asymptotics correspond to many functions; for example, to

$$n(t) = n_0 e^{-t/\tau} \tag{44}$$

$$n(t) = \frac{n_0}{1 + t/\tau} \tag{45}$$

The first can arise in the case of coalescence coupled with coagulation [14]; the second can arise for coalescence coupled with flocculation [29]. The discrimination between irreversible and reversible aggregations is only one component of emulsion dynamics modeling (EDM) and it is seen that without this discrimination, the difference in the prediction of the time necessary for a droplet concentration decrease; for example, 1000 times can be 7τ and 1000τ.

B. Perfection of Methods of Emulsion Stabilization (Destabilization) by Means of the Effect on Both Coalescence and Flocculation

Stability (instability) of an emulsion is caused by the coupling of coalescence and flocculation. Meanwhile, for emulsifiers (or demulsifiers), the elaboration of their influence on the elementary act of coalescence only is primarily taken into account.

The coupling of coalescence and flocculation is reflected in Eq. (18) and one concludes that it follows the multiplicativity rule and not the additivity rule. This means that the total result of the application of a stabilizer (destabilizer) depends very much on both flocculation and fragmentation.

The development of a more efficient technology for emulsion stabilization (destabilization) is possible by taking into account the joint effect on both the coalescence and the aggregation (disaggregation) processes.

The coalescence rate depends mainly on the thin (black) film stability and correspondingly on the short-range forces. The flocculation depends on the long-range surface forces. Due to this large difference, synergism in the dependence of these processes on the different factors can be absent.

The use of one surfactant only may not provide both the optimal fragmentation and optimal stability of an emulsion film. Probably the use of a binary surfactant mixture with one component which provides the film stability and a second one which prevents the flocculation may provide perfect emulsion stabilization. Naturally, their coadsorption is necessary. For such an investigation, a measurement method for both the doublet fragmentation time and the coalescence time is necessary.

C. Standardization of the Measurement of τ_c and τ_d

Direct investigation of the coalescence subprocess in emulsions is difficult. Instead, the entire destabilization process is usually investigated. Meanwhile, the rate of the destabilization process depends on the rates of both flocculation and disaggregation and on the floc structure as well. All these characteristics vary in a broad range. At a given unknown value for the time of the elementary act of coalescence τ_c, the different times can be measured for the integrated process and different evaluations of τ_c are possible.

Among the existing approaches, the one based on the theory of coupling of coagulation and coalescence is most correct [47]. The theory in Ref. 14 yields a time dependence of the decrease in the TDC due to coalescence within flocs by applying Smoluchowski's theory to the coagulation process. The entire destabilization process is sensitive to the coalescence time if it is longer than the Smoluchowski time. Slow coalescence occurs within rather large aggregates of different structures and dimensions if the aggregation is irreversible. The rate of coalescence in an aggregate essentially depends on the number of droplets within it and the packing type (i.e., on the number of films between the droplets). This complication is absent when considering the case of the SDE. Two important assumptions used in the model [14] are that the rate constants for both the coagulation and the coalescence are constant throughout the process. The second assumption is clearly an approximation because the lifetimes of thin films, which determine the coalescence rate constant, are known to be dependent on the radius of the droplets [48–51]. The validity and usefulness of the model clearly depend on how reasonable these approximations are.

Meanwhile, the dependence of the coalescence time (τ_c) on droplet dimension causes the main difficulty in the theory of coupled aggregation and coalescence. The possible advantage of τ_c measurement at SDE is in avoiding the difficulty caused by polydispersity of droplets appearing during preceding coalescence within large flocs. At the SDE, the initial stage of the entire coalescence process can be investigated when the narrow size distribution of an emulsion preserves.

The second advantage arises from the use of a low-density-contrast emulsion. The smallest droplets are distributed within a rather thick emulsion layer. As a result, their concentration near the microslide surface is low and their participation in the SDE can be neglected. This omittance provides a narrow distribution of droplet sizes participating in coalescence.

At SDE, the determination of the time dependence of the TDC is sufficient for the investigation of coalescence. In Refs. 27 and 28, this was accomplished through direct visual observation. By using video-enhanced microscopy and computerized image analysis the determination of the TDC can be automated. Such automated determination of the total droplet number in a dilute DCD-in-water emulsion at the SDE can be recommended as a standard method for the characterization of the elementary act of coalescence.

In parallel, the second important characteristic, namely the doublet fragmentation time, is determined by the substitution of τ_c, τ_{Sm} and measured τ into Eq. (18).

D. Experimental–Theoretical Emulsion Dynamics Modeling

Predicting the evolution of the droplet (floc) size distribution is the central problem in emulsion stability. It is possible in principle to predict the time dependence of the distribution of droplets (flocs) if information concerning the main subprocesses (flocculation, floc fragmentation, coalescence, creaming), constituting the whole phenomenon, is available. This prediction is based on consideration of the population balance equation (PBE).

The PBE concept was proposed by Smoluchowski. He specified this concept for suspensions and did not take into account the possibility of floc fragmentation. Even with this restriction, he succeeded in the analytical solution neglecting gravitational coagulation and creaming, and obtained the analytical time dependence for a number of aggregates n_i comprising i particles ($i = 2, 3 \ldots$).

The time dependence of the TDC in a dispersion, n_T, is of major importance when emulsions are under consideration. n_T comprises the number of individual droplets moving freely plus the number of droplets included in all kinds of aggregates:

$$n_T = \sum_{i=1}^{\infty} i n_i \qquad (46)$$

In the most general case, the equation for the evolution of the TDC takes into account the roles of aggregation, fragmentation, creaming, and coalescence. There is no attempt to propose an algorithm even for a numerical solution to such a problem.

The usual approach in the modeling of an extremely complicated process is the consideration of some extreme cases with further synthesis of the obtained results. The next three main simplifications are inherent to the current state of emulsion dynamics modeling: the neglect of the influence of the gravitational field (i.e., neglect of creaming/sedimentation); in first approximation, it is possible to consider either coagulation or flocculation, and the neglect of the rate constant dependence on droplet dimension.

1. Combined Approach in Investigations of Dilute and Concentrated Emulsions

The modeling of collective processes in concentrated emulsions is extremely complicated. Recently, the efficiency of computer simulation in the systematic study of aggregates, gels, and creams has been demonstrated [10]. Monte Carlo and Brownian dynamics are particularly suited to the simulation of concentrated emulsions. However, information about droplet pairwise interaction is necessary. The reliability of this information is very important in providing reasonable results concerning concentrated emulsions. In other words, the assumption concerning pairwise additive potentials for droplet–droplet interaction and the thin emulsion film stability must be experimentally confirmed. The extraction of this information from experiments with concentrated emulsions is very difficult. On the other hand, measurement of the doublet fragmentation time in dilute emulsions is a convenient method to obtain information about pairwise additive potentials.

Information about pairwise potentials and the elementary act of coalescence obtained in experiments with dilute emulsions preserves its significance for concentrated emulsions as well. One concludes that modeling of concentrated emulsions becomes possible by combining experimental investigation of the simplest emulsion model system with computer simulation accounting for the characteristics of a concentrated emulsion (high droplet volume fraction, etc.).

2. Kernel Determination Is the Main Task Which Must Be Solved to Transform the PBE in an Efficient Method for Emulsion Dynamics Modeling

The levels of knowledge concerning kernels describing different subprocesses differ strongly. There exists a possibility for quantification of kernels related to aggregation and fragmentation. On the other hand, the current state of knowledge is not sufficient for prediction of the thin-film disruption time.

The deficit in knowledge about thin-film stability makes pure theoretical modeling of emulsion dynamics impossible. As a result, a complex semitheoretical approach to EDM is necessary. The PBE equation is the main component of both the experimental and the theoretical stages of this approach. In the experimental stage, PBE provides the background for the determination of the coalescence kernels with the use of the experimental data.

For the determination of the coalescence kernels, the more complicated reverse task must be solved, namely their determination based on the comparison of experimental data about the emulsion evolution in time with the PBE solution. In the absence of an analytical solution, the reverse task is usually very difficult. The most efficient way to overcome this difficulty is the experimental realization with the use of the universally simplest conditions for emulsion time evolution, which can be described analytically.

3. Singlet–Doublet Quasiequilibrium with Slow Coalescence Within Doublets Is the Simplest Emulsion State for Which Investigation Can Provide Information About Coalescence

In distinction from the coupling of the two processes of coagulation and coalescence, the coupling of flocculation and coalescence is the combined manifestation of three processes. This, however, does not lead to additional difficulties in kinetic theory. On the contrary, floc fragmentation restricts the development of larger flocs. As a result, the simplest singlet–doublet emulsion can exist at singlet–doublet quasiequilibrium and slow coalescence within doublets. Its simplicity results in a very simple kinetic law for the entire kinetics of coupled flocculation and coalescence. Thus, singlet–doublet equilibrium provides the most convenient conditions for investigations of the elementary act of coalescence and the doublet fragmentation time.

The main condition for this important possibility for the determination of both emulsion thin-film lifetime and doublet fragmentation time is the elimination of the complications caused by gravity. The reliability of derived equations was justified [28] in experiments with emulsions of low-density contrast.

The main simplification in all existing models for emulsion dynamics is the neglect of the coalescence time dependence on droplet dimensions. This simplification is not justified and decreases the value of the prediction substantially, which can now be made with use of the PBE. For elimination of this unjustified simplification, it is necessary to determine the coalescence time for emulsion films between droplets of different dimensions i and j, namely τ_{cij}, similar to the existing analytical expressions for the doublet fragmentation time, π_{dij} [12]. The determination of a large set of τ_{cij} values by means of a comparison of experimental data obtained for an emulsion consisting of different multiplets and the PBE numerical solutions for it is impossible. On the contrary, this paramount experimental–theoretical task can be solved for a dilute emulsion at SDE and slow intradoublet coalescence.

4. Substitution of the Coalescence Kernels Makes the PBE Equation Definite and Ready for the Prediction of Emulsion Time Evolution with the Restriction of Low-Density Contrast and Without Account for Gravitational Coagulation and Creaming

With application of the scaling procedure for the representation of the kinetic rate constants for creaming and gravitational coagulation, the PBE is solved analytically in Ref. 52. This scaling theory creates a perspective for the incorporation of creaming in the emulsion dynamics model in parallel with coalescence, aggregation, and fragmentation.

V. DISCUSSION AND SUMMARY

In recent years, several research groups have improved the theoretical understanding of coalescence of droplets or bubbles significantly. The new results [53–57] together with results of earlier investigations [58–62] have clarified the role of double-layer interaction in the elementary act of coalescence.

The DLVO theory was applied [63,64] for the description of "spontaneous" and "forced" thinning of the liquid film separating the droplets. These experimental results and the DLVO theory were used [63] for the interpretation of the reported visual study of coalescence of macroscopic oil drops (70–140 μm in diameter) in water in a wide pH interval. In a comparison based on DLVO theory and these experimental data, the authors [63] concluded that "if the total interaction energy, V_T, is close to zero or has a positive slope in the critical thickness range, i.e., between 30 and 50 nm, the oil drops should be expected to coalesce." In the second paper [64], in which both ionic strength and pH effects were studied, coalescence was observed at constant pH values of 5.7 and 10.9 when the Debye thickness was less than 5 nm.

In Refs. 63 and 64, the surface potential was dependent on pH and no surfactant was used. In our experiments, the surface potential was dependent on the concentration of SDS. Nevertheless, the main trends in our experiments and in Refs. 63 and 64 are in accordance. It was difficult to establish the decrease in TND at NaCl concentrations lower than $5 \times 10^{-3}\,M$ (i.e., double-layer thickness larger than 5 nm).

An almost quantitative coincidence in the double-layer influence on coalescence established in this work for micrometer droplets and in Refs. 63 and 64 for droplets almost 100 times larger is important for the general knowledge about coalescence.

In contrast to the large success in industrial applications of emulsion surface chemistry, the potential of the physicochemical kinetics is almost not used in emulsion technology. This potential strength is the population balance equation (PBE) which enables the prediction of the evolution in time for the droplet size distribution (DSD) if the family of subprocesses including droplet aggregation, aggregate fragmentation, droplet coalescence, and droplet (floc) creaming are quantified. These subprocesses are characterized in PBE by means of the kinetic coefficients. The coupling of these four subprocesses, the droplet polydispersity, and immense variety in the droplet aggregate configurations cause extreme difficulty in EDM. Three subprocesses, namely aggregation, fragmentation, and creaming, can be quantified. In contrast, the experimental approach is only effective now for the accumulation of information concerning emulsion film stability and coalescence kernel quantification for EDM.

Correspondingly, EDM may be accomplished by combining experiment and theory: (1) The determination of coalescence and fragmentation kernels with the use of emulsion stability experiments at low-density contrast (LDC) and singlet–doublet equilibrium (SDE) because this enables the neglect of creaming and gravitational terms in PBE which simplifies it and makes the solution of the reverse task possible. (2) The prediction of the droplet size evolution with time by means of solution of PBE specified for the determined coalescence and fragmentation kernels. This mathematical model has to be based on PBE supplemented by terms accounting for the role of creaming and gravitational coagulation in the aggregation kinetics.

Emulsion dynamics modeling with experiments using LDC emulsions and SDE may result in the following: (1) The quantification of emulsion film stability, namely the establishment of the coalescence time dependence on the physicochemical specificity of the adsorption layer of a surfactant (polymer), its structure, and the droplet dimensions. This quantification can form a basis for the optimization of demulsifier and demulsifier selection and synthesis for emulsion technology applications instead of the current empirical level applied in this area. (2) The elaboration of a commercial device for coalescence time measurement, which in combination with EDM will represent a useful approach to the optimization of the emulsion technology with respect to stabilization and destabilization.

ACKNOWLEDGMENTS

This research has been sponsored by the Norwegian Research Council (NFR) by means of travel and subsistence grants. Also, the technology programme FLUCHA financed by NFR and the oil industry is acknowledged for partly financing this research.

REFERENCES

1. P Sherman, ed. Emulsion Science. New York: Academic Press, 1968.
2. P Becher, ed. Encyclopedia of Emulsion Technology, Vol. 1. New York: Marcel Dekker, 1983.
3. P Becher, ed. Encyclopedia of Emulsion Technology, Vol. 2. New York: Marcel Dekker, 1985.
4. TF Tadros, B Vincent. In: P Becher, ed. Encyclopedia of Emulsion Technology. New York: Marcel Dekker, 1983.
5. J Sjöblom, ed. Emulsions and Emulsion Stability. New York: Marcel Dekker, 1996.
6. SE Friberg. In: J Sjöblom, ed. Emulsions—A Fundamental and Practical Approach. Dordrecht: Kluwer Academic Publishers, 1992, p. 1.
7. IB Ivanov, PA Kralchevsky. Colloids Surf 128:155, 1997.
8. SS Dukhin, J Sjöblom. In: J Sjöblom, ed. Emulsions and Emulsion Stability. New York: Marcel Dekker, 1996, p. 41.
9. I Gregory. Crit Rev Environ Control 19:185 1989.
10. E Dickinson, SR Euston. Adv Colloid Interf Sci 42:89, 1992.
11. P Taylor. Adv Colloid Interf Sci 75:107, 1998.
12. HW Yarranton, YH Masliyah. J Colloid Interf Sci 196:157, 1997.
13. Ø Holt, Ø Sæther, J Sjöblom, SS Dukhin, NA Mishchuk. Colloids Surf 141:269, 1998.
14. RP Borwankar, LA Lobo, DT Wasan. Colloids Surf 69:135, 1992.
15. KD Danov, IB Ivanov, TD Gurkov, RP Borwankar. J Colloid Interf Sci 167:8, 1994.
16. GA Martynov, VM Muller. In: BV Derjaguin, ed. Surface Forces in Thin Films and Disperse Systems. Moscow: Nauka, 1972, p. 7.
17. GA Martynov, VM Muller. Kolloidn Zh 36:687, 1974.
18. VM Muller. Kolloidn Zh 40:885, 1978.
19. BV Derjaguin. Theory of Stability of Colloids and Thin Films. New York: Plenum Press, 1989.
20. JA Long, DW Osmond, B Vincent. J Colloid Interf Sci 42:545, 1973.
21. RM Cornell, JW Goodwin, RH Ottewill. J Colloid Interf Sci 71:254, 1979.
22. DYC Chan, B Halle. J Colloid Interf Sci 102:400, 1984.

23. GC Jeffrey, RH Ottewill. Colloid Polym Sci 268:179, 1988.
24. H Matsumura, K Watanabe, K Furusawa. Colloids Surf 98:175, 1995.
25. Ø Holt, Ø Sæther, J Sjöblom, SS Dukhin, NA Mishchuk. Colloids Surf 123–124: 195, 1997.
26. VM Muller. Colloid J 58:598, 1996.
27. Ø Sæther, J Sjöblom, SV Verbich, NA Mishchuk, SS Dukhin. Colloids Surf 142: 189, 1998.
28. Ø Sæther, J Sjöblom, SV Verbich, SS Dukhin. J Dispers Sci Technol 20:295, 1999.
29. SS Dukhin, J Sjöblom. J Dispers Sci Technol 19:311, 1998.
30. IM Elminyawi, S Gangopadhyay, CM Sorensen. J Colloid Interf Sci 144:315, 1991.
31. BJ McCoy, G Madras. J Colloid Interf Sci 201:200, 1998.
32. J Widmaier, E Pefferkorn. J Colloid Interf Sci 203:402, 1998.
33. M Smoluchowski. Phys Z 17:557, 1916.
34. M Van den Tempel. Rec Trav Chim 72:419, 1953; 72:433, 1953.
35. NA Mishchuk, J Sjöblom, SS Dukhin. Colloid J 57:785, 1995.
36. S Chandrasekhar. Rev Mod Phys 15:1, 1943.
37. WB Russel, DA Saville, WR Schowalter. Colloidal Dispersions. New York: Cambridge University Press, 1989.
38. JN Israelachvili. Intermolecular and Surface Forces. London: Academic Press, 1991.
39. J Lyklema. Fundamentals of Interface and Colloid Science, Vol. 1. London: Academic Press, 1993.
40. Ya I Rabinovich, NV Churaev. Kolloidn Zh 46:69, 1984; 52:309, 1990.
41. DJ Mitchell, P Richmond. J Colloid Interf Sci 46:128, 1974.
42. VN Gorelkin, VP Smilga. Kolloidn Zh 34:685, 1972.
43. VN Gorelkin, VP Smilga. In: BV Derjaguin, ed. Poverkhnostnye Sily v Tonkikh Plenkakh i Ustoichvost Kolloidov. Moscow: Nauka, 1974, p. 206.
44. J Sjöblom, DT Wasan, SS Dukhin. J Colloid Interf Sci (in press).
45. Ø Sæther, SS Dukhin, J Sjöblom, Ø Holt. Colloid J 57:836, 1995.
46. SV Verbich, SS Dukhin, A Tarovsky, Ø Holt, Ø Sæther, J Sjöblom. Colloids Surf 23:209, 1997.
47. LA Lobo, DT Wasan, M Ivanova. In: KL Mittal, DO Shah, eds. Surfactants in Solution. New York: Plenum Press, 1992, Vol. 11, p. 395.
48. DS Dimitrov, IB Ivanov. J Colloid Interf Sci 64:97, 1978.
49. AK Malhotra, DT Wasan. Chem Eng Commun 55:95, 1987.
50. OD Velev, TD Grukov, SvK Chakarova, BI Dimitrova, IB Ivanov, RP Borwankar. Colloids Surf 83:43, 1994.
51. OD Velev, GN Constantinides, DG Avraam, AC Payatakes, RP Borwankar. J Colloid Interf Sci 175:68, 1995.
52. SB Grant, C Poor, S Relle. Colloids Surf 107:155, 1996.
53. JD Chan, PS Hahn, JC Slattery. AIChE J 30:622, 1984.
54. JD Chan. J Colloid Interf Sci 107:209, 1985.
55. JD Chan, PS Hahn, JC Slattery. AIChE J 34:140, 1988.
56. SAK Jeelani, S Hartland. J Colloid Interf Sci 156:467, 1993.
57. RW Aul, WI Olbriht. J Colloid Interf Sci 115:478, 1991.
58. A Scheludko, D Exerova. Colloid J 168:24, 1960.
59. A Scheludko. Proc Konikal Ned Akad Wet Ser B 65:87, 1962.

60. D Platkanov, E Manev. In: Proceedings, 4th International Congress of Surface Active Substances. New York: Plenum Press, 1964, p. 1189.
61. KA Burrill, DR Woods. J Colloid Interf Sci 42:15, 1973.
62. KA Burrill, DR Woods. J Colloid Interf Sci 42:35, 1973.
63. SR Deshiikan, KD Papadopoulos. J Colloid Interf Sci 174:302, 1995.
64. SR Deshiikan, KD Papadopoulos. J Colloid Interf Sci 174:313, 1995.

4

Fat-Particle Structure Formation and Stability of Emulsions

K. Kumar, A. D. Nikolov, and Darsh T. Wasan
Illinois Institute of Technology, Chicago, Illinois

T. Tagawa
Mitsubishi Chemical Corporation, Yokohama, Japan

ABSTRACT

A digitized imaging technique is used to study the fat-particle structure formation in an aerated oil-in-water food emulsion. Using the imaging technique, we demonstrated the effect of surfactant–protein submicelles on the fat-particle structure formation and its impact on the stability of the emulsions. The fat particle–particle interaction results have been correlated with the creaming separation data. Experimental evidence is presented for layering of the surfactant–protein submicelles in the thin aqueous film between fat particles leading to fat-particle structure formation. The layering of surfactant–protein submicelles has been experimentally observed using the capillary force balance technique. We also measured the effect of shear, as encountered during whipping process, on the fat-particle packing structure in food emulsions. The foam lamella stability of the aerated food emulsion is also probed by studying the curved foam film rheology with the film rheometer.

I. INTRODUCTION

The development of a stable structure and a suitable product texture in whipped cream depends on interactions between fat particles and between fat particles

87

and air bubbles [1]. These interactions may control the rheological properties and the appearance of the product, as well as its physical instability, as reflected in changes of consistency and loss of homogeneity [2].

The fat particle–particle interactions and the stability of food emulsions is affected by many factors, such as surfactant concentration, protein concentration, surfactant–protein interactions, and surfactant composition. Koczo et al. [3] found a new stability mechanism for food emulsions due to the formation of microstructure and layering of sodium caseinate submicelles in the thin liquid films in the presence of fat particles. It was found that the existence of such submicelle layers in the thin liquid films led to the stabilization of the food emulsion.

Brooker et al. [4–6] used transmission and scanning electron microscopies to study the fat-particle packing structure inside whipped cream. They have reported qualitative information about the bulk structure of fat particles in their articles. So far, much information about the fat-particle packing structure has been mostly qualitative in nature due to the lack of measurement techniques. This prevents further understanding the role of fat particle–particle interactions and structure formation in stabilizing food emulsions. Xu et al. [7] used the Kossel diffraction technique to characterize the structure formation in concentrated food emulsions.

We have used a nondestructive technique of determining the fat-particle packing structure in food emulsions. The fat-particle packing structure is an important parameter in estimating the stability of a dispersed-phase system. The fat-particle packing structure is governed by the surfactant–protein submicellar interactions and can be important for the long-term stability of the food emulsion. In order to understand the role of the foam film dynamic properties, like film elasticity, in the stability of an aerated product, we studied the foam film rheology using the recently developed film rheometer [8]. The effect of surfactant and protein on the fat-particle structure formation and film rheology has been investigated. These results have been correlated with the creaming separation data. Overall, the objective of this research is to understand the role of fat particle–particle interactions and structure formation in the stability of the emulsion.

II. EXPERIMENTAL DETAILS

A. Materials

All food emulsion samples used in this research were prepared using an H-5000 homogenizer (manufactured by Microfluidics Intl. Corp.). The emulsion samples were prepared using the following ingredients: water (deionized water), fat (partially hydrogenated soybean oil), protein (skim milk powder), and surfactant (water-soluble sucrose esters). Sucrose esters are tasteless, odorless, nontoxic

Table 1 Ingredients of the Food Emulsion

Fat (partially hydrogenated soybean oil)
Lecithin
Water-soluble surfactants (sucrose stearate, sucrose
  oleate)
Proteins (skim milk powder)
Water (deionized water)

surfactants. Besides, they are easily digestible and are biodegradable. Thus, they are very suitable emulsifiers for food emulsions.

The composition of the prepared food emulsions is presented in Table 1. The sample preparation procedure was as follows. In gradients such as water and sucrose esters (surfactant) were stirred at 60°C to dissolve the surfactant. After the surfactant was dissolved, proteins were added to the surfactant solution and the solution was left stirring at 60°C for 24 h. The oil phase was prepared by mixing with lecithin at 70°C. After heating, the two phases were blended for 3 min in a kitchen blender to prepare a coarse emulsion. This coarse emulsion was then taken to a homogenizer. In the homogenizer, the coarse emulsion was homogenized at a pressure of 200 atm. After homogenization, the emulsion was cooled down to 3°C, and the final emulsion samples were produced. A schematic of the preparation of emulsion is shown in Figure 1. The fat-particle structure

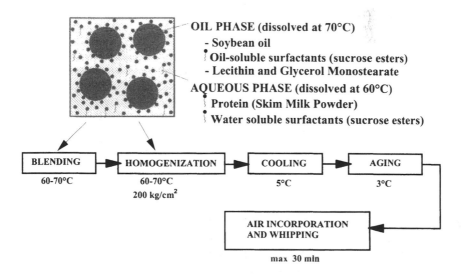

Figure 1 Procedure for preparation of the food emulsion.

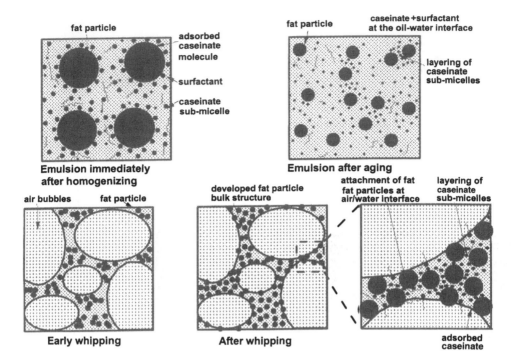

Figure 2   Fat-particle structure variation during a typical food emulsion and foam process.

variation during various processes, in the presence of surfactant and proteins, is shown in Figure 2. After preparing and cooling, a food emulsion sample was left to cream under gravity for several days. The creaming experiments were carried out at 3°C. The fat droplets, being lighter than the continuous phase (aqueous phase), begin to cream out as time progresses. This separation was recorded as a function of time.

B.   Digitized Imaging Technique

The digitized imaging technique (Fig. 3) was used to study the microstructure of the food emulsion. The cooled sample was taken under a microscope to record a microstructural image. This image was recorded using a video attached to the microscope. Using an imaging software (Image Pro), the microstructural image was magnified and then the analysis was done to measure the interparticle distance. This acquired data was processed in MATLAB to calculate radial distribu-

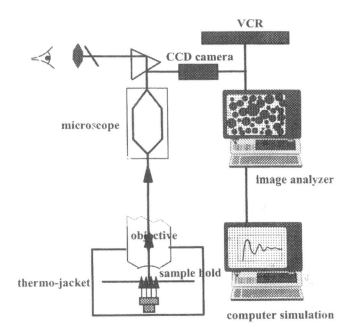

Figure 3 Digitized imaging technique.

tion function (RDF) and structure factor. The radial distribution function $g(r)$ measures the probability to find a fat-particle center at a distance r from a reference fat particle. It is oscillatory in nature and tends to unity as the distance from the reference fat particle tends to infinity, implying that the probability of finding a fat particle at infinity is the same as that in bulk. It typically has a maximum at a distance about one particle diameter for a monodisperse system.

$S(Q)$ is the static structure factor which describes the degree of particle packing structure inside a colloidal dispersion (IDC emulsion), and Q is the scattering vector that is defined by

$$Q = \frac{4\pi}{\lambda} \sin\left(\frac{\theta}{2}\right) \tag{1}$$

where θ is the scattering angle.

$$S(Q) = 1 + 4\pi\rho \int_0^\infty r^2[g(r) - 1]\left(\frac{\sin(Kr)}{Kr}\right) dr \tag{2}$$

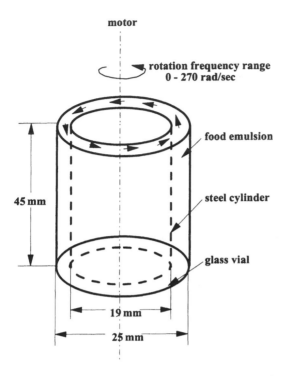

Figure 4 Experimental setup to study the effect of shear on food emulsion.

The structure factor is oscillatory in nature and tends to unity as the scattering vector increases. The structure factor is calculated using Eq. (2).

C. Fat-Particle Structure Under Shear

The effect of shear on the fat-particle packing structure as encountered during the whipping process was studied for the two surfactant systems. In this experiment, a concentric cylinder cell was used (Fig. 4) in which the inner cylinder rotated to produce a uniform steady-state shear rate on the food emulsion sample. The shear rate could be changed by controlling the rotation speed of the inner cylinder, and the sample temperature was kept at 25°C during the shear experiment.

D. Capillary Force Balance

A sketch of the experimental setup to study thin liquid films is shown in Figure 5. First, a biconcave drop was formed from the aqueous solution in a vertically

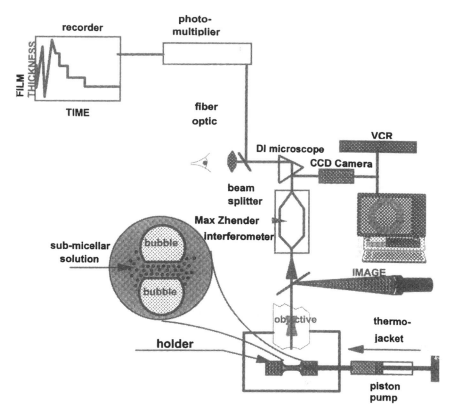

Figure 5   Capillary force balance.

oriented, cylindrical glass tube with hydrophilic inner walls inside a temperature-controlled glass cell. Then, liquid was slowly sucked out of the drop through a capillary orifice in the tube wall to create a horizontal flat film encircled by a biconcave liquid meniscus. The film diameter could be controlled in the 0.05–1.0-mm range. The film was illuminated by monochromatic light (546 nm wavelength) and observed by using a reflected-light microscope. The light rays which were reflected from the two film interfaces interfere and produce an interference pattern; the local film thickness can be estimated from the color intensity of the interference patterns. Further details of the instrument are given elsewhere [9].

E.   Curved Film Rheology

In the aerated product, when two air bubbles come close to each other, there is a thin aqueous film separating them, preventing them from coalescing. To study

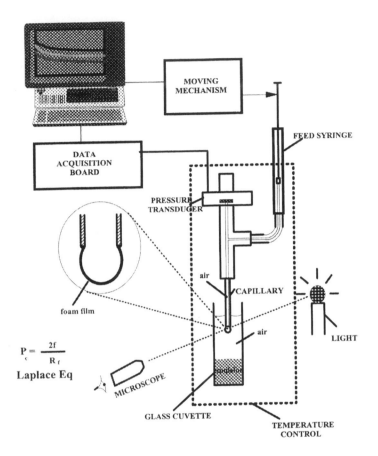

Figure 6 Film rheometer.

the film's rheological properties, we have simulated the process of film stress relaxation by forming a thin aqueous film on the tip of a glass capillary. The film rheometer (Fig. 6) was used to conduct the stress relaxation experiment. When the aqueous foam film is suddenly expanded, the concentration of the surfactant at the film interface decreases and this causes a jump in the film tension, which causes the capillary pressure to increase suddenly. A pressure transducer is used to record capillary pressure and the film radius is recorded using a microscope. After the expansion, surfactant molecules slowly diffuse to the interface and lower the film tension, which causes a relaxation in the capillary pressure. This experiment is used to calculate the Gibbs film elasticity which is defined as change of film tension against the increase of surface concentration (Γ). The

initial film tension after fast expansion in the film stress relaxation can be used
to measure the Gibbs film elasticity E_f [10]:

$$E_f = \frac{df}{d[\ln (A/A_0)]} \tag{3}$$

where f is the film tension and A/A_0 is the degree of expansion.

III. RESULTS AND DISCUSSION

A. Fat-Particle Structure Formation

Sucrose stearate hydrophilic lipophilic balance (HLB = 16) and sucrose oleate
(HLB = 15) were used to prepare two emulsion samples. The emulsion samples
had the same composition, except one had sucrose ester (0.1 wt%) as the water-
soluble surfactant and the other has sucrose oleate (0.1 wt%). The fat content
was 40 wt%, the protein content was 4 wt%, and the water content was 56 wt%.
These emulsion samples were kept for creaming at 3°C. The creaming experiment
(Fig. 7) shows that the sucrose oleate sample is more stable and has no creaming
separation even after a few days. In order to understand the role of fat particle–
particle interaction and fat-particle structure formation in emulsion stability, the

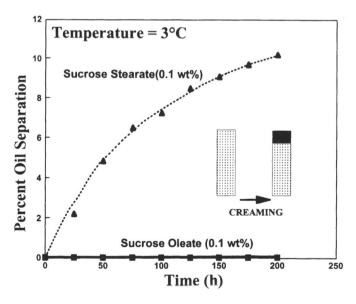

Figure 7 Effect of surfactant on oil-in-water emulsion stability.

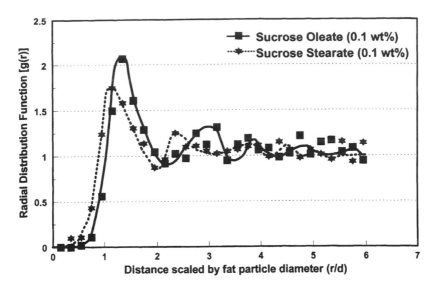

Figure 8 Effect of surfactant on radial distribution function of fat particles in oil-in-water emulsion.

digitized imaging technique was used to calculate the radial distribution function and structure factor. The above-mentioned emulsion samples were observed under a microscope to obtain the interparticle distances and particle size distribution. The radial distribution functions $g(r)$ for these samples are shown in Figure 8. The radial distribution function shows that the corresponding effective pair potential between fat particles is also oscillatory and depends on the polydispersity of the system and submicellar concentration. The periodicity of the curve is close to the diameter of the particles. The structure factor $S(Q)$ for these samples are calculated (Fig. 9). The first peak height of the structure factor $S(Q)$ of the sucrose oleate sample is higher, indicating that the addition of sucrose oleate facilitates the fat-particle structure formation. Thus, the fat-particle structure in the sucrose oleate sample is much better organized than the particle structure in sucrose stearate sample. This leads to higher fat-particle flocculation in sucrose stearate sample and the emulsion is less stable. This could be explained by the effect of polydispersity on the dispersion stability. The particle size distributions for these two samples are shown in Figure 10. The sucrose stearate sample is more polydisperse and has a larger mean size. The higher polydispersity in the sample leads to a less organized structure and destabilizes the emulsion [11] as fat particles flocculate to form a larger floc, which creams faster than a single fat particle.

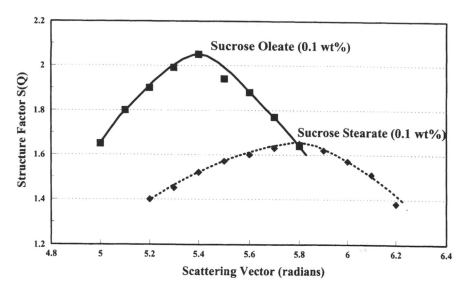

Figure 9   Effect of surfactant on structure factor of fat particles in oil-in-water emulsion.

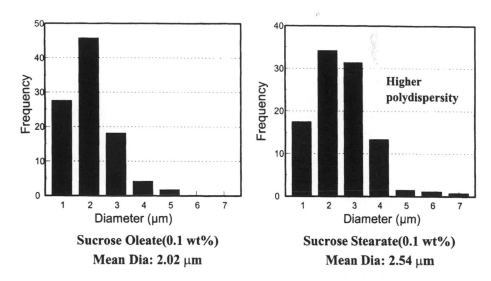

Figure 10   Effect of surfactant on fat-particle size distribution in oil-in-water emulsion.

Kumar et al.

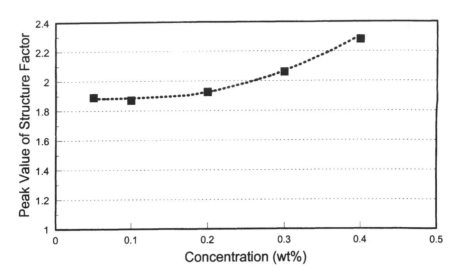

Figure 11 Structure factor variation versus concentration of sucrose stearate in oil-in-water emulsion.

B. Effect of Surfactant Concentration

To study the effect of surfactant concentration on fat-particle structure formation, samples with different amounts of sucrose stearate and sucrose oleate were prepared. The samples had 30 wt% fat, 3 wt% proteins, and 67 wt% water.

The effect of sucrose stearate concentration on the first peak height of the structure factor is shown in Figure 11. It is observed that first peak height of the structure factor $S(Q)$ increases with surfactant concentration. The same effect is observed for sucrose oleate (Fig. 12). Thus, we can conclude that the fat-particle structure becomes more ordered as the concentration of surfactant increases; this is obvious from the increase in the value of first peak height of the structure factor. This ordering of the fat-particle structure factor leads to an increase in the energy barrier which the fat particles have to overcome in order to destabilize the emulsion by aggregation [11]. Thus, the emulsion becomes more stable.

The capillary force balance technique was used to probe the surfactant–protein submicellar structure in the thin aqueous film between fat particles (Fig. 13). It was found that in the presence of surfactant–protein submicelles the film thinned in a stepwise manner. The principle behind this mechanism is that the submicelles existing inside the film can form a microlayering structure parallel to the film [3]. Due to the layered structure inside the film, the structural dis-

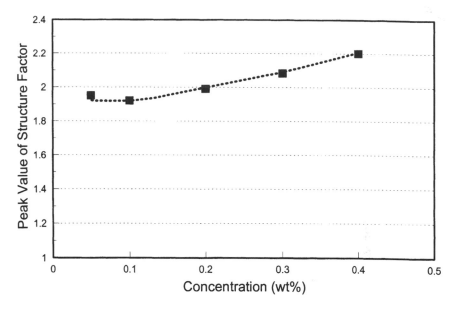

Figure 12 Structure factor variation versus concentration of sucrose oleate in oil-in-water emulsion.

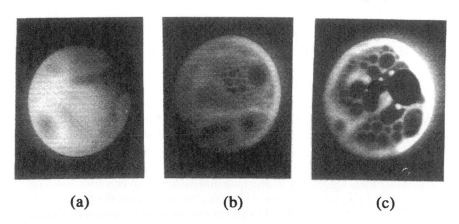

(a)   **(b)**   **(c)**

Figure 13 Photomicrographs of a foam film containing 4 wt% proteins and 0.1 wt% sucrose oleate at 25°C showing various stages of drainage.

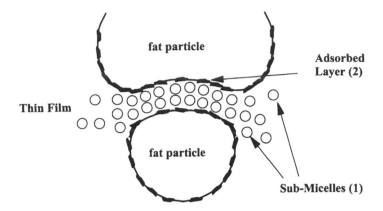

Figure 14 Schematic of surfactant stabilization mechanisms.

joining pressure becomes oscillatory. Nikolov and Wasan [12] have experimentally shown that number of thickness transitions is a strong function of micellar concentration. It was found that the emulsion film, and thereby the emulsion, was very stable if the draining emulsion film (small size film) contained layers of surfactant micelles [13,14]. Chu et al. [15] have shown by Monte Carlo simulation that the structural disjoining pressure increases with micellar concentration. This shows that an increase in submicellar concentration leads to an increased stability because the thin liquid film between particles drains slowly. Based on the above experimental data, it is concluded that submicellar concentration plays an important role in emulsion stability. The emulsion is stabilized by the layering of surfactant micelles in the thin liquid film (Fig. 14). In addition, the adsorbed surfactant and protein layers form a sort of "protective skin" around the droplets, which acts as a mechanical barrier to aggregation and thus stabilizes the emulsion.

C. Effect of Shear

In order to investigate the fat-particle structure variation during whipping, the effect of shear on fat-particle structure was studied by the imaging technique (Fig. 15). The two samples were made with sucrose oleate and sucrose stearate as surfactants. The fat-particle packing structure for both samples changes and there is a decrease in the first peak height of the structure factor. This is because fat particles form aggregates under shear and the structure becomes less ordered. Therefore, the stability of the samples is lowered by shearing the emulsion. However, the change in the first peak height of the structure factor after shearing for the sucrose oleate sample was less than that for the sucrose stearate sample. This could be explained by the microlayered structure of sucrose stearate micelles

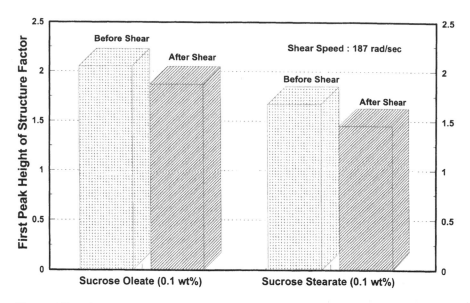

Figure 15 Effect of shear on fat-particle structure in oil-in-water emulsion.

around fat particles, which increased the resistance force of fat particles against deformation during shearing.

D. Curved Film Rheology

The film between the air bubbles plays an important role in stabilizing an aerated food emulsion. In order to investigate the role of the film rheology in stabilizing the emulsion, a stress relaxation experiment was carried out (Fig. 16) at different degrees of expansion, for both sucrose stearate and sucrose oleate emulsions. The Gibbs elasticity for the two samples was calculated from the slopes of the curves shown in Figure 17. It is observed that the sucrose oleate sample (more stable sample) has higher elasticity. It is also observed that the sucrose oleate foam film has a lower surface tension than the foam film of sucrose oleate, due to its higher surface activity. The higher elasticity of the film implies higher interfacial tension gradients (the Gibbs–Marangoni effect) and, thus, the film will drain more slowly due to the Gibbs–Marangoni effect.

IV. CONCLUSIONS

A nondestructive method of the digitized imaging technique was used to study the fat-particle structure formation in an oil-in-water food emulsion. The role of

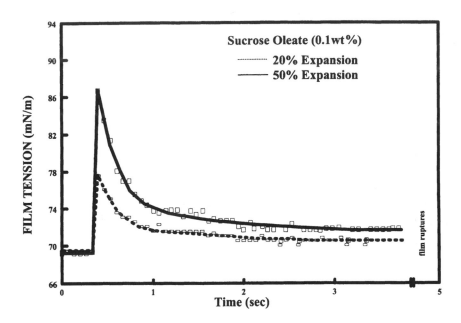

Figure 16 Stress relaxation curve for foam film of aerated food product.

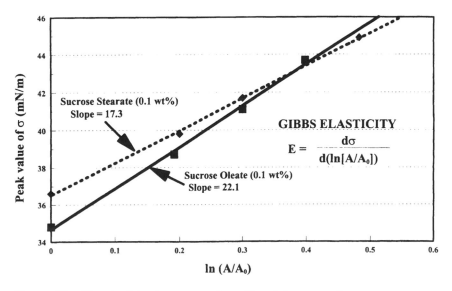

Figure 17 Film elasticity for aqueous foam film of the two surfactants.

fat particle–particle interactions in structure formation leading to stabilization of food emulsions was investigated by studying the effect of surfactant–protein submicelles on emulsion stability. The emulsion was found to be more stable when sucrose oleate was used as the surfactant compared to sucrose stearate. It was found that due to surfactant–protein interactions, the sucrose oleate emulsion had a more organized fat-particle structure. The capillary force balance technique was used to probe the surfactant–protein submicellar structure in the thin liquid film. The surfactant–protein submicelles formed a layered structure in the film and the film was found to thin in a stepwise manner. This stepwise thinning increased the film lifetime and stabilized the emulsion. On increasing the surfactant concentration, the fat-particle structure became more organized and the emulsion stability increased. The emulsion with sucrose oleate was found to be more stable during shearing due to the better fat-particle packing structure. The foam film of the sucrose oleate emulsion was found to have a higher Gibbs elasticity, which contributes to the stabilization of an aerated food product.

REFERENCES

1. W Buchheim, NM Barlod, N Krog. Food Microstruct 4:221, 1985.
2. P Walstra. In: E Dickinson, P Walstra, eds. Food Colloids and Polymers: Stability and Mechanical Properties. London: The Royal Society of Chemistry, 1993, pp. 3–15.
3. K Koczo, AD Nikolov, DT Wasan, RP Borwankar, AJ Gonsalves. J Colloid Interf Sci 178:694, 1996.
4. BE Brooker, M Anderson, AT Andrews. Food Microstruct 5:277, 1986.
5. BE Brooker. Food Microstruct 9:223, 1990.
6. BE Brooker. Food Microstruct 12:115, 1993.
7. W Xu, AD Nikolov, DT Wasan, AJ Gonsalves, RP Borwankar. J Food Sci 63:183, 1998.
8. J Soos, K Koczo, E Erdos, DT Wasan. Rev Sci Instrum 65:3555, 1994.
9. AD Nikolov, DT Wasan. J Colloid Interf Sci 133:1, 1989.
10. YH Kim, K Koczo, DT Wasan. J Colloid Interf Sci 187:29, 1997.
11. W Xu, AD Nikolov, DT Wasan, AJ Gonsalves, RP Borwankar. J Colloid Interf Sci 191:471, 1997.
12. AD Nikolov, DT Wasan. Langmuir 8:2985, 1992.
13. DT Wasan, AD Nikolov. Proceedings of the First World Congress on Emulsions, Paris, 1993, Vol. 4, p. 93.
14. ED Manev, SV Sazdanova, DT Wasan. J Dispers Sci Technol 5:111, 1984.
15. XL Chu, AD Nikolov, DT Wasan. J Chem Phys 103:6653, 1995.

5

Spontaneous Emulsification in Oil–Water–Surfactant Systems

Clarence A. Miller and Moon-Jeong Rang*
Rice University, Houston, Texas

ABSTRACT

An emulsion can form spontaneously when oil and aqueous phases are brought into contact without mixing if diffusion produces supersaturated compositions within a phase. This behavior can occur even if both initial compositions are in single-phase regions, provided that the phase diagram and initial compositions are suitable. Examples are given of spontaneous emulsification when oil drops are contacted with dilute surfactant solutions, a situation of possible interest in detergency. In other cases, a drop of oil containing a surfactant and an alcohol can emulsify spontaneously when injected into water. This behavior is sometimes called self-emulsification and is important in applications such as drug delivery and use of emulsifiable concentrates for agricultural chemicals. Whether the surfactant is initially in the aqueous or oil phase, complete emulsification to form small oil droplets occurs only when phase behavior is such that the injected oil is completely converted to another phase such as a microemulsion or the lamellar liquid crystal and when this phase subsequently becomes supersaturated in oil.

---

* *Current affiliation*: LG Chemical Ltd., Taejon, Korea.

I. INTRODUCTION

Emulsions have interested Darsh Wasan throughout his career and he has published extensively on emulsion stability and related topics. It is a pleasure to contribute this review on some aspects of spontaneous emulsification to the volume dedicated to Wasan.

Many emulsions are made by mechanical mixing of oil and water phases. Generally, vigorous mixing is required to obtain small drop sizes. However, in some cases, such mixing is impossible or undesirable, and spontaneous emulsion formation is preferred; that is, compositions of the initial oil and water phases are not in equilibrium and are chosen such that small drops form spontaneously when the phases are brought into contact.

Because it is an intriguing phenomenon, an extensive literature exists on spontaneous emulsification. Various mechanisms have been suggested, for instance, spontaneous expansion and breakup of an oil–water interface produced by transient local negative values of interfacial tension. Davies and Rideal [1] reviewed the older literature on this mechanism and put forward their own ideas. Gopal [2] presented an analysis of interfacial instability produced by negative tensions. Granek et al. [3] recently developed an improved model of how such an instability could arise. Experiments in which an oil containing some dissolved alcohol was contacted with a dilute surfactant solution have exhibited spontaneous emulsification for conditions in which *extrapolation* of measured interfacial tensions suggested that negative values could occur [1,4]. However, tensions below 0.02 mN/m were not measured, even in the more detailed study [4]. Interfacial tensions of this magnitude and lower are common in microemulsion systems. In the absence of information on phase behavior, it is not clear that negative interfacial tensions, in fact, occurred in these experiments. It may be that a small amount of a microemulsion phase formed and/or that the diffusion process produced local supersaturation, a mechanism of emulsification discussed below.

Vigorous Marangoni flow leading to breakoff of drops from an interface has also been proposed as a mechanism of spontaneous emulsification. However, although both phenomena occur simultaneously in numerous systems, conclusive evidence is lacking that the former causes the latter. Indeed, in some systems, it is clear that emulsification stems from the local supersaturation mechanism discussed next but that Marangoni flow greatly increases the rate of emulsification by increasing rates of mass transport near the interface [1,5].

This chapter deals with emulsification involving formation of drops of one phase in another that has become locally supersaturated as a result of diffusion. Davies and Rideal [1] recognized that supersaturation was involved in many cases of spontaneous emulsification and called this mechanism "diffusion and stranding." Ruschak and Miller [6] clarified the mechanism and presented a theoretical

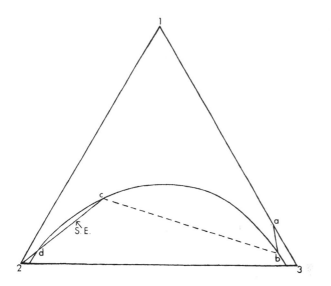

Figure 1   Diffusion path in a typical alcohol (1)–water (2)–oil (3) system indicating spontaneous emulsification (SE) in the aqueous phase.

analysis. They considered the solution to the diffusion equations in a ternary system for semi-infinite phases brought into contact without mixing. Because a similarity solution exists when certain plausible assumptions are made, it is readily shown that the set of compositions in the system is independent of time and can be plotted directly on the ternary phase diagram, the so-called "diffusion path."

Ruschak and Miller found that in some cases, the diffusion path in one or both phases passed through the two-phase region of the ternary diagram. For example, Figure 1 shows a schematic oil–water–alcohol ternary diagram and diffusion path for an oil–alcohol mixture having a composition brought into contact with pure water. Segments ab and cd2 represent compositions in the oil and water phases, respectively. They are straight lines as drawn whenever all diffusion coefficients in each phase are equal; otherwise, they are curved. The dotted line bc is an equilibrium tie line connecting compositions at the interface. Because cd lies within the two-phase region and ab does not, spontaneous emulsification is predicted in the aqueous phase but not in the oil. This method was able to predict successfully when spontaneous emulsification would occur and in which phase(s) for several oil–water–alcohol systems [5,6]. Videomicroscopy experiments [5] confirmed that emulsification in a typical system of this type was, in fact, the result of drop formation within a bulk liquid phase near the interface and not of drops breaking off the interface. As this example illustrates, knowing when local supersaturation can be expected requires knowledge both of equilibrium-phase behavior and of the diffusion processes which take place.

II. SURFACTANT SOLUTIONS IN CONTACT WITH OIL

Consideration of Figure 1 in light of the above discussion reveals that local supersaturation is more likely to occur when the solute (e.g., alcohol) diffuses into the phase in which it is more soluble—in this case, water. Similarly, experiments in oil–water–surfactant systems showed that spontaneous emulsification was favored when the surfactant diffused from the aqueous phase into an oil phase in which it was more soluble. For nonionic surfactants, this situation occurs when the system is above its phase inversion temperature (PIT) (i.e., the temperature at which hydrophilic and lipophilic properties of the surfactant films separating oil and water are balanced). Extensive emulsification of water in the oil phase was, in fact, observed when mixtures of the pure nonionic surfactant $C_{12}E_5$ and water and of $C_{12}E_4$ and water were brought into contact with n-hexadecane at temperatures above their respective PITs ([7–9]; Fig. 2). Not surprisingly, the emulsions formed were more stable than those in oil–water–alcohol systems.

   The general location and arrangement of single-phase and multiphase regions on the phase diagrams are known for these systems but not precise phase boundaries. As a result, the basic form of the diffusion paths could be determined but not the exact compositions along them. Nevertheless, it was clear from the

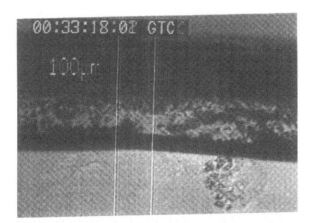

Figure 2 Video frame showing spontaneous emulsification of water in the oil phase observed 18 min after initial contact in the $C_{12}E_5$–water–n-tetradecane system at 60°C.

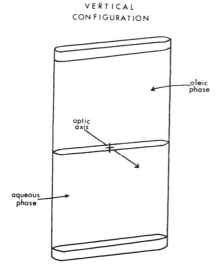

VERTICAL
CONFIGURATION

oleic
phase

optic
axis

aqueous
phase

Figure 3 Rectangular glass cell in vertical orientation used in the contacting experiments.

approximate diffusion paths why local supersaturation and spontaneous emulsification in the oil phase occurred for temperatures above the PIT while a very different behavior was seen for temperatures below the PIT, where phase behavior was different.

The corresponding situation for ionic surfactants is contact of a surfactant–salt–water mixture, frequently containing an alcohol cosurfactant as well, with oil above the optimal salinity of the system. Videomicroscopy experiments with anionic surfactant–alcohol–NaCl–water mixtures confirmed the occurrence of spontaneous emulsification of brine in the oil phase for these conditions [10]. Quantitative diffusion paths predicting emulsification were determined using hypothetical but plausible phase behavior for such a system [11].

The discussion so far has dealt with local supersaturation which occurs upon initial contact of oil and water phases. The similarity solution of diffusion-path theory requires that the composition in each phase approach its initial value at increasing distances from the surface of contact. This situation was achieved in the videomicroscopy experiments by gently placing the oil phase on top of the aqueous phase in a vertically oriented rectangular glass cell ([5,7], Fig. 3). Cells ranging from 200 to 400 μm in thickness have been used. This technique was employed in obtaining the video frame of Figure 2.

If, in contrast, a small drop of oil some 50–100 μm in diameter is injected

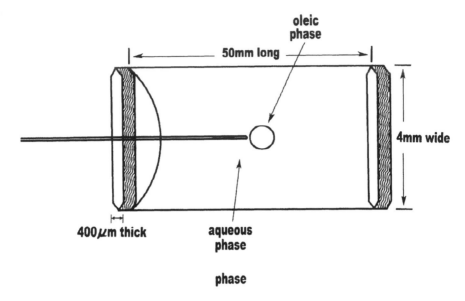

Figure 4 Schematic illustration of contacting experiment in which a small oil drop is injected into an aqueous surfactant solution.

into such a cell placed on a conventional horizontal microscope stage (Fig. 4) and nearly filled with an aqueous surfactant solution, composition at the center of the drop soon departs from its initial value. The time for diffusion to occur within the drop and produce significant composition changes is of the order r^2/D, where r is the drop radius and D is a characteristic diffusion coefficient. For $r = 50$ µm and $D = 5 \times 10^{-10}$ m²/s, the time required is only 5 s. As this time is typically much less than the duration of the experiment, the behavior of such a drop can be analyzed using a quasi-steady-state approach [9,12]. According to this analysis, the drop's composition is uniform but time dependent and follows the coexistence curve on the phase diagram of an oil-rich phase which is in equilibrium with the aqueous phase.

Consider, for example, injection of a drop of a hydrocarbon–alcohol mixture into a dilute aqueous micellar solution. Our interest is in situations in which the drop is above the relevant PIT on initial contact (i.e., where the alcohol content exceeds that of the excess oil phase in equilibrium with a balanced microemulsion and water at the experimental temperature). For long-chain hydrocarbons and alcohols which have negligible solubility in water (e.g., n-hexadecane/oleyl alcohol mixtures), the drop loses little oil following injection but swells and becomes a microemulsion by taking up surfactant and water from the micellar

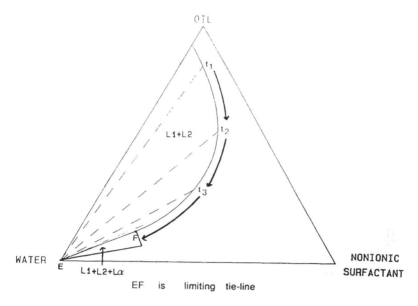

Figure 5 Schematic pseudoternary phase diagram of nonionic surfactant–water–hydro-carbon–oleyl alcohol mixture showing the change in the drop's composition with time.

solution. The rate-limiting step for this process is the diffusion of surfactant in the aqueous phase. This behavior is shown schematically in Figure 5, where the drop's composition follows the coexistence curve in the direction of the arrows as the microemulsion becomes ever richer in water. Eventually, composition of the drop reaches point F, which is one end of the limiting tie line separating the L_1–L_2 region from the L_1–L_2–L_α region, where L_α is the lamellar liquid-crystalline phase. The videomicroscopy experiments show that at this point, the lamellar phase begins to grow at the interface in the form of small myelinic figures [12]. Although the myelinic figures appear to be rather flexible, they do not break up into drops. In fact, no spontaneous emulsification was observed except when the initial drop composition was very near to that of the PIT, in which case a few oil drops formed within the microemulsion.

Emulsification is typically seen, however, when the alcohol has a shorter chain length and, hence, appreciable solubility in water. In this case, both the diffusion of surfactant into the drop and of alcohol out of the drop act to make it more hydrophilic. Thus, the situation differs from that of Figure 5 in that the alcohol/hydrocarbon ratio of the drop decreases during the experiment instead of remaining constant. As a result, the phase behavior for a series of alcohol/hydrocarbon ratios is required to follow drop composition as it moves along the L_2 coexistence surface with L_1 and becomes a microemulsion ever richer in water.

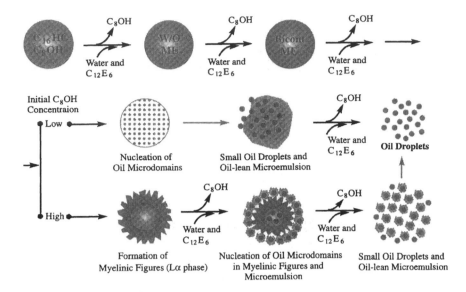

Figure 6 Schematic behavior showing the spontaneous emulsification process for drops of hydrocarbon/alcohol contacting dilute surfactant solutions. The behavior is different at high- and low-octanol contents in *n*-hexadecane drops contacting $C_{12}E_6$ solutions as indicated.

Such phase behavior was determined for the $C_{12}E_6$–*n*-hexadecane–*n*-octanol–water system [13]. When a hexadecane–octanol drop containing between 10 wt% and 30 wt% alcohol was injected into a 0.05 wt% solution of $C_{12}E_6$, the drop followed the coexistence surface and eventually reached a limiting tie line of the L_1–ME (microemulsion) region at the boundary of the L_1–ME–oil region. As the microemulsion continued to lose alcohol, it became supersaturated in oil and spontaneous emulsification ensued. Further loss of alcohol caused the microemulsion to become miscible with water, leaving an oil-in-water emulsion [13]. A schematic diagram of this process is shown in the low-alcohol path of Figure 6.

For drops with initial alcohol contents exceeding 30 wt% behavior was different in that the limiting tie-line reached was bounded by the L_1–ME–L_α region. Consequently, myelinic figures grew which were smaller and more fluid than those for the long-chain alcohol systems discussed above. With further loss of alcohol the microemulsion and lamellar phases first became supersaturated in oil, so that spontaneous emulsification occurred, and later became miscible with water [13]. A schematic diagram of this behavior is shown in the high-alcohol path of Figure 6.

The behavior illustrated by the high-alcohol path of Figure 6 was also seen for drops consisting of suitable mixtures of n-decane–n-heptanol or n-decane–n-octanol injected into 0.05 wt% solutions of $C_{14}DMAO$, tetradecyldimethylamine oxide [14]. In this case, the process sometimes occurred so rapidly that the emulsification resembled an explosion. Equilibrium-phase behavior studies confirmed that a lamellar intermediate phase was expected in these systems for initial drop compositions above the PIT, and that as alcohol content continued to decrease, a four-phase region of L_1–ME–L_α–oil was encountered. Emulsification would be expected to start when the drop's composition entered this four-phase region.

Spontaneous formation of oil droplets also occurred when drops of triolein–oleyl alcohol were injected into dilute nonionic surfactant solutions [15,16]. Here, too, the drop became a microemulsion as it took up surfactant and water, but in this case, the L_3 or sponge phase formed as an intermediate phase. Then, alcohol was able to leave the microemulsion by transferring to this phase instead of to the surfactant solution, as in the preceding examples with shorter-chain alcohols. As a result, the microemulsion became supersaturated in oil, leading to the spontaneous formation of oil droplets. This behavior occurred, for example, when a drop containing 75 wt% oleyl alcohol and 25 wt% triolein was injected into 0.05 wt% $C_{12}E_6$ at 40°C [15]. Emulsification within a similar drop initially placed on a small polyester fiber immersed in a mixture of water and a different nonionic surfactant caused almost all of the drop to be removed from the fiber and dispersed into the aqueous phase [16]. Although experiments were not conducted for drops containing oleic acid instead of oleyl alcohol, a similar behavior would be expected in view of the similar phase behavior found with oleic acid. Thus, spontaneous emulsification seems of potential use in improving detergency for situations where the phase behavior for the relevant surfactant–soil combination is such that the soil can be converted to a microemulsion, which then becomes supersaturated with oil.

III. SELF-EMULSIFICATION

The term ''spontaneous emulsification''is usually limited to situations where no external energy of agitation is supplied. ''Self-emulsification'' generally refers to situations where a small amount of energy is supplied to achieve gentle mixing. Sometimes (e.g., when interfacial tension is very low), such mixing may be sufficient to produce drops and form an emulsion. In other cases, the drops form spontaneously and mixing simply serves to disperse them throughout a large volume and to bring together portions of the oil and water phases which were not near the initial surface of contact.

Self-emulsification of an oil phase in water is of interest, for example, in the use of ''emulsifiable concentrates'' of agricultural chemicals [17,18] and in some methods for drug delivery [19,20]. Basically, a surfactant or surfactant mixture is added to a mixture of a pesticide or drug and a suitable organic solvent to enable it to disperse when added to water under conditions where gentle mixing occurs. The mechanism of such self-emulsification is not well understood. Lee and Tadros [21] found that development of low interfacial tensions could not fully explain the emulsification phenomena they observed. Sometimes, self-emulsification has been attributed to the formation of liquid-crystalline phases near the surface of contact between the two phases [20,22] although the precise role of the liquid crystal was not made clear by these authors. Accordingly, the selection of surfactants for self-emulsification and the determination of the minimum surfactant concentration needed has been largely empirical.

Recently, a systematic study of spontaneous emulsification of hydrocarbon–alcohol–surfactant mixtures in water was carried out at 30°C using the drop-injection technique of Figure 4. Two of these systems, in which the alcohols had appreciable solubility in water, were the n-hexadecane–n-octanol–$C_{12}E_6$ and n-decane–n-heptanol–$C_{14}DMAO$ systems mentioned previously [13]. An example in which complete emulsification occurred to form uniformly small droplets a few micrometers in diameter is shown in Figure 7 for an oil drop initially containing 20 wt% $C_{12}E_6$ and 80 wt% of a 90/10 mixture by weight of hexadecane and octanol. Such behavior was observed only when initial surfactant and octanol concentrations in the injected drop exceeded approximately 15 wt% and 10 wt%, respectively. Phase-behavior studies demonstrated that the latter constraint amounts to a requirement that alcohol content exceed that of the excess oil phase in equilibrium with a balanced microemulsion and water at the experimental temperature (i.e., that the system be above the PIT as in the similar experiments described earlier where the surfactant was initially in the aqueous phase).

Indeed, the mechanism of emulsification is closely related to that described earlier for the same system. As earlier, water is transferred to the drop and alcohol to the water. Little surfactant leaves the drop because its solubility in water is basically the critical micelle concentration (cmc), which is very low, for conditions above the PIT. Also, as earlier, the drop becomes a microemulsion which is made ever more hydrophilic as alcohol diffuses into the water phase. Eventually, it becomes supersaturated in oil, initiating spontaneous emulsification. Later, it becomes miscible with water, the emulsification continuing as the solubilization capacity for oil decreases. A schematic description of this behavior is shown in Figure 8.

The behavior was similar for n-decane–n-heptanol–$C_{14}DMAO$ drops injected into water. Here, too, the initial alcohol content of the drop had to be large enough for the system to be above the PIT. However, the phase diagram dictated that the first intermediate phase formed was the lamellar liquid crystal. Nevertheless, as alcohol continued to diffuse into the water, both the liquid crystal and microemulsion became supersaturated in oil, leading to spontaneous emulsifica-

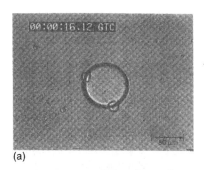

(a)

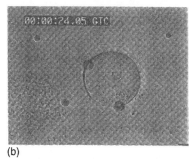

(b)

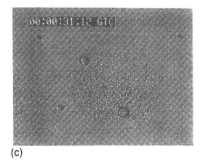

(c)

Figure 7   Video frames showing spontaneous emulsification of an oil drop with an initial composition of 20 wt% $C_{12}E_6$ and 80 wt% of a 90/10 mixture by weight of hexadecane and octanol. The behavior at approximately 16, 24, and 31 s after drop injection is shown.

tion. As in the nonionic surfactant system, these phases ultimately became miscible with water, so that only an oil-in-water emulsion was present at the end of the experiment. The behavior was basically that shown for the high-alcohol path of Figure 6, except that no surfactant was initially present in the aqueous phase to diffuse into the drop.

In the $C_{14}DMAO$ system, it was found that complete emulsification forming

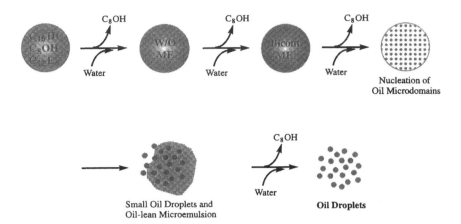

Figure 8 Schematic behavior showing the spontaneous emulsification process for drops of hydrocarbon–alcohol–surfactant contacting water.

only small oil droplets occurred in several cases with only 5 wt% surfactant present in the injected drop, whereas 15 wt% was required for the nonionic surfactant. The reason was rapid coalescence of the spontaneously generated oil droplets to form larger drops in the latter system at low surfactant concentrations. Factors likely responsible for the minimal coalescence observed in the $C_{14}DMAO$ system were the presence of liquid crystal and the higher water content of the microemulsion when it became supersaturated with oil, the latter effect causing the nucleated droplets to be farther apart.

Self-emulsification was also investigated systematically for the system n-hexadecane–oleyl alcohol–$C_{12}E_6$–water [23], where the alcohol is nearly insoluble in water. Complete emulsification to form small oil droplets was seen only when the hexadecane–oleyl alcohol ratio in the injected drop was between 95/5 and 90/10 by weight and when the surfactant concentration was at least 20 wt%. Figure 9 illustrates the effect of surfactant concentration on the size of droplets formed for the 90/10 case. Phase-behavior studies indicated that the higher surfactant concentrations were necessary in order for the drop to be completely converted to the lamellar phase as it took up water following injection. As yet more water entered the drop, the lamellar phase was partially or completely converted to a microemulsion, according to the phase diagram. Then, these phases became supersaturated in oil, and spontaneous emulsification ensued, as shown schematically in Figure 10. Complete conversion of the original oil to the lamellar phase assured that all oil droplets in the final state would be formed by nucleation (i.e., that no larger drops of the original oil phase remained).

The condition cited in the preceding paragraph on the hexadecane–oleyl

$[(C_{16}HC/C_{18:1}OH = 90/10)/C_{12}E_6 = 95/5\ ]$ $[(C_{16}HC/C_{18:1}OH = 90/10)/C_{12}E_6 = 85/15]$

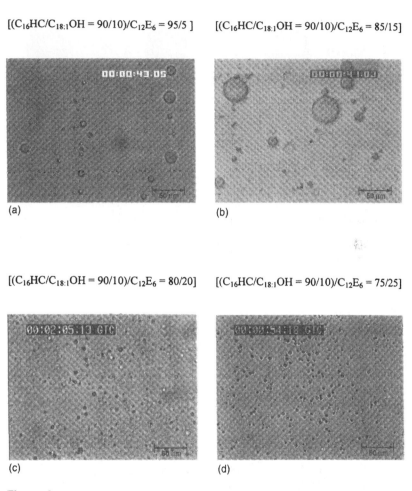

$[(C_{16}HC/C_{18:1}OH = 90/10)/C_{12}E_6 = 80/20]$ $[(C_{16}HC/C_{18:1}OH = 90/10)/C_{12}E_6 = 75/25]$

Figure 9 Video frames showing droplets formed due to spontaneous emulsification of oil drops initially having a hexadecane–oleyl alcohol ratio of 90/10 by weight and (a) 5 wt%, (b) 15 wt%, (c) 20 wt%, and (d) 25 wt% $C_{12}E_6$.

alcohol ratio is basically a requirement that the ratio be near that corresponding to the PIT. Generally speaking, the lamellar phase extends to the lowest surfactant concentrations near the PIT.

    Emulsification occurred more quickly for this system than for those with shorter-chain alcohols discussed previously. In fact, it was not possible to observe details of the emulsification process as in the previous systems. The reason is that the process was controlled by diffusion in the small injected drop instead

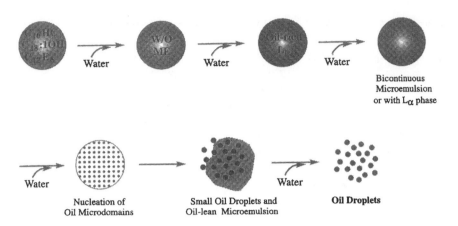

Figure 10 Schematic behavior showing spontaneous emulsification process for drops of *n*-hexadecane–oleyl alcohol–$C_{12}E_6$ contacting water.

of by diffusion in the surrounding water, as was the case for the shorter-chain alcohols.

The stability of the emulsions so formed was studied using turbidity measurements in a gently stirred system [23]. The most stable emulsion was found for an initial drop composition where the equilibrium-phase behavior indicated that the lamellar phase was present to form a protective coating around the small oil droplets formed by the mechanism described earlier.

In all these studies with different surfactants and with alcohols of various chain lengths, it was found that spontaneous emulsification yielding only small droplets (of the order 1 µm in diameter) required that the injected drop be (1) completely converted to a phase such as a microemulsion or the lamellar liquid crystal which (2) subsequently became supersaturated in oil. Although the studies all involved rather hydrophilic surfactants and lipophilic alcohols, these criteria should also apply for mixtures of hydrophilic and lipophilic surfactants. Indeed, videomicroscopy experiments demonstrated that spontaneous emulsification to small droplets occurred when drops containing a 65/35 mixture by weight of *n*-hexadecane and the commercial nonionic surfactant Neodol 25-7 (a Shell product) were injected into water. This surfactant is a mixture of linear alcohol ethoxylates with hydrocarbon chain lengths between 12 and 15 and with an average ethylene oxide number of 7.3. Preliminary phase-behavior experiments confirmed that the lamellar phase formed upon the addition of approximately 10 wt% water to the 65/35 mixture.

The systems satisfying the above criteria and thus exhibiting self-emulsification of oil drops had at least four components. It was found that ternary systems consisting of water, hydrocarbon, and a pure nonionic surfactant could not satisfy both criteria simultaneously [13].

Recently, formation of silicone oil-in-water emulsions with small droplets was reported to take place when a mixture of the oil and a surfactant was stirred gently with an appropriate mixture of ethanol and water [24]. The ethanol concentration was chosen to assure that the oil passed through the single-phase region of the lamellar liquid crystal during the early stages of dilution. Later, this phase became supersaturated with oil, and the final state was a water-continuous emulsion. The same authors also made emulsions having small drops using various oils in systems where a substantial amount of glycerol or similar compound was present instead of ethanol. However, local supersaturation was apparently not a major factor because an oil–lamellar phase dispersion having the consistency of a gel was prepared and diluted with water while stirring. The liquid crystal coated the oil drops to provide stability.

It is noteworthy that local supersaturation can also be produced by changing temperature (e.g., by cooling a bicontinuous microemulsion or the lamellar phase of a system containing a single nonionic surfactant or a mixture of such surfactants). Oil-in-water emulsions with small drops have been produced in this way [25,26]. A recent study showed that growth of oil droplets after their initial formation is a concern that must be addressed in such processes [27]. Highly concentrated water-in-oil emulsions with polyhedral drops were produced by a similar scheme, except that the temperature was raised instead of lowered [28]. Some initial theoretical considerations involving nucleation of a new phase in a microemulsion have been put forward by Vollmer et al. [29].

IV. SUMMARY

Spontaneous emulsification can be produced by local supersaturation both in oil–water–alcohol and in oil–water–surfactant systems. In oil–water–surfactant–alcohol systems, examples of possible interest in detergency and self-emulsification of oils are presented. In either case, emulsification of an oil to form only small droplets requires that diffusion first convert the oil to a microemulsion and/or the lamellar liquid-crystalline phase and then cause this phase (or these phases) to subsequently become supersaturated in oil. The presence of some liquid crystal during the emulsification process apparently helps stabilize the droplets against coalescence. In any case, knowledge of the equilibrium-phase behavior of the system of interest is of great value in choosing suitable conditions for emulsification.

ACKNOWLEDGMENT

Michael Unton performed the experiments with Neodol 25-7.

REFERENCES

1. JT Davies, EK Rideal. Interfacial Phenomena. 2nd ed. New York: Academic Press, 1963.
2. ESR Gopal. In: P Sherman, ed. Rheology of Emulsions. Oxford: Pergamon Press, 1963, pp. 15ff.
3. R Granek, RC Ball, ME Cates. J Phys II (France) 3:829, 1993.
4. M Vermeulen, P Joos, L Ghosh. J Colloid Interf Sci 140:41, 1990.
5. CA Miller. Colloids Surf 29:89, 1988.
6. KJ Ruschak, CA Miller. Ind Eng Chem Fundam 11:534, 1972.
7. WJ Benton, KH Raney, CA Miller. J Colloid Interf Sci 110:363, 1986.
8. KH Raney, WJ Benton, CA Miller. J Colloid Interf Sci 117:282, 1987.
9. CA Miller, KH Raney. Colloids Surf A 74:169, 1993.
10. KH Raney, WJ Benton, CA Miller. In: DO Shah, ed. Macro- and Microemulsions. ACS Symposium Series 272. Washington, DC: American Chemical Society, 1985, Chap. 14.
11. KH Raney, CA Miller. AIChE J 33:1791, 1987.
12. J-C Lim, CA Miller. Langmuir 7:2021, 1991.
13. M-J Rang, CA Miller. Prog Colloid Polym Sci 109:101, 1998.
14. M-J Rang, CA Miller, HH Hoffmann, C Thunig. Ind Eng Chem Res 35:3233, 1996.
15. T Tungsubutra, CA Miller. In: S Friberg, B Lindman, eds. Organized Solutions. New York: Marcel Dekker, 1992, Chap. 7.
16. T Tungsubutra, CA Miller. J Am Oil Chem Soc 71:65, 1994.
17. DZ Becher. In: P Becher, ed. Encyclopedia of Emulsion Technology, Vol. 2. New York: Marcel Dekker, 1985, pp. 239ff.
18. ThF Tadros. In: ThF Tadros, ed. Surfactants in Agrochemicals. New York: Marcel Dekker, 1995, pp. 63ff.
19. PP Constantinides. Pharm Res 12:1561, 1995.
20. MG Wakerly, CW Pouton, BJ Meakin, FS Morton. In: JF Scamehorn, ed. Phenomena in Mixed Surfactant Systems. ACS Symposium Series 311. Washington, DC: American Chemical Society, 1986, pp. 242ff.
21. GWJ Lee, ThF Tadros. Colloids Surf 5:105, 117, 129, 1982.
22. MJ Groves. Chem Ind 417, June 17, 1978.
23. M-J Rang, CA Miller. J Colloid Interf Sci 209:179, 1999.
24. T Suzuki. In: H Ohshima, K Furusawa, eds. Electrical Phenomena at Interfaces. New York: Marcel Dekker, 1998, Chap. 29.
25. H Sagitani. In: S Friberg, B Lindman, eds. Organized Solutions. New York: Marcel Dekker, 1992, pp. 259ff.
26. T Förster, W von Rybinski, A Wadle. Adv Colloid Interf Sci 58:119, 1995.
27. L Taisne, B Cabane. Langmuir 14:4744, 1998.
28. H Kunieda, Y Fukui, H Uchiyama, C Solans. Langmuir 12:2136, 1996.
29. D Vollmer, R Strey, J Vollmer. J Chem Phys 107:3619, 3627, 1997.

6

A New Technique for Drug Release Using Multiple Emulsion Technology

Shuming Zheng*
Hampton University, Hampton, Virginia

R. L. Beissinger
Illinois Institute of Technology, Chicago, Illinois

Lakshman R. Sehgal
Evanston Northwestern Healthcare, Evanston, Illinois

Darsh T. Wasan
Illinois Institute of Technology, Chicago, Illinois

ABSTRACT

Ketamine [2-(chlorophenyl)-2-(methylamino) cyclohexanone, $C_{13}H_{16}ClNO$, anesthetic agent], the cyclohexylamine, is used as a short-acting anesthetic in humans and in some animal species. Ketamine is poorly bound to plasma proteins and has a half-life of approximately 4 h following intravenous injection. Ketamine leaves the blood very rapidly to be distributed into the tissues with high lipid solubility. The recommended dosage of intravenous ketamine is 2.5–20 mg/kg. The LD_{50} dose injected intraperitoneally in mice and rats is 100 times the intravenous dose and 30 times the intramuscular

* *Current affiliation*: Harvard University, Cambridge, Massachusetts.

121

dose used in humans. The objective of this study was to formulate a ketamine multiple emulsion, which has high porosity and lower viscosity at 37°C for sustained anesthesia release. The results in this study demonstrated that the ketamine–O/W multiple emulsion had a number of desirable properties, which suggests that it presents an appropriate mechanism for the development of an emulsion-based drug delivery system.

I. INTRODUCTION

A large number of investigations have been made with oil-in-water-in-oil (O/W/O) and water-in-oil-in-water (W/O/W) systems from different standpoints of emulsion science and technology. Multiple-emulsion technology provides a mechanism for the encapsulation and in vivo delivery of drugs and other materials which would otherwise be degraded, released rapidly, or be toxic to the host. During recent years, a number of approaches have been used in the attempt to develop drug delivery systems using the liquid membrane technology: the study of antigen adjuvants and other drug systems [1,2], and other biochemical and biomedical applications [3–6], such as immobilization of enzymes [7], treatment of chronic uremia [8], and as an oxygen delivery system [9,10]. A wide variety of oils and surfactants can be used in making W/O/W multiple emulsions, but only a few of them can be used in the food and biomedical fields. The choice of materials in a multiple-emulsion system is dependent on the application of the product. For intravenous administration in humans, the range of oils and emulsifiers is severely limited. Purified paraffin oils (mineral oils) are used for intramuscular use, as are vegetable oils. Vegetable oils are used almost exclusively for intravenous purposes as nutritional aids. Oils derived from vegetable sources are biodegradable, whereas purified mineral oils are not [11]. Surfactants are used in the membrane or middle phase and outer phase to form and stabilize the multiple-emulsion droplets. The choice of the optimum surfactants in the primary and secondary emulsions is often based on the hydrophile–lipophile balance (HLB).

In vitro studies were performed to evaluate a prototype multiple emulsion as a stable ketamine delivery system. Studies have been conducted in this ketamine/O/W multiple-emulsion system to improve the stability of ketamine encapsulation in primary emulsion, the porosity of ketamine release in secondary emulsion, and satisfactory rheological properties.

II. MATERIALS AND METHODS

A. Preparation of Oil Phase

Purified mineral oil (carnation oil, Witco Chemical Co., New York, NY) and vegetable oil (soybean oil, Croda Co., Edison City, NY) were used to prepare the liquid membrane phase of the ketamine multiple emulsion. These oils were chosen based on their ability to achieve high loading of the inner aqueous phase in the primary emulsion and for their ability to achieve reasonable porosity of the liquid membrane in secondary emulsion, as well as for their low viscosity. Emulsifying agents for the primary emulsion were dissolved in the oil to promote the formation of a ketamine aqueous solution-in-oil (primary) emulsion.

B. Preparation of Outer Aqueous Phase

Phosphate-buffered saline (PBS) solution of pH 7.4 was used in the outer phase of the ketamine multiple emulsion to make the solution isotonic. Emulsifiers for the secondary emulsion were dissolved in the outer aqueous phase to improve the overall dispersion of the multiple-emulsion system.

C. Preparation of Inner Aqueous Phase

Ketamine (Sigma Chemical Co., St. Louis, MO) solution was prepared and stored at room temperature at a concentration up to 100 mg/mL in PBS solution (pH = 7.3–7.4).

D. Emulsifiers in Oil and Outer Phase

The emulsifying agents used in the outer phase were either 0.3 wt% Tween-60 or 0.3 wt% Pluronic F68 (ICI Americas, Inc., Wilmington, DE) as the emulsifying agent for secondary emulsion in the carnation oil or soybean oil multiple-emulsion system, or 5.0 wt% Brij 93 (ICI Americas, Inc., Wilmington, DE) in carnation oil primary emulsion.

The emulsifying agents used in the oil phase for the preparation of primary emulsion were a combination of 5.0% wt% Brij 93 with Pluronic L101 at 5.0 wt% (BASF, Wyandotte Corp., Parsippany, MI) in the ketamine/soybean oil primary emulsion.

E. Preparation of the Ketamine/O/W Multiple Emulsion

In order to encapsulate large quantities of concentrated ketamine solution, it is necessary to follow a procedure as described by Zheng et al. [10]. The surfactants 5 wt% Brij 93 and 5 wt% Pluronic L101 were dissolved in soybean oil to prepare the ketamine/oil primary emulsion. The ketamine solution was added slowly to the oil phase and stirred vigorously on a magnetic stirring table. A 60 vol% loading of ketamine solution was generally used. In some of these studies, in order to increase the stability of the ketamine/oil primary emulsion and to enable the subsequent formation of smaller multiple-emulsion droplets, a high-energy method was used to further emulsify the primary emulsion. This employed either

Microfluidizer™ M110 or H-5000A Homogenizer (preferred) (HMZ, Microfluidics Corp., Newton, MA) at a pressure drop range across the interaction chamber of about 2000 psig. The resulting primary emulsion was then stored at room temperature until injection.

F. Characterization of the Ketamine/O/W Multiple Emulsion

In order to test whether the stability of the ketamine in the soybean oil system could be enhanced, the HMZ processed ketamine/soybean oil primary emulsion was dispersed into an oil phase (carnation oil with surfactants) at different volume ratios (11–1.5 preferred) and mixed for 30 min using a magnetic stirrer. The resulting emulsion was evaluated for stability, porosity, and viscosity. The results showed that the use of HMZ following the resuspension step to form the emulsion resulted in higher stability of the soybean oil primary emulsion system.

G. Determination of Ketamine Encapsulation Efficiency in Primary Emulsion During Storage

The method used to measure the encapsulation efficiency of ketamine in the primary emulsion during storage at room temperature was developed in this study. The W/O emulsion was centrifuged ($150 \times g$) and the ketamine in the water phase (W_k) was collected. The ketamine concentration during storage was analyzed using a spectrophotometer at 261 nm wavelength; the solvent used for the ultraviolet (UV) measurement was 10 mM NaOH in 95 vol% methanol (CH_3OH). Because the organic solvent must be 95% CH_3OH with 10 mM NaOH, which indicates that the W_k could be only 5% in the overall sample (i.e., 5 mL), W_k was dissolved in 95 mL 100% CH_3OH with 10 mM NaOH. In order to achieve this concentration, 100% CH_3OH with 10 mM NaOH was prepared as the solvent for the measurement, and ketamine released from the inner water phase was dissolved in the organic solvent to make the volume ratio of 5/95 prior to the UV measurement.

H. Drop-Size Analysis

The ketamine emulsion droplet sizes were examined by dispersion of the multiple emulsion into a gelatin solution (0.1 g gelatin per 100 mL of aqueous outer phase); a homogeneous dispersion was obtained [12]. The diluted sample was placed in a Howard cell (Rascher and Betzold, Inc., Chicago, IL) and was examined under a microscope at 100× magnification.

I. Calibration Curve

By preparing solutions with the concentration of ketamine at 0.000, 0.0625, 0.125, 0.25, 0.5000, 0.825, and 1.65 mg/mL and measuring the optical density (OD) at 261 nm, the calibration curve was obtained. The volume ratio of W_k to methanol solvent used was 5/95 as mentioned earlier.

J. Multiple-Emulsion Viscosity and Shear Stability

The steady shear viscosity of the multiple emulsion was measured with a Wells–Brookfield Syncro-Lectric Microviscometer (Model LVT) using a cone-and-plate system with a 0.80° cone Model CP-40 with a temperature controller (Brookfield Engineering Laboratories, Inc., Soughton, MA). The sample to be tested was contained within the conical gap, and the lower conical platen was rotated at various angular speeds. The viscosity was measured following the exposure of the ketamine/O/W multiple emulsion to clinically relevant shear rates.

   The shear-sensitive stability studies involved exposure of the ketamine multiple emulsion to clinically relevant shear-rate (up to 450 s^{-1}) levels and measurement of ketamine release as a function of time. The release of ketamine from either carnation-oil- or soybean-oil-based ketamine/O/W multiple-emulsion droplets was evaluated using the spectrophotometer at 261 nm. The samples were sheared at 37°C for 30 min.

III. RESULTS AND DISCUSSION

A. Stability of Primary Emulsion

The release of ketamine from the primary emulsion was measured as a function of storage time. No significant release of ketamine was observed during storage at room temperature for up to 6 months for the ketamine/carnation oil/W multiple emulsion, for up to 20 days for the ketamine/soybean oil/W multiple emulsion, and 35 days for the ketamine/soybean oil/carnation oil/W multiple emulsion (see Fig. 1), although 50% of the ketamine leaked out after 2 days at 4°C from all the emulsions tested. The release at 4°C may be due to the crystallization of ketamine in the inner phase, which breaks the liquid membrane by its new molecular crystal structure at 4°C. The experimental results also showed that the use of Pluronic L101 along with Brij 93 as emulsifiers in the oil phase, and the resuspension procedure (i.e., dispersion of the microfluidization-processed primary emulsion in carnation oil at a volume ratio of 11/1.5) increased the stability of the ketamine/soybean oil emulsion.

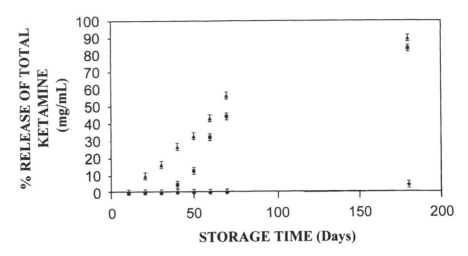

Figure 1   Stability of ketamine in oil primary emulsion during storage at room temperature. ◆: Ketamine/carnation oil/W; ■: ketamine/soybean oil/carnation oil/W; ▲:
ketamine/soybean oil/W.

B.   Drop Sizes of Primary and Secondary Emulsions

The particle size range of the ketamine primary emulsion appears to be from as
small as 0.1 to no larger than 1.0 μm, and most of the droplets in the secondary
emulsion are about 3.0 μm prior to filtration. Both Tween-60 and Pluronic F68
were found to be effective surfactants for making a secondary emulsion. Tween-
60 resulted in smaller-sized (Fig. 2) droplets and gave a better sustained ketamine
release (Fig. 3). We chose to use Pluronic F68, as Pluronic F68 is widely used
in the medical field as a fat reducer and a protective agent for hemolysis during
extracorporeal circulation and is shown to be nontoxic at low concentrations [13].

C.   Multiple-Emulsion Viscosity

Steady shear viscosity results were obtained for the ketamine multiple emulsion.
The shear viscosities of the multiple emulsion for a volume fraction ratio of
droplets to the outer aqueous phase of 1/1 under shear rates from 45 to 450 s^{-1}
were measured at a temperature of 37°C. The shear viscosity of the multiple
emulsion varied from 4.1 to 3.3 cP (±0.2) for ketamine/carnation oil/W and
from 4.6 to 2.4 cP (±0.2) for the ketamine/soybean oil/W multiple emulsion as
the shear rate was varied from 45 to 450 s^{-1} at a temperature of 37°C (see
Fig. 4).

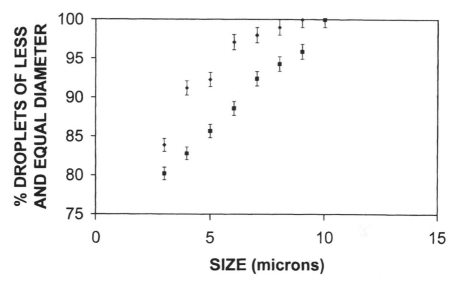

Figure 2   Effect of surfactant in outer phase on size distribution of the ketamine/soybean oil/W multiple emulsion. ◆: Tween-60; ■: Pluronic F68.

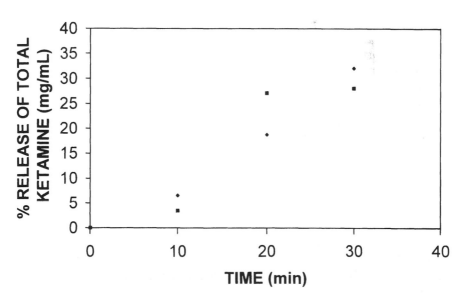

Figure 3   Effect of surfactant in outer phase on ketamine release from ketamine/carnation oil/W emulsion with shearing. ◆: Tween-60; ■: Pluronic F68.

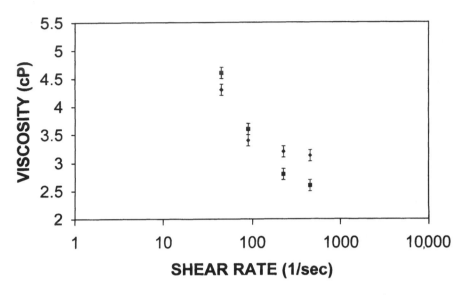

Figure 4 Shear viscosity of ketamine/O/W multiple emulsions at 37°C. ◆: Ketamine/carnation oil/W; ■: ketamine/soybean oil/W.

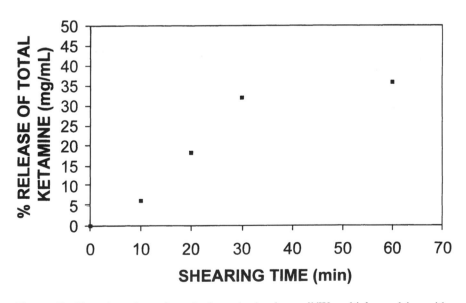

Figure 5 Ketamine release from the ketamine/soybean oil/W multiple emulsion with shearing.

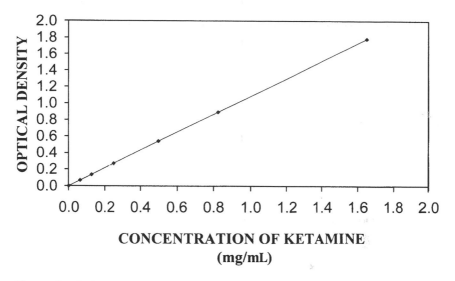

Figure 6 Calibration curve for ketamine concentration measurement.

D. Ketamine Release from Secondary Emulsion with Shearing

The release of the encapsulated ketamine in multiple-emulsion droplets with shearing at 450 s^{-1} is shown in Figure 5. Figure 6 shows a calibration curve of the optical density versus ketamine concentration at 261 nm wavelength. The results showed that the ketamine (100 mg/mL in inner phase) was released at 8.2% at 10 min, 67.0% at 30 min, and 95.5% at 60 min from the ketamine/soybean oil/W multiple emulsion in a well-controlled manner. The efficacy and toxicity of the ketamine/O/W multiple emulsion will be evaluated in the near future.

ACKNOWLEDGMENT

The authors acknowledge the Microfluidics Corporation for aid in adapting the Microfluidizer™ M110 equipment for their purpose.

REFERENCES

1. AF Brodin, DR Kavaliunas, SG Frank. Acta Pharm Suecica 15:1–9, 1978.
2. WJ Herbert. Lancet 771:16–18, 1965.

3. JW Frandenfeld, WJ Acher, NN Li. Recent Developments in Separation Science, Vol. 4. West Palm Beach, FL: CRC Press, 1978, pp. 39–50.

4. S Zheng, RL Beissinger, DT Wasan. J Colloid Interf Sci 144:72–83, 1991.

5. S Zheng, RL Beissinger, DT Wasan, DL McCormick. Biochim Biophys Acta 1158: 65–74, 1993.

6. C Chiang, GC Fuller, JW Frandenfeld, CT Rhodes. J Pharm Sci 67:63–66, 1978.

7. SW May, NN Li. Biochem Biophys Res Commun 47:1179–1182, 1975.

8. WJ Asher, KC Bovee, JW Frandenfeld, RW Hamilton, LW Henderson, PG Holtzapple, NN Li. Kidney Int 7:s409–s412, 1975.

9. S Zheng, RL Beissinger, DT Wasan, LR Sehgal, AL Rosen. U.S. Patent 5,438,041, 1995.

10. S Zheng, RL Beissinger, DT Wasan, LR Sehgal, AL Rosen. U.S. Patent 5,217,648, 1993.

11. TA Davis, WJ Asher, HW Wallace. Appl Biochem Biotechnol 10:12–21, 1984.

12. S Zheng, RL Beissinger. Biomater Artif Cells Immobil Biotechnol 22:487–501, 1994.

13. K Yokoyama, K Yamanouchi, M Watanabe, T Matsumoto, R Murashima, T Daimoto, T Hamano, H Okamoto, T Suyama, R Watanabe, R Naito. Fed Proc 34:1478–1484, 1975.

7

Surface Rheology of a Nonaqueous System and Its Relationship with Foam Stability

Mojahedul Islam
Witco Corporation, Dublin, Ohio

Anita I. Bailey
Imperial College of Science, Technology, and Medicine, London, England

ABSTRACT

Surface rheological properties of a surfactant system are important parameters in determining the degree of foam stability. The investigation described in this chapter deals with the determination of surface rheological parameters of a nonaqueous system and its foam stability. The system investigated is a mixture of *n*-decane/decalin (50/50 wt%) and a nonionic fluorocarbon surfactant, FC740. Shear viscosity measurements were carried out using a deep-channel viscometer. The surface dilatational properties were obtained from dynamic surface tension measurements using the stress relaxation technique. The results revealed that the FC740 was capable of forming a gellike structure at the air–oil interface and the surface tension relaxation process was governed by two major processes. The fast process is mainly attributed to hindered diffusion, whereas the slow process is attributed to reorientation of the adsorbed molecules. The correlation between these surface rheological parameters and foam stability is also reported.

I. INTRODUCTION

The stability and breaking of a foam is a subject of great importance because foams occur both as a desirable product and as an undesirable entity in many industrial processes, such as firefighting, froth flotation, foam fractionation, food products (cream, sponges), personal care products, foamed latex, expanded polymers, and so forth. In all these applications, stable foams of varying degrees are necessary [1–7]. Foams can also be undesirable in many industrial processes. Examples are their occurrence in the refining of crude oil, sugar processing, steel production, paper manufacturing, and so forth [8–13].

The various factors that determine foam stability are foam drainage, surface elasticity (dynamic surface tension), surface shear viscosity, and electrical double-layer repulsion. For very stable foams, interbubble diffusion of the gas also plays an important role. A considerable amount of work has been carried out on aqueous surfactant systems [14–21]. However, with respect to nonaqueous systems, investigation has been limited [22].

The investigation described in this chapter deals with the determination of foam stability and the surface rheological properties of a nonaqueous system. The surface dilatational properties were obtained from dynamic surface tension measurements using the stress relaxation technique [23,24], and surface shear viscosity was measured using a deep-channel viscometer [25].

II. EXPERIMENT

A. Materials

1. Surfactant

An oil-soluble polymeric fluorocarbon surfactant (FC740) was used as the foam-stabilizing agent. This surfactant is manufactured by 3M, USA. It is nonionic in nature and was obtained as 50% active, containing an aromatic solvent. It was used as received without any further purification.

2. Solvents

Solvents used were an equiweight mixture of n-decane and decalin. The purpose of using this mixture was to simplistically model a crude oil on the basis of the equivalent alkane carbon atom number (EACN) [26,27]. They were obtained from a chemical vendor and were of analytical grade.

B. Methods

1. Foam Stability

Foam stability was measured using a column foam cell [12], and nitrogen was used as the gaseous medium. In this method, a steady column of foam height was obtained by bubbling nitrogen through the sintered disk of the column containing 5 mL of the desired surfactant solution. After achieving a steady height, nitrogen flow was shut off and foam collapse was measured as a function of time. The average foam lifetime, L_f, was calculated from the experimental collapse profile curve using

$$L_f = \frac{1}{h_0} \int_0^{t_c} h \, dt \tag{1}$$

where h_0 is the initial or steady foam height, h is the foam height at time t, and t_c is the total collapse time.

2. Surface Shear Viscosity

Surface shear viscosity was measured using a deep-channel viscometer [25]. Figure 1 is a schematic diagram of the apparatus used. The instrument consists of

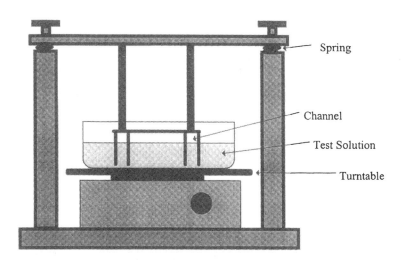

Figure 1 Schematic diagram of a deep-channel viscometer. The apparatus consists of circular channel (formed by two concentric cylinders), a Pyrex dish, a turntable and a coil (not shown) immersed into the solution for temperature control.

a channel formed by two concentric cylinders mounted on a support. This support was, in turn, spring mounted on three metal posts, thus enabling the channel to be leveled and the gap between the channel and the moving floor to be adjusted to any desired value. A Pyrex dish was used as the container for the solution. The depth of the solution was measured by means of a cathetometer. Throughout the investigation, the volume (250 mL) of the solution and the gap between the channel and the moving floor were kept constant. The temperature of the solution was controlled by circulating water through a coil immersed into the solution. The turntable could be rotated at various angular velocities ranging from 0.217 to 0.101 rad/s.

Prior to charging the vessel, the solution was equilibrated to a desired temperature. After pouring the solution into the vessel, a small Teflon particle was placed on the surface of the solution inside the channel. In all experiments, the particle size was kept as constant as possible. Due to the concave nature of the meniscus, the particle remained in the center of the channel. Thus, when the turntable was rotated, the centerline velocity, v_c, of the surface could be evaluated by measuring the time, t_p, taken for the Teflon particle to complete one revolution. The relative surface shear viscosity η_s^r was then calculated using [25]

$$\eta_s^r = \frac{v_c - v_c^*}{v_c^*} = \frac{t_p - t_p^*}{t_p^*} \tag{2}$$

where t_p^* is the particle revolution time without any surfactant. v_c^* is the centerline velocity of the surface without any surfactant.

3. Dilatational Rheology

The dilatational rheological properties is a measure of the response of a surface subjected to a dilatation and is described by a surface modulus, ε, given by

$$\varepsilon = \frac{d\gamma}{d \ln A_s} \tag{3}$$

where γ is the surface tension and A_s is the area of the surface element considered. The magnitude of ε, in general, depends on the extent and rate of the expansion. The modulus is a complex function of the frequency and can be written

$$\varepsilon(i\,\omega) = \varepsilon'(\omega) + i\varepsilon''(\omega) \tag{4}$$

where ε' and ε'' are the real and imaginary parts of the complex modulus, respectively, $i = \sqrt{-1}$, and ω is the angular frequency. The real part $\varepsilon'(\omega)$ accounts for the elastic energy stored in the system and is known as the dilatational elasticity. The imaginary part $\varepsilon''(\omega)$ accounts for energy dissipation in the system and may be expressed in terms of surface dilatational viscosity, η_d,

$$\varepsilon'' = \omega\eta_d \tag{5}$$

The surface dilatational properties were measured using the stress relaxation method [23,24]. In this method, a "surface strain" is applied by expanding the surface in a single step and the corresponding change in surface tension $\Delta\gamma$ and its decay with time $\Delta\gamma(t)$ are measured, thus obtaining the stress relaxation function. The dilatational modulus as a function of frequency was then calculated from the Fourier transformation (FT) of stress relaxation function from time to frequency domain as

$$\varepsilon = \frac{FT(\Delta\gamma(t))}{FT(\Delta A_s(t)/A_s)} \tag{6}$$

where the Fourier transformation of a function $f(t)$ is defined by

$$FT(f(t)) = \int_0^\infty f(t)e^{-i\omega t}\, dt \tag{7}$$

Because $\Delta A_s(t)/A_s$ can be assumed to be a step function [i.e., $\Delta A_s(t)/A_s = 0$ for $t < 0$ and $\Delta A_s(t)/A_s =$ constant for $t \geq 0$], the FT was obtained analytically. The Fourier transformation of the surface tension decay, $\Delta\gamma(t)$, was obtained by first fitting an equation to the data, followed by transformation of the decay curve from time to frequency domain. The equation was fitted on the basis that the relaxation was due to multiple processes occurring simultaneously. For a given process, one may represent it by an exponential decay; thus, for the collective processes, it may be represented by a summation of exponentials. Therefore, the surface tension relaxation may be represented by

$$\Delta\gamma(t) = \sum_{k=1}^{n} A_k e^{-B_k t} \tag{8}$$

where B_k is the characteristic frequency of the kth process and $A_k/\sum A_k$ is the fractional contribution of that process. The above function was fitted to the experimental data using a nonlinear routine [28]. On substituting for $\Delta\gamma(t)$ in Eq. (6), this leads to

$$\varepsilon = \frac{FT\left(\sum_{k=1}^{n} A_k e^{-B_k t}\right)}{FT(\Delta A_s(t)/A_s)} \tag{9}$$

Solving the above equation gives

$$\varepsilon' \frac{1}{F_A} \sum_{k=1}^{n} \frac{A_k \omega^2}{B_k^2 + \omega^2} \tag{10a}$$

where $F_A = \Delta A_s/A_s$ is the fractional change in area.

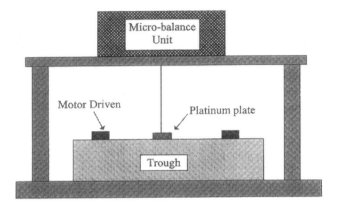

Figure 2 Schematic diagram of the apparatus for measuring the dynamic surface tension. The temperature of the solution was controlled by a coil (not shown) immersed into the solution.

and

$$\varepsilon'' = \frac{1}{F_A} \sum_{k=1}^{n} \frac{A_k B_k \omega}{B_k^2 + \omega^2} \tag{10b}$$

Using Eqs. (10a), (10b), and (5), both the surface dilatational elasticity and viscosity were calculated.

The apparatus used in this experiment is shown in Figure 2; it consists of a Langmuir trough with motor-driven barriers and a microbalance system (Perkin–Elmer microbalance and a Perkin–Elmer TGS-2 balance control unit) connected to a Lenseis x–t chart recorder. The surface tension of the liquid in the trough was measured using the Wilhelmy plate technique. The solution of desired surfactant concentration was poured into the trough and was allowed to equilibrate (temperature and surface tension). The temperature of the solution was controlled by means of water circulating from a bath through the coils inserted into the solution in the trough. The surface of the liquid bounded between the barriers was then quickly expanded by simultaneously moving both barriers. The platinum plate used for measuring surface tension was placed in the center of the trough and perpendicular to the barriers—this minimizes the resistance to surface flow during expansion. The change in surface tension and its decay with time was then continuously measured and recorded on the chart recorder. The characteristic frequency of the relaxation process, the surface dilatational elasticity, and the dilatational viscosity were evaluated using the above equations.

4. Surface Tension

The surface tension of all the solutions used was measured using the Wilhelmy plate technique. A platinum plate was used as the probe. Prior to every use, the plate was cleaned by flaming it to red hot.

III. RESULTS AND DISCUSSION

A. Surface Shear Viscosity

Preliminary experiments were carried out using solution containing 1×10^{-4} g/mL FC740 in order to ascertain the shear rate to be used for measuring the viscosity. During this investigation, it was observed that the surface viscosity increased with time as it was sheared. Initially, it was thought to be due to the effect of shearing. To confirm this, an experiment was carried out in which the surface was initially sheared for a given time, and then allowed to rest, followed by resumption of shearing. The data obtained from this experiment are shown in Figure 3. It can be seen that increase in viscosity continued irrespective of shearing of the surface. In addition, the exponential shape of the curve suggests that it is not only due to adsorption of surfactant molecules. If this effect were due

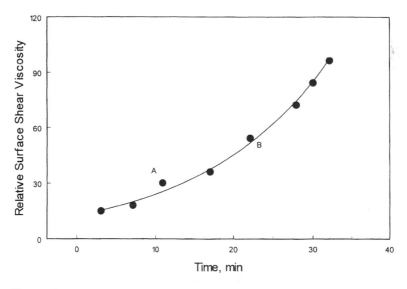

Figure 3 Relative surface shear viscosity as a function of time. The surfactant (FC740) concentration was 1×10^{-4} g/mL and temperature was $-20°C$. At point A, shearing was stopped, and at point B, shearing was resumed.

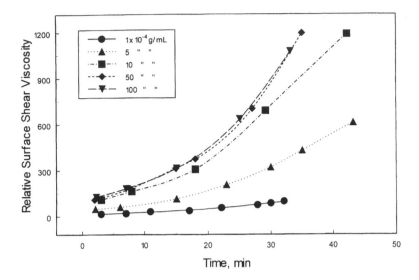

Figure 4 Relative surface shear viscosity as a function of time at 20°C for different concentrations of FC740.

to surfactant adsorption, one would expect an initial rapid increase in surface viscosity approaching a limiting value. Because the surfactant molecule is polymeric in nature, it is possible that this is due to the formation of a structured layer resulting from entanglement of the molecules, which would be expected to have a profound effect on viscosity. Thus, it was decided that in addition to temperature and concentration, the effect of time should also be investigated. It was also decided to measure viscosities at a given shear rate, namely 0.159 s^{-1}, for all experiments conducted.

Figures 4 and 5 show the data for relative surface shear viscosity, η_s^r, as a function of time for different concentrations of FC740 at 20°C and 60°C, respectively. From Figure 4, it can be seen that the exponential behavior is even more marked at a higher concentration. The shape of the curves implies that at a higher concentration, not only does the surfactant molecules have greater interaction resulting in a more entangled surface layer and hence higher viscosity, but the rate of entanglement is faster. This is to be expected, as a higher bulk concentration means a greater number of molecules at the surface, which will increase the probability of intermolecular collision and, thus, greater rate of entanglement. Another feature is that at a higher concentration, the η_s^r versus time plots are very similar, which suggests that at higher concentrations, viscosities are similar for a given time. Figure 6 shows the variation of surface shear viscosity with the

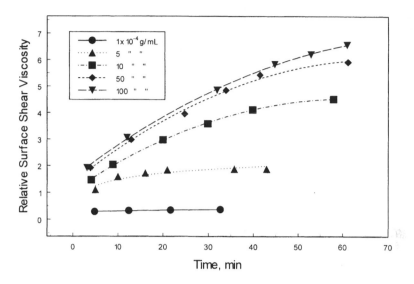

Figure 5 Relative surface shear viscosity as a function of time at 60°C for different concentrations of FC740.

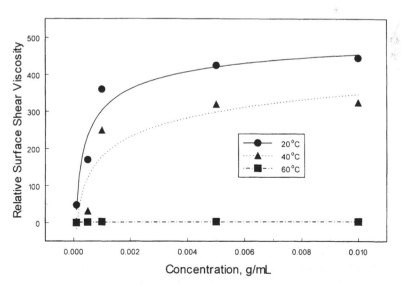

Figure 6 Relative surface shear viscosity as a function concentration at time $t = 20$ min for different temperatures.

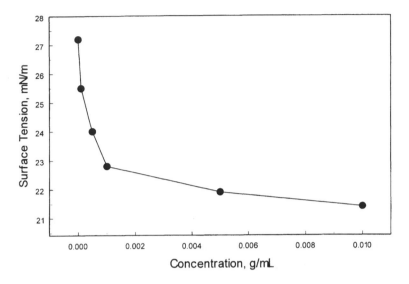

Figure 7   Surface tension versus concentration of FC740 at 20°C.

concentration of FC740 at time $t = 20$ min for different temperatures. Indeed, η_s^r reaches a limiting value at a concentration greater than 1×10^{-3} g/mL, but it increases rapidly up to this concentration. This can be attributed to the surface concentration, which increases rapidly but soon reaches a saturation value at concentrations greater than 1×10^{-3} g/mL, as indicated by the γ versus concentration plot (Fig. 7).

The effect of temperature on surface shear viscosity is shown in Figures 4–6. An examination of these graphs shows that as the temperature increases, the surface shear viscosity decreases dramatically. In addition, at a higher temperature, 60°C, the exponential behavior of η_s^r versus time is no longer present (Fig. 5). These observations can be attributed to the greater kinetic energy of surfactant molecules, which breaks up the structured semirigid layer, hence the absence of exponential behavior.

B.   Surface Dilatational Rheology

Dynamic surface tension measurements were made using the stress relaxation experimental method at four different concentrations and three different temperatures. The concentrations and temperature investigated were 5×10^{-2}, 1×10^{-2}, 1×10^{-3}, and 1×10^{-4} g/mL and at 20°C, 40°C, and 60°C. The surface tension decay data $\Delta\gamma(t)$ were fitted to an analytical expression as per Eq. (8). A compari-

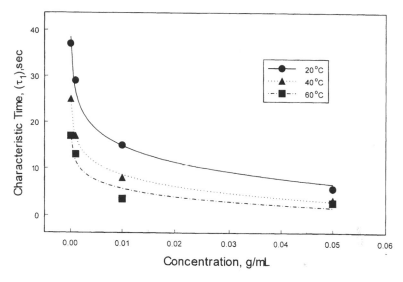

Figure 8 Effect of concentration of FC740 on the characteristic time (τ_1) of the surface tension relaxation at different temperatures.

son between experimental and fitted lines shows that the fitting is in good agreement, thus confirming the validity of the equations used. Using the A and B parameters of Eq. (8) one can then obtain the viscoelastic spectrum (i.e., ε' and ε'' at different frequencies). At a very high frequency, the dilatational elasticity (ε') reaches a limiting value (ε_0) and is known as the limiting elasticity.

It was also observed that for all experimental data, an adequate fit was obtainable using two exponentials. This indicates that at least two dominant processes were responsible for the relaxation. The characteristic times τ_1 and τ_2 ranges from 1.7 to 62 s and from 55 to 2267 s, respectively, depending on the concentration and temperature. The variation of these characteristic times as a function of concentration at different temperatures are shown in Figures 8 and 9. The fast process is most probably due to bulk diffusion, whereas the slow process could be attributed to the reorientation of the surfactant molecules on the surface. The time scale of the fast process is in the range of 2–60 s, which is rather large for the diffusion relaxation time, even for a polymeric surfactant. The surface shear viscosity data (discussed earlier) indicate that FC740 forms a structured semirigid layer. Therefore, it is likely that this monolayer could hinder adsorption, thus resulting in much longer characteristic times.

From Figures 8 and 9, it can be seen that as the concentration increases, the characteristic times τ_1 and τ_2 both decrease sharply up to bulk concentration

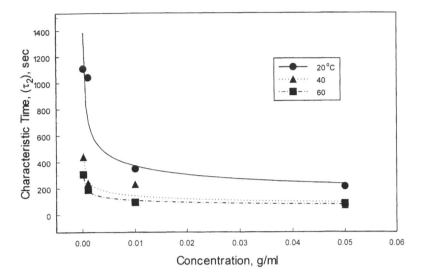

Figure 9   Effect of concentration of FC740 on the characteristic time (τ_2) of the surface tension relaxation at different temperatures.

of 1×10^{-3} g/mL; beyond this, the decrease is much less dramatic. For τ_1, the profile can be partly attributed to an increase in the rate of adsorption due to the increase in monomer concentration. Although one would expect that as the concentration increases, the barrier to adsorption also increases, as indicated by the shear viscosity data (i.e., formation of a more structured semirigid film). On the other hand, this implies that as the surface is expanded, there will be more spaces of less dense areas because the surface diffusion of the molecules to the newly created space will be more hindered. This suggests that the adsorption barrier may be even less than what it was at the lower bulk concentration. Therefore, a combination of a greater diffusion rate due to increase in monomer concentration as well as a decrease in adsorption barrier will result in an increase in τ_1.

Similarly, there is now a greater entropy difference between the newly created space which has been partially filled by the approaching molecules from the bulk and the semirigid structured domain of the surface. This implies that molecular rearrangement between these areas will occur at a faster rate than at a lower bulk concentration, and as a result, τ_2 will also increase. This phenomenon is shown schematically in Figure 10.

From Figures 8 and 9, it can also be seen that at a given concentration, both relaxation times τ_1 and τ_2 decrease as the temperature increases. This can

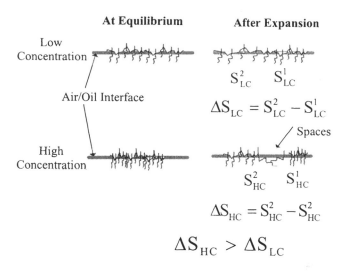

At Equilibrium   After Expansion

$$\Delta S_{LC} = S_{LC}^{2} - S_{LC}^{1}$$

$$\Delta S_{HC} = S_{HC}^{2} - S_{HC}^{2}$$

$$\Delta S_{HC} > \Delta S_{LC}$$

Figure 10 Schematic representation of the possible behavior of FC740 at an oil–air interface at low and high concentrations. S is the entropy, subscripts LC and HC are low and high concentrations, respectively, and superscript 1 and 2 are domains of different surface concentrations of the molecules immediately after expansion.

be attributed to the higher kinetic energy of the molecules, which will result in both a greater adsorption rate as well as a faster reorientation of the molecules.

If a film is deformed very quickly, such that the time scale is less than any of the characteristic times of the relaxation processes, then the monolayer will behave as though it were insoluble [29]. The degree of resistance to such a film will depend on the nature of the film (i.e., liquid or condensed state) and is given by the limiting elasticity, ε_0. Figure 11 shows the variation of ε_0 as a function of concentration at different temperatures, and it can be seen that ε_0 increases very rapidly within a very narrow range of concentrations and reaches a limiting value. This indicates that the resistance of the film increases with increasing concentration, implying that the film becomes more rigid, which is in accord with the results of the shear viscosity study (discussed earlier). In addition, from surface tension measurements it has been evaluated that the saturation adsorption is reached around 1×10^{-3} g/mL, which corresponds well with ε_0 values, which reaches its limiting value around a similar concentration. With increasing temperature as the film becomes fluid, as indicated by the shear viscosity measurements, the ε_0 value also decreases.

The resistance to deformation of a film due to its viscous contribution is given by the dilatational viscosity, η_d. This is also a measure of the film's ability

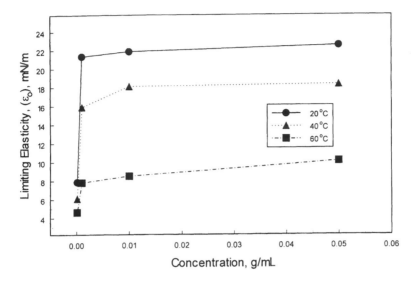

Figure 11 Variation of the limiting elasticity (ε_0) of the adsorbed layer of FC740 on the air–water interface, as a function of concentration at different temperatures.

to absorb external disturbance (e.g., mechanical shock). As Eq. (5) implies, the values of η_d will depend on the frequency at which it has been evaluated. Because an external disturbance can occur at many ranges of frequencies, it is not practical to evaluate η_d at all frequencies; thus, an arbitrarily chosen frequency of 0.1 Hz was used to calculate the dilatational viscosities. A plot of dilatational viscosity versus concentration is shown in Figure 12, and it can be seen that with increasing concentration, η_d increases sharply and reaches a pseudoequilibrium value. At a higher temperature (60°C), the film exhibits low values of surface dilatational viscosity at all concentrations.

C. Foam Stability

The foaming ability of FC740 and the decane/decalin system was measured at different concentrations and temperatures. Figure 13 shows the foam stability data as a function of the concentration of FC740. As the concentration increases, the foam stability increases sharply up to 1×10^{-3} g/mL; a further increase in concentration only increases the stability slightly. This can be attributed to surface tension reduction of the solution; as indicated in Figure 7, it almost reaches a limiting value around this concentration.

Figure 14 shows the variation of foam stability with relative shear viscosity.

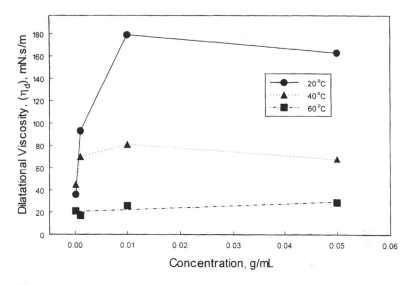

Figure 12 Variation of the dilatational viscosity (η_d) of the adsorbed layer of FC740 on the air–water interface, as a function of concentration at different temperatures. Frequency is set at 0.1 Hz.

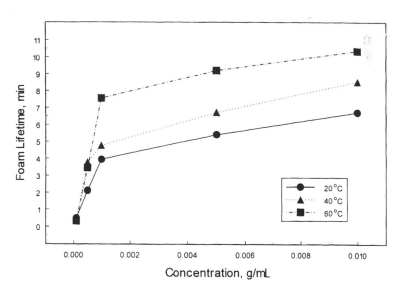

Figure 13 Foam stability (foam lifetime) of the FC740–decane/decalin system at different surfactant concentrations.

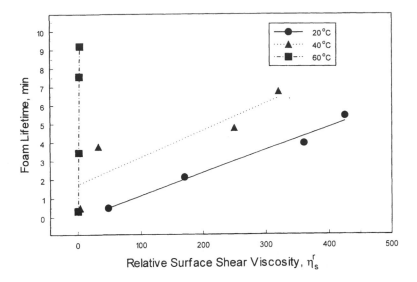

Figure 14 Correlation between foam stability and surface shear viscosity at different temperatures.

For the isothermal curves, the foam stability increases with increasing η_s^r. It can be explained in the following manner. First, the increase in η_s^r retards the drainage of the liquid from the lamellae of the bubbles [14], thus increasing the foam lifetime. Second, high values of η_s^r enable the foam films to absorb more shocks. Higher foam stability resulting from higher surface viscosities has also been observed by other investigators [30,31]. At the highest temperature (60°C), it seems that a slight increase in viscosity increases the foam life dramatically. Because viscosity is very low at this temperature, this increase of foam stability is mainly due to the dilatational properties.

The effect of characteristic times τ_1 and τ_2 of the surface tension relaxation function on foam stability are shown in Figures 15 and 16, respectively. It can be seen that at a given temperature, the foam lifetime decreases as the relaxation time increases. A longer relaxation time suggests that the surface film will take longer to restore to its equilibrium condition if subjected to an external disturbance. Thus, the foam lamellae will remain at the unsteady condition for a longer period of time, which may result in film rupture. At a higher temperature, the decline in foam stability is sharper and can be correlated with low values of shear viscosities encountered at higher temperature. It has been observed by other investigators [32] that low values of η_s^r are associated with foams which are less

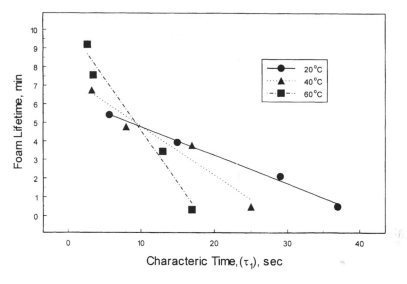

Figure 15 Effect of characteristic time (τ_1) of the fast process of the surface tension relaxation on foam stability at different temperatures.

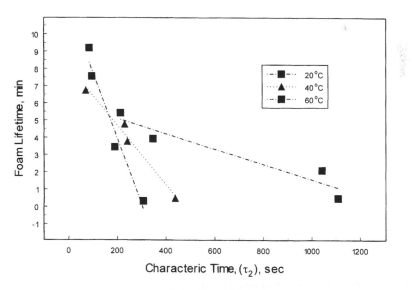

Figure 16 Effect of characteristic time (τ_2) of the slow process of the surface tension relaxation on foam stability at different temperatures.

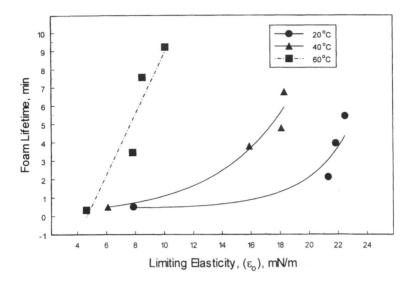

Figure 17 Correlation between foam stability and limiting elasticity (ε_0) at different temperatures.

stable. Thus, if a system approaches having low values of η_s^r and long relaxation times, the foam stability is expected to decrease sharply.

The relationship between other dilatational surface rheological properties, limiting elasticity, ε_0, and dilatational viscosity, η_d, are shown in Figures 17 and 18, respectively. In general, higher values of ε_0 favors high foam stability. This effect is more marked in the higher temperature range. In the case of dilatational viscosity, η_d, it can be seen that for a given temperature, higher values of η_d also favor high foam stability. A high η_d means that the film has a greater ability to resist surface deformation, hence promoting higher foam stability. A similar finding has been reported for crude oil foam: High values of η_d were associated with high foam stability [12,13].

IV. CONCLUSION

Surface rheological properties and foam stability of a nonaqueous system, a fluorocarbon surfactant (FC740), and a decane/decalin mixture were investigated. The results of shear viscosity measurements revealed that the FC740 was capable of forming a gellike structure at the air–oil interface. Furthermore, the values of relative surface shear viscosity, η_s^r, were found to be time dependent and this is attributed to entanglement between surfactant molecules.

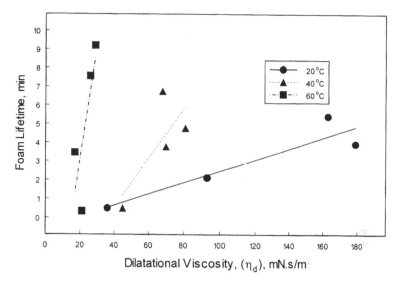

Figure 18 Correlation between foam stability and dilatational viscosity (η_d) at different temperatures.

The surface dilatational properties were obtained from dynamic surface tension measurements using the stress relaxation technique. The results indicated that the relaxation process of the FC740 system was governed by two major processes having characteristic times τ_1 and τ_2 ranging from 1.7 to 62 and from 55 to 2267 s, respectively. The fast process is mainly attributed to a hindered diffusion process, whereas the slow process to reorientation of the adsorbed molecules.

The correlation between foam lifetime and the surface properties measured showed that the relatively high foam stability will be attained by a system having the following properties:

1. Low values of both τ_1 and τ_2
2. High values of limiting elasticity and dilatational viscosity
3. High values of η_s^r

REFERENCES

1. JG Corrie. In: RJ Akers, ed. Foams, Proceedings of a Symposium Organized by the Society of Chemical Industries, Colloid & Surface Chemistry Group, London. Academic Press, 1976, p. 195.

2. JJ Bikerman. Foams, New York: Springer-Verlag, 1973.
3. JW Mansvelt. In: RJ Akers, ed. Foams, Proceedings of a Symposium Organized by the Society of Chemical Industries, Colloid & Surface Chemistry Group, London. Academic Press, 1976, p. 283.
4. TF Cooke. In: RK Prud'homme, SA Khan, eds. Foams. Surfactant Science Series 57. New York: Marcel Dekker, 1995, p. 339.
5. M Rieger. In: RK Prud'homme, SA Khan, eds. Foams. Surfactant Science Series 57. New York: Marcel Dekker, 1995, p. 381.
6. T Briggs. In: RK Prud'homme, SA Khan, eds. Foams. Surfactant Science Series 57. New York: Marcel Dekker, 1995, p. 465.
7. R Prud'homme, GG Warr. In: RK Prud'homme, SA Khan, eds. Foams. Surfactant Science Series 57. New York: Marcel Dekker, 1995, p. 511.
8. AD Barber, S Hartland. Trans Inst Chem Eng 53:106, 1975.
9. AD Barber, EF Wijn. Inst Chem Eng 56:3.1/15–3.1/31, 1979.
10. WL Bolles. Chem Eng Prog 63(9):48, 1967.
11. VA Kuznetsov, IG Chufarova. Zh Prikl Khim 29:688, 1956.
12. IC Callaghan, EL Nuestadter. Chem Ind (London), No. 2:53, 1981.
13. IC Callaghan, CM Gould, RJ Hamilton, EL Nuestadter. Colloids Surf 8:17, 1983.
14. JA Kitchner, CF Cooper. Quart Rev Chem Soc 13:71, 1959.
15. JT Davies, EK Rideal. Interfacial Phenomena. New York: Academic Press, 1963.
16. S Ross. Chem Eng Prog 63:41, 1967.
17. JA Wingrave, TP Matson. J Am Oil Chem Soc 58:347A, 1981.
18. RH Ottewill, DL Segal, RC Watkins. Chem Ind (London), No. 2:57, 1981.
19. DT Wasan, SP Christiano. In: KS Birdi, ed. Handbook of Surface and Colloid Chemistry. Boca Raton, FL: CRC Press 1997, p. 179.
20. A Bhakta, E Ruckenstein. Adv Colloid Interf Sci 70:1, 1997.
21. A Colin, J Giermanska-Kahn, D Langevin, B Desbat. Langmuir 13:2953, 1997.
22. DL Schmidt. In: RK Prud'homme, SA Khan, eds. Foams. Surfactant Science Series 57. New York: Marcel Dekker, 1995, p. 287.
23. G Loglio, V Tesei, R Cini. Ber Bunsenges Phys Chem 81:541, 1977.
24. G Loglio, V Tesei, R Cini. J Colloid Interf Sci 71:316, 1979.
25. RA Burton, RJ Mannheimer. Adv Chem Ser 63:315, 1967.
26. RL Cash, JL Cayias, RG Fournier, JK Jacobson, T Schares, RS Schechter, WH Wade. Improved Oil Recovery Symposium of the Society of Petroleum Engineers of AIME, Tulsa, Oklahoma, March 22–24, 1976. Paper number SPE 5813.
27. RL Cash, JL Cayias, RG Fournier, JK Jacobson, T Schares, RS Schechter, WH Wade. J Colloid Interf Sci 59:39, 1977.
28. PG Guest. Numerical Methods of Curve Fitting. Cambridge: Cambridge University Press, 1961.
29. EH Lucassen-Reynders, J Lucassen, PR Garrett, D Giles, F Hollway. Advances in Chemistry Series 144, 272 (1975).
30. AG Brown, WC Thuman, JW McBain. J Colloid Sci 8:491, 509, 1953.
31. AK Malhotra, DT Wasan. In: IB Ivanov, ed. Thin Liquid Films. New York: Marcel Dekker, 1988, p. 294.
32. DO Shah, NF Djabbarah, DT Wasan. Colloid Polym Sci 256:1002, 1978.

8

Stratification of Foam Films Made from Polyelectrolyte Solutions

Dominique Langevin
Université Paris Sud, Orsay, France

ABSTRACT

Studies of thin liquid films, made from semidilute polyelectrolyte solutions, are presented. The disjoining pressure variation with film thickness exhibits oscillations, corresponding to film stratification. The oscillations become sharper as the polymer concentration, c, increases, and disappear when salt is added. The period of the oscillations scales as $c^{-1/2}$. The observed stratification is related to the polymer network and the size of the steps to the mesh size, ξ.

I. INTRODUCTION

The behavior of polymers near surfaces and in confined environments has been the subject of many studies, both fundamental and applied [1]. Indeed, it is important to understand this behavior for the control of separation processes (filtration, gel chromatography, osmosis, etc.) and oil-recovery processes (drilling muds with added polymers, viscosity control of injected water), among others. From the fundamental point of view, it is interesting to describe how the polymer coils or the polymer networks are affected by the confinement. When the polymers are confined between two surfaces and when the surfaces are close enough, surface forces arise. Their origin is steric most of the time, caused by the repulsion due

151

to the overlap of chains adsorbed on each surface. The knowledge of these forces is important to control colloid stability or flocculation. Numerous measurements with the surface-force apparatus (SFA) have been reported in the literature for polymers adsorbed onto mica surfaces [2]. In addition to the usual steric repulsion, small attractive forces have been evidenced when the chains are in a "poor" solvent. These attractive forces are due to monomer–monomer interactions and, in part, to bridging of the surfaces, when the polymer surface coverage is low. Little is known for the case of nonadsorbed polymers, where depletion layers and depletion forces are expected. Depletion forces have been measured in solutions of polymer like micelles that do not adsorb at the mica surfaces (coated with surfactant of the same charge sign as that of the micelles) [3]. Oscillatory forces have been recently observed with an atomic force microscope in polyelectrolyte solutions. These forces were attributed to the formation of a crystal of coils stabilized by long-range electrostatic forces [4], although they are more likely related to the polymer network, as will be seen in the following with measurements on freely suspended polymer films.

II. DRAINAGE OF SURFACTANT SOLUTIONS

When a foam is first formed, the amount of liquid in the foam films is large, and these films are thick. Afterward, they drain under the influence of gravity and capillary pressure: for example, the Plateau borders connecting the foam films are curved in such a way that the liquid pressure is smaller there than in the flat parts of the foam films. The drainage of foam films has been extensively studied, in particular in horizontal films with devices in which the pressure can be controlled more easily; for example, Scheludko cells, and porous plates [5,6]. The easiest way to estimate the velocity of drainage is to assume that the film is flat parallel, that its surfaces are immobile, and that the fluid flows regularly from the center toward the Plateau borders. In such a simple case, an expression derived by Reynolds for the flow between two rigid plates brought together can be used [7]:

$$V = -\frac{dh}{dt} = \frac{h^3}{3\eta R^2}\Delta P \tag{1}$$

where h is the film thickness, R is the film radius, η is the fluid viscosity, and ΔP is the difference in pressure between the film center and border. In practice, the film surfaces are not solid, and surface flow can arise, this flow being controlled by the surface rheology. Numerical calculations done in the simple case of a flat parallel film show that the influence of surface compression elasticity can be important for thick films [8].

Sometimes, "dimpling" instabilities appear during the drainage of large circular horizontal films, in which the film is pinched off around its boundary and is thicker in the center [9]. When the surface elasticity is moderate, a second instability can take place in which the dimple loses its circular shape and is sucked away rapidly into the Plateau border. During this fast process, the film frequently ruptures.

When the surfactant solutions contain aggregates, the drainage process becomes more complex. At the beginning of the century, Johnnott and Perrin reported that the drainage of the foam films occurred stepwise [10]. Later, it was shown that this was due to the presence of surfactant micelles which form layered structures below the film surfaces, and that the drainage proceeds by expulsion of the layers, one after the other, into the Plateau border [11–13]. The expulsion of a layer of micelles begins by the nucleation of one or more holes, which expand in different ways.

When the film is thick and contains several layers, the radius of the holes increases as the square root of time. This has been modeled by Nikolov et al. by assuming that the micelle layer flows regularly through the Plateau border [14].

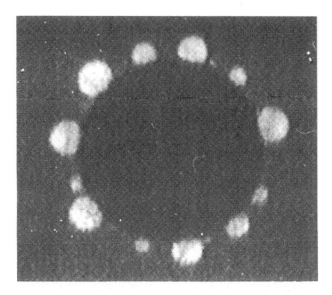

Figure 1   Image of a film made with a dodecyltrimethylammonium bromide solution above the critical micelle concentration. The darker regions are thinner, and when they expand, they push a rim that can fracture into tiny little drops visible in the picture. (After Ref. 13.)

It has also been noted that the problem is analogous to the spreading of stratified drops on solids, where an equivalent diffusion coefficient can be defined: $R_{hole} \sim \sqrt{D_{eff}t}$, with

$$D_{eff} = -\frac{h^3}{3\eta} \frac{\partial \Pi_d}{\partial h} \tag{2}$$

where Π_d is the "disjoining pressure" (i.e., the force per unit area exerted between the two film surfaces) [15]. When this force is mainly a van der Waals interaction, $D_{eff} \sim A/6\pi\eta h$, where A is the Hamaker constant. This expression gives an order of magnitude of 10^{-8} cm^2/s for D_{eff}, close to the measured values [13]. When the films are thinner, holes open in a micelle layer; the holes are bordered by a liquid rim, which can rupture via a mechanism similar to the Rayleigh instability of a liquid jet (fig. 1) [16]. The radius of the hole then increases linearly with time. This is similar to dewetting processes [17] and, although less rapid, to the bursting of foam films [18,19].

Very similar features were observed in mixed polyelectrolyte–surfactant solutions, at concentrations where micelles were not present [20,21]. As in the case of micellar solutions [12], this stratification is associated with oscillatory surface forces, as will be discussed later.

III. SURFACE FORCES

If the foam drainage has been smooth enough, so that no early rupture has occurred, a regime where interactions between the two sides of the film become significant can be attained; this is in a thickness region of about 50 nm and less. Typical interactions are van der Waals forces (attractive) and electrostatic forces (repulsive). Short-range forces (steric, hydration) are also present [22]. The disjoining pressure is the sum of the corresponding terms:

$$\Pi_{vdw} \sim -\frac{A}{6\pi h^3}, \qquad \Pi_{elect} \sim Be^{-\kappa h}, \qquad \Pi_{steric, hydration} \sim Ce^{-h/\lambda} \tag{3}$$

where B and C are constants, κ^{-1} is the Debye–Hückel screening length, and λ is the range of the short-range forces, typically a few tens of nanometers. When the electrostatic repulsion is strong enough, an energy minimum can be found for thicknesses of about 10 nm. Figure 2 is a schematic diagram for the disjoining pressure in which short-range repulsive forces have also been taken into account. If a lateral pressure ΔP is applied to the film (gravity, Laplace pressure), the thickness of the film decreases down to h_1 (see Fig. 2); the corresponding equilibrium film is called "common black film." If the pressure ΔP is larger than the

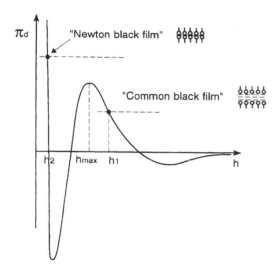

Figure 2 Schematic representation of the variation of the disjoining pressure Π_d with film thickness h. The dotted lines correspond to different applied pressures ΔP and show the final equilibrium thicknesses of the film.

electrostatic barrier, one reaches a very small film thickness after drainage, where the water layer thickness is of order λ; this is the so-called "Newton black film."

We have studied films with the porous plate technique formed from mixed aqueous solutions of surfactants (DTAB, dodecyltrimethylammonium bromide; $C_{10}E_5$, pentaethylene glycol decyl ether) and various polyelectrolytes:

- PAMPS, a random copolymer of acrylamide AM and acrylamidomethylpropane sulfonate (AMPS), with a fraction of charged monomers of 25% (each AMPS monomer bears one elementary charge, AM being neutral); two molecular weights were used; 2.2×10^6 and 4×10^5 [23]
- PSS (polystyrene sulfonate), 100% sulfonated, molecular weight of 7×10^5 [24]
- Xanthan, a natural polysaccharide, molecular weight of the order of 10^7 [25]

The surfactant concentrations used were well below the concentration at which mixed polymer–surfactant aggregates form in the bulk. Part of the surfactant is concentrated at the film surfaces where it forms a mixed monolayer with the polymer when the surfactant is ionic [23,26]. This surface layer confers a good stability to the thin film; without the polymer, the DTAB film breaks almost immediately. At the polymer concentrations used, the polymer alone does not adsorb at the film surfaces (surface tension equal to that of water). Because the polyelectrolyte chains dimensions are large, the solutions are expected to be in the semidilute range. We have checked that this was the case by measuring the viscosity of the polymer solutions. The reduced viscosity $\eta_r = (\eta - \eta_0)/\eta_0 c$ varies as $c^{-1/2}$ for both PAMPS and PSS, as usual for semidilute polyelectrolyte solutions [27]. For xanthan, it varies as c^2, as expected for solutions of rigid rods [28].

A typical disjoining pressure variation with film thickness is shown in Figure 3. At small thicknesses, a repulsive force is observed which can be fitted with an exponential form exp $-\kappa h$, as expected for screened electrostatic repulsion; κ is close to the calculated inverse Debye–Hückel length in the solution:

$$\kappa^2 = 4\pi l_B \sum Z_i^2 n_i \quad \text{with } l_B = \frac{e^2}{4\pi\epsilon kT} \text{ and } n_i = \frac{Nc_i}{M_i} \tag{4}$$

where l_B is the Bjerrum length, n_i is the number concentration of the ionic species i, c_i is its concentration by weight, M_i is its molecular weight, Z_i is its valency, N is Avogadro's number, e is the electron charge, ϵ is the dielectric constant of the solution, and kT is the thermal energy. For water at room temperature, $l_B \sim$ 0.7 nm.

When the polymer concentration is increased, additional branches appear [21,24,25]. It should be noted that only repulsive forces can be measured with the porous plate technique, at the difference of the SFA. When ΔP increases, h decreases and when the top of a given branch is reached, the thickness jumps to a smaller value on the next branch. When ΔP is decreased, the different branches

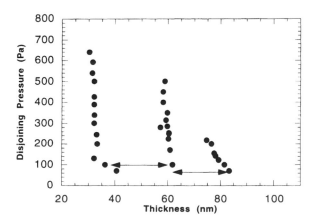

Figure 3 Disjoining pressure versus film thickness for PAMPS; $c = 1900$ ppm. The distance d is taken as the distance between arrows (distance between branches in the limit of vanishing disjoining pressures). (After Ref. 21.)

are obtained back (although not the jumps). At the pressures where jumps occur, spots of different thicknesses can be observed on the film. The number of spots depends on the pressure jump used to nucleate the thinner spots on the film. When the number of spots is small, the spots expand with a radius proportional to the square root of time. This is as observed for micellar solutions [11,13]. Further work is currently under way to extract information about the viscosity in the polymer films from this time evolution.

The distance d between the branches of the curves of Figure 3 is the same (when more than two branches are observed) and does not depend on the polymer's molecular weight. This distance d varies with the polymer concentration as $d \sim c^{-1/2}$ (Table 1). Numerically, d is about four times larger than the Debye length κ^{-1} and is close to the expected mesh size of the polymer network. It does not depend on the type of surfactant used to stabilize the film; for PSS, similar oscillations have been observed with both DTAB and $C_{10}E_5$. Finally, we have checked the effect of added salt (sodium chloride). The effect of increasing the salt concentration is similar to the effect of decreasing polymer concentration: The outer branches disappear and one is left with a single branch consistent with

Table 1 Comparison Between the Measured Distance d Between Force Branches and Calculated Mesh Sizes ξ for the Polymer Network as a Function of Polymer Concentration

Polymer	c (mM)	Measured d (nm)	Calculated ξ (nm)
PAMPS	2.72	70	46
	6.78	40	29
	8.73	29	26
	17.2	23	18
	36.2	19	13
PSS	4.85	43	37
	9.71	31	26
	19.4	22	19
	29.1	18	15
	38.8	15	13
Xanthan	0.8	57	59
	1	46	52
	1.25	41	47
	1.5	37.5	43
	1.9	33.6	38
	2.95	30	31

the electrostatic repulsion $\exp(-\kappa h)$ when the salt concentration becomes comparable to the polymer counterion concentration.

Let us recall briefly the present state of understanding of semidilute solutions of flexible polyelectrolytes [29]. If the distance between charges on the chain is smaller than the Bjerrum length l_B, there is a counterion condensation (Manning condensation). This is the case for PSS; for PAMPS, the polymer is weakly charged, and no condensation is expected to occur [30]. These polyelectrolyte chains are flexible enough; the intrinsic persistence length (the persistence length when the electric charges are fully screened, for instance by adding excess salt) is comparable to the monomer size. In such a case, the chain starts to coil until the electrostatic repulsion energy becomes larger than kT. The coil size is then ξ_e. If the chain is larger, it starts to form a new coil or "blob" and the whole chain can be viewed as a rodlike succession of electrostatic blobs of size ξ_e; this picture holds until the rod length reaches a value comparable to the (total) persistence length l_p, after which, a coil with larger dimensions (units l_p) is formed. In the semidilute regime, the rods form a mesh of size ξ, the average distance between overlap points of the rods. For rigid polyelectrolytes such as xanthan, the persistence length is large (up to 100 nm) and there are no electrostatic blobs; the chain itself is rodlike.

Small-angle x-ray and neutron-scattering experiments have evidenced the presence of a peak at a wave vector q^* corresponding to a characteristic distance that varies with polymer concentration as $c^{-1/2}$ and that has been identified with the mesh size [30–32]. The peak disappears when salt is added around a concentration comparable to that of the polymer ions. The scattering data and, in particular, the structure factor can be Fourier transformed to obtain the pair correlation function $g(r)$. A peak in the structure factor leads to oscillations in $g(r)$. When the polymer is confined between two surfaces and when the distance h is comparable to the period of the oscillations of $g(r)$, oscillatory forces between the surfaces are expected. These oscillations have been observed in simple fluids made of spherical molecules in the SFA apparatus [22]. They do not arise because the molecules tend to structure into semiordered layers at surfaces, but because of the disruption or change of this ordering during the approach of the second surface. A similar interpretation has been given for the oscillatory forces in micellar solutions [33]. Recently, they have been predicted to exist for polyelectrolyte solutions [34]. Clearly, our data confirm that such structural forces do exist in semidilute polyelectrolyte solutions.

We have compared the distance between branches d to the theoretical mesh size ξ [29]:

$$\xi = \frac{\alpha}{c^{1/2}} \tag{5}$$

with $\alpha = a^{-1/2}M^{1/2}N^{-1/2}(l_B/a)^{-1/7}f^{-2/7}$ for a flexible polyelectrolyte without Manning condensation, where a is the monomer size and f is the fraction of charged monomers. With Manning condensation, α is the distance between charges. The agreement between d and ξ is good (Table 1). The data are also consistent with the disappearance of the scattering peak by the addition of salt, with its sharpening with increasing polymer concentration, and with the absence of its dependence on polymer molecular weight.

Studies of charged wormlike micelles led to similar results: The disjoining pressure is oscillatory, with a period equal to the mesh size of the network formed by these micelles, as obtained from neutron-scattering data [35]. The behavior observed is, therefore, remarkably universal.

IV. CONCLUSION

Thin liquid films made from semidilute polyelectrolyte solutions can be stabilized with small amounts of surfactants. A surfactant monolayer is formed at the two free surfaces of the film, eventually mixed with the polymer when both have opposite charges, but no mixed polymer–surfactant aggregates are present in the bulk. A film stratification is observed, with a stratum thickness corresponding to the mesh size ξ of the polymeric network (i.e., the distance between overlap points of two polymer chains). The oscillatory forces are particular to polyelectrolytes and disappear when the electrostatic forces are screened with salt. The study of freely suspended films gives new useful insights into the structure of semidilute polymer solutions, which are presently the object of numerous speculations. They allow one to probe concentration regimes inaccessible to scattering experiments because of poor contrast conditions, but which are the topic of most theories.

ACKNOWLEDGMENTS

This short article is a review of the work performed at the Laboratoire de Physique de l'Ecole Normale Supericure, Paris and at the Centre de Recherche Paul Pascal, Bordeaux with many colleagues: A. Sonin, A. Asnacios, J. F. Argillier, V. Bergeron, A. Espert, R. v. Klitzing, and A. Colin. They are all gratefully acknowledged.

REFERENCES

1. GJ Fleer, JMHM Scheutjens, MA Cohen-Stuart, T Cosgrove, B Vincent. Polymers at Interfaces. London: Chapman & Hall, 1993.

2. SS Patel, M Tirrell. Annu Rev Phys Chem 40:597, 1989.

3. P Kékicheff, F Nallet, P. Richetti. J Phys II (France) 4, 735, 1994.

4. AJ Milling. J Phys Chem 100:8986, 1996.

5. A Scheludko. Adv. Colloid Interf Sci 1:391, 1967.

6. KJ Mysels, MN Jones. Discuss Faraday Soc 42:42, 1966.

7. I Ivanov, DS Dimitrov. In: I Ivanov, ed. Thin Liquid Films. New York: Marcel Dekker, 1988, p. 379.

8. A Sonin, A Bonfillon, D Langevin. J Colloid Interf Sci 162:323, 1994.

9. JL Joye, G Hirasaki, CA Miller. Langmuir 8, 3085, 1992; 10:3174, 1994.

10. ES Johnnott. Philos Mag 11, 746, 1906; J Perrin. Ann Phys 10:160, 1918.

11. AD Nikolov, DT Wasan. J Colloid Interf Sci 133:1, 1989; AD Nikolov, DT Wasan, ND Denkov, PA Kralchevski, IB Ivanov. Prog Colloid Polym Sci 82:87, 1990.

12. V Bergeron, C Radke. Langmuir 8:3020, 1992.

13. A Sonin, D Langevin. Europhys Lett 22:271, 1993.

14. AD Nikolov, PA Kralchevski, IB Ivanov, DT Wasan. J Colloid Interf Sci 133:13, 1989; PA Kralchevski, AD Nikolov, DT Wasan, IB Ivanov. Langmuir 6:1180, 1990.

15. PG de Gennes, AM Cazabat. CR Acad Sci Ser II 310:1601, 1990.

16. A Jimenez-Laguna, C Radke. Langmuir 8:3027, 1992.

17. F Brochard-Wyart, C Redon. Langmuir 8:2324, 1992.

18. K Mysels, K Shinoda, S Frankel. Soap Films. Elmsford, NY: Pergamon, 1959.

19. PG de Gennes. In J Meunier, D Langevin, N Boccara, eds. Physics of Amphiphilic Layers. New York: Springer-Verlag, 1987, p. 64.

20. V Bergeron, A Asnacios, D Langevin. Langmuir 12:1550, 1996.

21. A Asnacios, A Espert, A Colin, D Langevin. Phys Rev Lett 78:4974, 1997.

22. JN Israelachvili. Intermolecular and Surface Forces. London: Academic Press, 1985.

23. A Asnacios, D Langevin, JF Argillier. Macromolecules 29:7412, 1996.

24. R v Klitzing, A Espert, A Asnacios, T Hellweg, A Colin, D Langevin. Colloids Surfaces A 149:131, 1999.

25. R v Klitzing, A Espert, A Colin, D Langevin. Colloids Surfaces A (in press).

26. A Asnacios, R Klitzing, D Langevin. Colloids Surfaces A (in press).

27. S Förster, M Schmidt. Adv Polym Sci 120:51, 1995.

28. M Doï, SF Edwards. The Theory of Polymer Dynamics. Oxford: Oxford University Press, 1986.

29. JL Barrat, JF Joanny. Adv Chem Phys 94:1, 1996.

30. W. Essafi, Thesis, University of Paris, 1997.

31. M Nierlich, CE Williams, F Boué, JP Cotton, M Daoud, B Farnoux, G Janninck, C Picot, M Moan, C Wolf, M Rinaudo, PG de Gennes. J Phys 40:701, 1979.

32. M Milas, M Rinaudo, R Duplessix, R Borsali, P Lindner. Macromolecules 28:3119, 1995.

33. ML Pollard, CJ Radke. J Chem Phys 101:6979, 1994.

34. X Châtellier, JF Joanny. J Phys II (France) 6:1669, 1996.

35. A Espert, R v Klitzing, P Poulin, A Colin, R Zana, D Langevin. Langmuir 14:4251, 1998.

9
Preparation of Novel Silicone-Based Antifoams Having a High Defoaming Performance

Takamitsu Tamura, Motohiro Kageyama, Yukihiro Kaneko, and Masanori Nikaido
Lion Corporation, Tokyo, Japan

ABSTRACT

It is well known that the stability of the pseudoemulsion films formed on the surface of many antifoamers is important during defoaming action. We have compared, by using laser microscopic techniques, the bursting behavior of a two-dimensional pseudoemulsion film from an aqueous surfactant solution, in which four types of antifoam particles have been dispersed on a glass plate. The antifoamers used were a silicone oil, a mixed-type antifoamer (silicone oil with hydrophobic silica particles), hydrophobic silica, and a silicone-based solid antifoam (prepared by interfacial polymerization with oil and water). The pseudoemulsion film breaking was observed at a film thickness less than 0.1 μm for both the silicone oil and the hydrophobic silica. On the other hand, the pseudoemulsion film on the top of the particles can be easily broken by convex parts of the surface for the silicone-based antifoam. Furthermore, it was discovered that the acute top of the solid particles in the mixed-type antifoam ran out of oil and broke the liquid film when the film thickness decreased to a diameter comparable to the antifoam particle (ca. 8μm).

Based on these results, we have prepared novel silicone-based solid antifoams having a high defoaming performance by using interfacial polymerization with oil and water. These solid particles had a number of sharp-pointed needles on the surface and the roughness was estimated quantitatively as the ten point height of irregularities (R_Z) by analyzing the laser microscopic patterns. Defoaming performance was studied by monitoring the foam volume measured by the cylinder shaking test and the drum-type laundry washing machine. It was found that the defoaming efficiency was proportional to the value of R_Z, an indicator of surface roughness.

I. INTRODUCTION

A number of investigations about the breakdown of foams by dispersed insoluble oils [1–6], hydrophobic solid particles [7–15], and oils and particles in combination [16–18] as antifoams have been reported. The mixed-type antifoams are very effective; thus, they are widely employed in many industries. Several theories have been presented in the literature on the mechanisms of defoaming by these three types of antifoams [17,18]. The oil lens–bridge theory was suggested by Garrett [19] and Frye and Berg [9]. The oil drop first enters one of the foam-film surfaces and forms a lens. On further thinning of the film, the lens enters the opposite film surface and an oil bridge is formed. The bridge is unstable because the capillary forces dewet the film from the bridge, and the film ruptures. The theory of foam film bridging by a solid particle was suggested by Garrett [14]. When the hydrophobic particle bridges with the film, the capillary pressure dewets the particle, and the film ruptures. On the contrary, the hydrophilic particle's capillary pressure acts in the opposite direction and tends to increase the film thickness. The mechanism for the mixed-type anitifoam was suggested by Wasan and co-workers [18,20]. They proposed that the antifoaming efficiency strongly depended on the stability of the pseudoemulsion film. The oil drop containing solid particles collects and gets trapped in the thinning Plateau border. The pseudoemulsion film breaks, and a drop enters and forms a solid plus oil lens. This lens gets trapped in a later stage of thinning and then bridges the film at the Plateau border and the bridge ruptures. In these terms, it is qualitatively explained that the hydrophobic solid particles destabilized the pseudoemulsion film.

We have adopted the measurement techniques using scanning laser microscope (SLM) to reproduce the pseudoemulsion films for the several kinds of antifoam particles on a slide glass plate. In the case of particles with rough edges (the mixed-type antifoam and the silicone-based solid antifoam), it was found that the pseudoemulsion film was momentarily ruptured as soon as the distortion took place on the top of the particles. Especially, the acute top of the solid particles in the mixed-type antifoams runs out of silicone oil and causes an instantaneous break of the thick pseudoemulsion film. These effects clearly affect the defoaming performance of the antifoams.

The aim of the present study was to clarify the defoaming mechanism when the foam film contained four types of antifoaming particles during the defoaming process. The antifoams were a silicone oil, a mixed-type antifoamer (silicone oil with hydrophobic silica particles), a hydrophobic silica, and a silicone-based solid antifoam (prepared by interfacial polymerization with oil and water). We have compared the bursting behavior using laser microscopic techniques of a two-dimensional pseudoemulsion film from an aqueous surfactant solution, in which the four types of antifoam particles were dispersed on a glass plate.

We have also investigated the behavior of a novel solid antifoam by observing the defoaming performance and the physical properties of the solid particles. Hydrophobic silicone resins having the required physical properties for the optimum defoaming conditions were prepared by the interfacial polymerization via the hydrolysis reaction of trichloromethylsilane. In this method, polymerization proceeded at the oil–water interface. The silicone resin particles with various surface shapes could be obtained by this method.

II. EXPERIMENT

A. Materials

Sodium dodecyl sulfate (SDS) was obtained from Nacalai Tesque Inc. (Kyoto) in a specially prepared reagent, and poly(vinyl alcohol) (PVA) with average monomer units of 2500 was obtained from Wako (Tokyo) in extrapure grade. A standard stable foam was generated from an aqueous 60 mM SDS/0.5 wt% PVA solution as the model foaming solution. Hexane, chloroform, and methyl alcohol were obtained from Kanto Chemical Co. (Tokyo) in extrapure grade. Trichloromethyl silane was obtained from Tokyo Kasei Co. (Tokyo) in extrapure grade. The silanizing agent used to render the silica and silicone resin surfaces hydrophobic was trimethylsilychloride (TMSCl) which was obtained from Tokyo Kasei Co. (Tokyo) in extrapure grade. All surfactant solutions were prepared with purified water that had been passed through a cell membrane, NANO pure II (Brebsted Co.), after distillation.

B. Antifoams

Commercially available antifoams were used for the oil-type and mixed-type antifoams. Dimethylsilicone with a viscosity of 1000 cSt (KF 96) was obtained from Shinetsu Chemical Co. The mixed-type antifoam (F-16) was obtained from

Dow-Corning Asia Co. The hydrophobic silica was used after silanization of the spherical silica (H51) from Asahi Glass Co. (Tokyo) The surface treatment was carried out as follows. One part of the H51 was treated with the silanizing solution of 1.5 parts of TMSCl in chloroform below 50°C and then treated with a supersonic wave for 3 h using the Ultra Sonic Automatic Washer [Model US-3 (Iuchi Seieido Co., Tokyo)], washed with an excess of a methanol–chloroform solvent and dried. The specific gravity and the contact angle of the model foaming solution on this particle was found to be 2.1 g/cm^2 and 130°, respectively. A hydrophobic silicone resin with a rough surface shape was obtained by the following interfacial polymerization via the hydrolysis reaction of trichloromethylsilane [21]. A 20-mL aliquot of trichloromethylsilane was dropped by a pipette (5 mL) into 120 mL of n-tetradecane and 120 mL of water in a screw-capped glass container (500 mL) at 50°C. The reaction mixture was immediately capped after the addition and was stored for 24 h without stirring at 50°C. After the completion of the reaction, unreacted trichloromethylsilane was treated with methanol and the mixtures were filtered. The filtrate was given a final rinse with 500 mL of methanol with water three times and dried under reduced pressure at 50°C. The dried solid was powdered using a mixer [Trio Blender (Trio Science Co., Tokyo)] and the surface treatment by TMSCl was carried out in the same manner as already described. The specific gravity and the contact angle of the model foaming solution on this particle were found to be 1.1 g/cm^2 and 130°, respectively. Laser microscopic photographs of these four types of antifoamers are shown in Figures 1 and 2.

The following parameters were varied to prepare the various kinds of silicone-based solid antifoams:

1. The reaction temperature varied from 20°C to 60°C.
2. The number of carbon atom in the hydrocarbon used as the oil phase varied (6, 8, 10, 12, 14, and 16). In the case of silicone oil, its viscosity varied (10, 30, and 100 cSt).
3. As the water phase, H_2O and 10% NH_3 aqueous solution was chosen.
4. As the substrates, $MeSiCl_3$, $MeSi(OMe)_3$, and the mixture of $MeSiCl_3$ and Me_2SiCl_2 were chosen.

C. Procedures

1. Scanning Laser Microscopy

A scanning laser microscope (SLM) [Model 2LM31 (Lasertech Co., Tokyo)] was used to observe the phenomenon during foam breakdown by the antifoams. A single He–Ne laser beam ($\lambda = 633$ nm) was directed on the sample situated on a slide glass plate, and the reflective interference fringes that appeared around

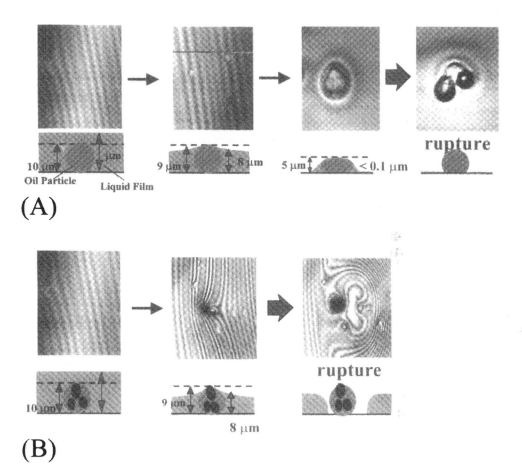

(A)

(B)

Figure 1 Scanning laser microscopic reflection features of pseudoemulsion films from the model foaming solution (60 mM SDS/0.5 wt% PVA) for the silicone oil antifoams (A series) and the mixed-type antifoams (B series). Estimated images of vertical sectional views are shown under these features.

the sample were observed by a charge-coupled device (CCD) camera. A sample solution was dropped on the slide glass fixed on a movable stage and the reflective fringe patterns around the antifoam particles were recorded during the spontaneous spreading processes. The interval of each interference fringe is always regular and corresponds to 0.3 μm in the direction of thickness when using monochromic light. Furthermore, by focusing on the top and bottom of the sample, any change in the particle form can be observed with a high resolution (>0.1 μm).

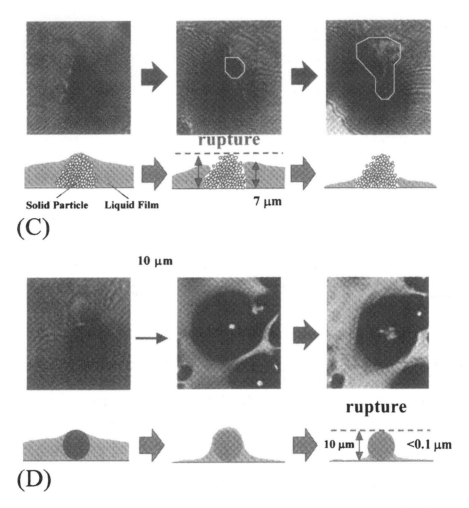

Figure 2 Scanning laser microscopic reflection features of pseudoemulsion films from the model foaming solution (60 m*M* SDS/0.5 wt% PVA) for the hydrophobic silica antifoams (C series) and the silicone resin antifoams (D series). Estimated images of vertical sectional views are shown under these features.

D. Surface Roughness Measurements

The surface shape of the particle could be observed using a scanning electron microscope (SEM). For the quantitative value, we adopted the method known as the 10- point height irregularities (R_z). Five points which are above the standard line of the sample surface in order of distance are chosen; also, five points below the standard line are chosen. R_Z is defined by dividing the sum of the absolute values of these numbers. It gives the value of surface roughness of the SLM image of a solid particle. Figure 3 shows the SLM image of the silicone resin particle with a rough surface. First, the sample of a solid particle was scanned in the z-axis direction using the SLM. The obtained image was processed three dimensionally; then the R_Z of the particles was measured five times by scanning a 2.25-mm^2 area. The average of these 25 values was adopted as R_Z. By using the R_Z, the surface roughness of a solid particle can be quantitatively determined as a physical property.

E. Specific Gravity and Contact Angle Measurements

The specific gravity of the solid particles was measured at 25°C using a picnometer according to the standard method [22]. The solid particles were immersed with methanol in the picnometer and treated with a supersonic wave until no further generation of bubbles from the particles could be observed.

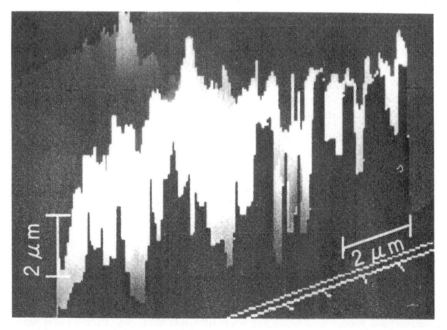

Figure 3 Three-dimensional image using a scanning laser microscope on the surface of developed silicone resin antifoam.

The contact angles of the surfactant solution with solid antifoaming agents were measured using a contact-angle goniometer [Model CA-Z (Kyowa Interface Science Co., Tokyo)]. A small amount of the solid samples was rubbed on both faces of 18-mm-wide piece of adhesive tape [Type CW-18 (Nichiban Co., Tokyo)] on a slide glass and the remaining particles were removed by air-blowing. A microscopic observation was carried out before measurements to clarify that no openings larger than the particle diameters were observed. A Hamilton microsyringe was used to apply solution drops to the sample surfaces at 20°C.

F. Determination of Defoaming Effect

Sample solutions were prepared by mechanically mixing the model foaming solution with 200 ppm of antifoam using a blender [ULTRA-DISPERSER Model LK-22 (Yamato Kagaku Co., Tokyo)] at 8000 rpm for 15 min. The average diameters of these dispersed particles were very uniform in the 5–10-µm range.

The defoaming effect of the antifoams was estimated by the glass cylinder shaking test. Samples (20 mL) of aqueous surfactant solution, both with and without antifoam, were shaken by hand in a 100-mL stoppered measuring cylinder in a reproducible way 10 times for 10 s. The foam volume was then monitored over an appropriate period.

For application, the defoaming effect of the antifoams was also estimated by the drum-type laundry washing machine. One kilogram of cotton shirts were washed in the drum-type laundry washing machine [ES-E60 (Sharp Co.)] with 24 g of laundry detergent, ["Super Top^R" (Lion Co.)] mainly containing α-sulfo fatty acid methylester (α-SF). In this case, hot water (53°C) was first poured and the temperature of the washing machine was set at 60°C. Then, 0.3 wt%, based on the detergent weight, of antifoam was added and the defoaming performance was observed.

III. RESULTS AND DISCUSSION

A. Rupture of Pseudoemulsion Foam Film on Liquid Containing Antifoams

The defoaming efficiency strongly depends on the stability of the pseudoemulsion film. The stability of the pseudoemulsion film was directly studied by Wasan and co-workers, forming such a film from a surfactant solution on the tip of a capillary [18]. They pointed out that the edges of the particles penetrating into an aqueous phase could pierce and break the pseudoemulsion film. Furthermore, the particle concentration required to break the foam film was very low, because it is sufficient that the film be pierced at only one point to rupture. Garrett [14], Frye and Berg [9], and Aveyard et al. [7] reported the stability behavior of a pseudoemulsion film studied from the viewpoint of the contact angle on the solid particles. We have tried to directly clarify the mechanism of breakdown of such films and to correlate it to the defoaming efficiency of different types of defoaming particles using laser microscopic techniques.

Figure 1 illustrates the SLM reflection features of the pseudoemulsion films from the model foaming solution formed on the top of the silicone oil antifoams (A series) and the mixed-type antifoams (B series). When the interval of interference fringe in the image is wide, the liquid film has a gentle surface incline; when the interval is narrow, it has a steep incline. The estimated images of the vertical sectional views are shown under these SLM features. It is seen that widespread fringes were observed for both antifoams when the thickness of the surfactant solutions was greater than the diameter of the antifoam droplets, indicating gentle surface inclines (A-1 and B-1). Considerable differences in the features appeared as the drainage of the solution progressed. In the case of the silicone oil, the pseudoemulsion film was formed around the top of the droplets, which was confirmed from the appearance of circular fringes (A-2). Although changes in the shape of the droplets and aggregation of such particles occurred with additional drainage, the pseudoemulsion film persisted until the film thickness became less than 0.1 μm (A-3). The foam-film rupture occurred around the droplets after the removal of most of the surfactant solution (A-4). On the other hand, when the liquid film thickness decreased and became slightly less than the diameter of the mixed-type antifoam, the acute top of solid particles in this droplet ran out of silicone oil and distorted the pseudoemulsion film (B-2). As soon as the distortion took place on the top of the antifoam droplet, the pseudoemulsion film could be instantaneously ruptured (B-3). A large area of the slide glass surface can be seen in the direction to the thinner part of the liquid film after the rupture. Thus, the liquid film rupture occurred when the film thickness decreased to a diameter comparable to the antifoam particle (~8 μm).

B. Rupture of Pseudoemulsion Foam Film on Solid Antifoams

Figure 2 illustrates the SLM reflection features of the pseudoemulsion films from the model foaming solution formed on the top of the hydrophobic silica antifoamers (C series) and the silicone resin antifoams (D series). Estimated images of the vertical sectional views are also shown under these SLM features. It is seen that the pseudoemulsion films were present on the top of their particles when the thickness of the surfactant solutions was slightly less than the diameter of the antifoam particles (C-1 and D-1). Considerable differences in the features appeared as the drainage of the solution progressed similar to the liquid antifoams. In the case of the hydrophobic silica particles, the pseudoemulsion film completely surrounded the particles with drainage and the film persisted until the film thickness became less than 0.1 μm (C-2). The foam-film rupture occurred only on the surface of the particles after the removal of most of the surfactant solution (C-3). On the other hand, when the liquid film thickness decreased and became slightly less than the diameter of the silicone resin particles, the acute top of solid particles distorted the pseudoemulsion film. As soon as the distortion took place on the top of the antifoamer droplet, the pseudoemulsion film was momentarily ruptured and part of the particle appeared from the liquid film as shown in Fig. 2C (surrounded by the white line). The surface of the particle gradually appeared, depending on the thinning of the liquid film. These results clearly show that the rupture of the pseudoemulsion film was caused by the acute top of the silicone resin antifoam, although instantaneous rupture did not occur as for the case of the mixed-type antifoam.

C. Defoaming Effect of Antifoamers

The decay curves of the foam volume as a function of time with and without the four types of antifoams are shown in Figure 4. It is seen that the fairly high initial foam volumes were observed for both the silicone oil and the silica particles. Although the silicone oil has a slightly higher defoaming effect than the silica particles, only slight defoaming took place in a similar manner for both foam films. Compared with these agents, the initial foam volumes were reduced to one-half its blank solution for both the mixed-type and the silicone resin systems. Furthermore, an effective defoaming performance was observed within a short time; particularly, only a small amount of foam could be observed in the mixed-

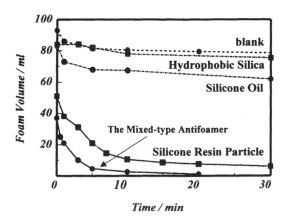

Figure 4 Effects of different antifoams on the rate of breakdown of foams from the model foaming solution (60 mM SDS/0.5 wt% PVA). Glass cylinder shaking test conditions are described in the text.

type system. These defoaming efficiencies were consistent with the results reported by Wasan and co-workers [19,23]. They also pointed out that the oil alone had a low foam-breaking efficiency, the hydrophobic particles alone were more effective, and the mixed-type agent was a much better antifoam.

As shown in Figure 1, we could not see any differences in the droplet shape between the silicone oil and the mixed-type antifoams in the model foaming solution. The diameter (d) of the dispersed silicone oil was about 10 µm, and a similar size of silicone oil droplets, in which some silica particles ($d < 0.1$ µm) existed as aggregates ($d = 2–3$ µm), was observed in the mixed-type antifoams. Thus, there is a clear dependence of defoaming efficiency on the amount of solid particles. On the other hand, a large difference in defoaming efficiency was also observed between the hydrophobic silica and the silicone resin, although there was a slight difference in their particle diameters, as seen in Figure 4. Because the silica particle has a higher specific gravity, it might be readily swept from the foam film and hence reduce the defoaming efficiency. However, Frye and Berg reported the effects of silica particle geometry on the foam stability [8]. The polyhedral silica particles had a higher defoaming efficiency than the spherical silica particles. Therefore, our results suggest that the significant factor for foam-breaking could be derived from the surface roughness of the silicone resin antifoam.

D. Relationship Between Physical Properties and Deforming Performance

Various kinds of silicone resin particles with different values for the R_Z (0.82–1.87) and the contact angle (110°–135°) were prepared with different reaction conditions (temperature, oil type, and silylchloride type). It gives particles with different surface roughnesses when treated at different temperatures [24].

Figure 5a shows the relationship between the contact angle and the defoaming performance in the cylinder shaking test. The contact angle indicates the hydrophobicity of the solid surface. We chose those silicone resin particles that satisfied the requirements of the physical properties (a contact angle greater than 90°) and evaluated their defoaming performances. No significant correlation was found between the defoaming performance and the hydrophobicity of the silicone resin particle. However, the defoaming effect is highly correlated to the R_Z values, as shown in Figure 5b. Also, it appeared that the solid particles of lower specific gravity tended to have a higher defoaming effect.

E. Defoaming Performance in the Drum-Type Laundry Washing Machine

Figure 6 shows the defoaming performance of the silicone resin particle and the mixed-type antifoams. The height of the foam was measured from the bottom of

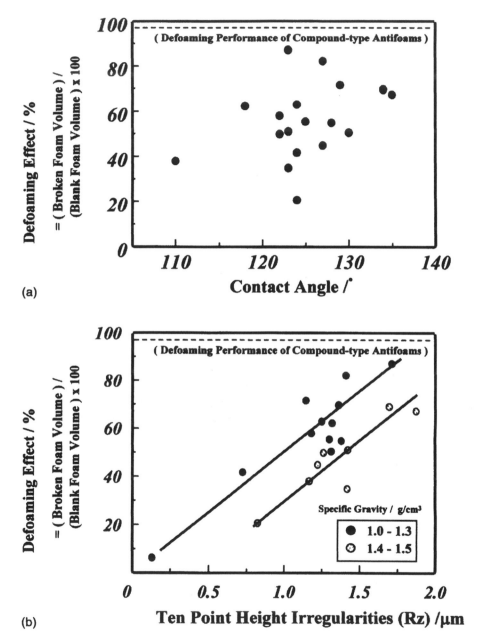

Figure 5   Relationship between defoaming performance evaluated by the cylinder shaking test and the contact angle (a), as well as the 10-point height of irregularity (R_Z) of various kinds of the developed silicone resin antifoam (b).

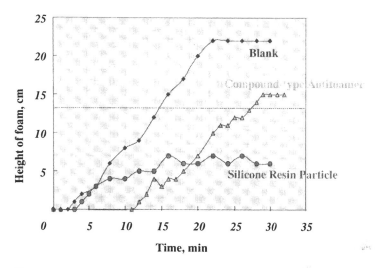

Figure 6 Effects of defoaming performance of several kinds of antifoams evaluated by the drum-type laundry washing machine. Conditions are given in the text.

the front window of the machine. In the case of no antifoam, foam filled the inside of the machine after 20 min. When the mixed-type antifoam was added, little foam was observed in the first 10 min, but after 10 min, foam gradually appeared, and finally increased to a height of 15 cm. This phenomenon suggested that the defoaming performance of the compound-type antifoam decreased with time or was influenced by the washing temperature. On the other hand, when the silicone resin particle was added, foam was observed in a few minutes; however, it did not increase with time. It shows that this particle has a high defoaming performance and it is lasting.

IV. CONCLUSIONS

The antifoaming efficiency of antifoams strongly depends on the stability of the pseudoemulsion films. The antifoamer droplets enter the air–solution interface, forming a lens, with the particles destabilizing the pseudoemulsion film during the early stage of foam breakdown. We adopted the measurement techniques using SLM to reproduce the pseudoemulsion films for the four types of antifoam particles on a slide glass plate. We found that the film stability was largely affected by surface roughness for both the liquid and the solid antifoams; that is to say, in the case of smooth particles (the silicone oil antifoam and the hydrophobic silica), the drainage caused liquid to flow away from the droplet or the particle and form a very thin liquid film on the surface. The film then attained a critical thickness (<0.1 μm) and caused dewetting on the antifoam surface. However, in the case of particles with rough edges (the mixed-type antifoam and the silicone-based antifoam), it was found that the pseudoemulsion film was momentarily ruptured as soon as the distortion of the pseudoemultion film took place on the top of the particles. Especially, the acute top of the solid particles in the mixed-type antifoams runs out of silicone oil and causes an instantaneous break of the thick pseudoemulsion film. These effects clearly affect the defoaming performances of the antifoams. We also took notice of the solid particle itself. Thus, the various silicone-based solid antifoams, not including any silicone oil, were prepared and their defoaming performance was measured. The silicone-based solid antifoams were prepared by interfacial polymerization with oil and water. These solid particles had many sharp needles on the surface, and the surface roughness of the solid particles was quantitatively evaluated based on the R_Z analyzed by the SLM. The defoaming performance of these solid particles was highly correlated to the R_Z value, indicating surface roughness, although the defoaming performance was not correlated to the contact angle, indicating hydrophobicity.

REFERENCES

1. DT Wasan, AD Nikolov, LA Lobo, K Koczo, DA Edwards. Prog Surface Sci 39: 119, 1992.
2. LA Lobo, DT Wasan. Langmuir 9:1668, 1993.
3. LA Lobo, AD Nikolov, DT Wasan. J Disper Sci Technol 10:143, 1989.
4. K Koczo, LA Lobo, DT Wasan. J Colloid Interf Sci 150:492, 1992.
5. R Aveyard, JH Clint. J Chem Soc Faraday Trans 91:2681, 1995.
6. MP Aronson. Langmuir 2:653, 1986.
7. R Aveyard, BP Binks, PDI Fletcher, CE Rutherford. J Disper Sci Technol 15:251, 1994.
8. GC Frye, JC Berg. J Colloid Interf Sci 127:222, 1989.
9. GC Frye, JC Berg. J Colloid Interf Sci 130:54, 1989.
10. A Dippenaar. Int J Miner Process 9:1, 1982.
11. A Dippenaar. Int J Miner Process 9:15, 1982.
12. G Johansson, RJ Pugh. Int J Miner Process 34:1, 1992.
13. F-Q Tang, Z Xiao, J-A Tang, L Jiang. J Colloid Interf Sci 131:498, 1989.
14. PR Garrett. J Colloid Interf Sci 69:107, 1979.
15. RD Kulkarni, ED Goddard, B Kanner. Ind Eng Chem Fundam 16:472, 1977.
16. R Aveyard, P Cooper, PDI Fletcher, CE Rutherford. Langmuir 9:604, 1993.
17. PR Garrett, J Davis, HM Rendall. Colloids Surfaces A 85:159, 1994.
18. G Racz, K Koczo, DT Wasan. J Colloid Interf Sci 181:124, 1996.

19. PR Garrett. In: Defoaming: Theory and Industrial Applications: New York: Wiley–Inerscience, 1992.
20. K Koczo, KJ Koczone, DT Wasan. J Colloid Interf Sci 166:225, 1994.
21. T Tamura, M Kageyama, K Kaneko, T Kishino, M Nikaido. J Colloid Interf Sci 213:179, 1999.
22. Japanese Patent JIS K 0061.
23. DT Wasan, K Koczo, AD Nikolov. In: Foams: Fundamentals and Applications in the Petroleum Industry. Washington, DC: American Chemical Society, 1994, p. 47.
24. Lion Co. Japan patent pending, JN98278690.

10

Preliminary Observations Concerning the Use of Perfluoroalkyl Alkanes in Hydrocarbon-Based Antifoams

Richard John Curtis, P. R. Garrett, M. Nicholls, and John W. H. Yorke
Port Sunlight Laboratory, Unilever Research, Wirral, Merseyside, England

ABSTRACT

Perfluoroalkyl alkanes are known to effectively lower the air-oil surface tensions of hydrocarbons. This reduction in air-oil surface tension produces a concomitant increase in the effectiveness of hydrocarbon/hydrophobed particle antifoams. However, the resulting increase in effectiveness declines relatively rapidly during continuous foam generation. This is probably associated with transient spreading at the air-water interface, induced by low values of the air-oil surface tension. Such spreading may lead to disproportionation, suboptimal antifoam droplet sizes and diminished antifoam effectiveness. Addition of a hydrocarbon polymer to antifoams prepared from these mixtures of perfluoroalkyl alkanes and hydrocarbons results in increased viscosities and a diminished rate of disproportionation. Antifoam effectiveness therefore declines less rapidly during continuous foam generation.

I. INTRODUCTION

Effective antifoams for aqueous systems usually consist of mixtures of hydrophobic oils and hydrophobic particles [1]. Here, the particles adhere to oil–water interfaces and rupture the aqueous films between oil droplets and air bubbles. The oil droplets then emerge into the air–water surface. The behavior of the oil droplet in that surface is determined by the spreading coefficient S defined as

$$S = \gamma_{AW} - \gamma_{AO} - \gamma_{OW} \tag{1}$$

where γ_{AW} and γ_{OW} are respectively the air–water and oil–water surface tensions and γ_{AO} is the air–oil surface tension. If $S < 0$, the oil droplet will form oil lenses. In turn, these may bridge aqueous foam films. Such bridging lenses are necessarily unstable provided the condition

$$B = (\gamma_{AW}^2 - \gamma_{OA}^2 + \gamma_{OW}^2) > 0 \tag{2}$$

is satisfied, where B is the bridging coefficient [1–3]. When this inequality is satisfied, the bridging oil droplet must cause a capillary pressure in the foam film near the droplet leading to enhanced drainage rates and, according to Frye and Berg [4], to film collapse by "film pinch-off." The process is shown schematically in Figure 1. Here, θ is the angle formed between the air–water and oil–water interfaces at the three-phase contact line. If $B > 0$, then it is easy to show that $\theta > \pi/2$ [1].

For such antifoams, the properties of the oil are of critical importance. Thus, the oil should not be solubilized in surfactant micelles and should not be volatilized at the temperature of the foaming solution. Equation (2) indicates that it should also possess an oil–air surface tension lower than the air–water surface tension of the foaming surfactant solution (if we make the reasonable assumption that γ_{OW} is small). In general, the lower the value of γ_{AO}, the higher the capillary pressure due to the bridging droplet and the greater the resultant enhancement of foam-film drainage rates [3]. Such enhancement, albeit generated by changes to γ_{AW} with constant γ_{AO}, has been shown to correlate with increased antifoam effectiveness [3]. Low air–oil surface tensions should then also correlate with enhanced antifoam effectiveness.

A complicating factor concerns the possibility that at sufficiently low air–oil surface tensions, $S > 0$ and the oil spreads over the air–water surface. Then, B has no meaning and $\theta = \pi$. A bridging oil droplet with $S > 0$ would produce enhanced foam film drainage as a result of *both* a shear force derived from gradients in the spreading pressure *and* a capillary force due to the shape of the oil–water interface [5]. It seems likely that film collapse by such bridging droplets would be accompanied by disproportionation of the droplet as shown in Fig. 1.

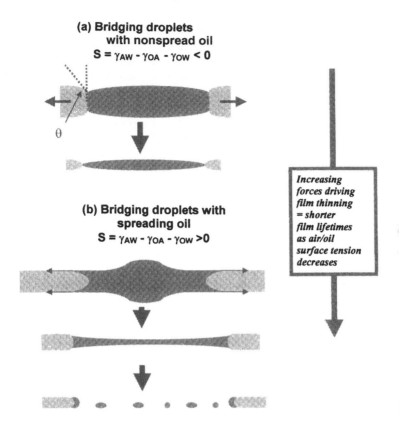

Figure 1 Mechanism of antifoam action by bridging oil lens where (a) $S < 0$ and $B > 0$ and (b) $S \geq 0$. (See Refs. 1, 4, and 5.)

Situations where $S > 0$ are necessarily transient. They arise because the relevant interfaces are not at mutual equilibrium. Thus, if initially $S > 0$, then we find, after equilibration, formation of either a stable duplex film (where $S = 0$ at equilibrium) or a thin liquid film in equilibrium with an oil lens (where $S < 0$ at equilibrium). This causes some ambiguity in assessing the relevance of transient spreading in antifoam action. The process will depend on the relevance of the pre-equilibrated interfacial tensions. Irrespective of such ambiguity, it seems probable, however, that reduction of oil–air surface tensions so that $S > 0$ will, if anything, produce enhanced antifoam effectiveness relative to situations where $S < 0$.

Typically, antifoam oils are either hydrocarbons or polydimethylsiloxanes. Hydrocarbon oil–air surface tensions are in the region of 25–30 mN/m, at ambient temperatures, whereas polydimethylsiloxane oil–air surface tensions exhibit little variability and lie in the region of 20–22 mN/m. Many practical aqueous foam control situations involve hydrocarbon-chain surfactants, which dissolve to produce air–water surface tensions within the range exhibited by the air–oil surface tensions of hydrocarbon oils. The use of polydimethylsiloxane-based antifoams may then become necessary if condition Eq. (2) is to be satisfied.

Not only are there marked differences in air–oil surface tensions between antifoam polydimethylsiloxanes and hydrocarbons, but there are also marked differences in viscosity. Typically, the polydimethylsiloxanes used in commercial antifoams have viscosities > 1000 mPa · s, whereas those of hydrocarbons are < 100 mPa · s. The role of oil viscosity in determining antifoam effectiveness does not appear to be well established. It does, however, seem likely to concern oil droplet disproportionation and the existence of optimum droplet sizes [5,6].

It seems probable that differences in performance of polydimethylsiloxane- and hydrocarbon-based oil/particle antifoams derive, at least in part, from differences in air–oil surface tensions and viscosities of the oils. An improved understanding of the role of these physical properties in determining antifoam effectiveness would clearly lead to improved antifoam design. In achieving such understanding, it is desirable to select a model system where viscosity and air–oil surface tension can be separately varied. Here, we describe preliminary work concerned with the selection of a suitable system.

Alternatives to polydimethylsiloxanes as low-surface-tension antifoam components are not readily available. Nonvolatile oils with air–oil surface tensions as low as those of polydimethylsiloxanes and which do not dissolve in micellar surfactant solutions are not common. Arguably, the only possibilities concern various perfluorinated materials in general and perfluorinated alkanes in particular. Surface tensions in the region of 20 mN/m are readily achieved with the latter. However, the viscosities of perfluorinated alkanes tend to be low.

Low-surface-tension oils may be prepared by mixing comparatively high-surface-tension hydrocarbon oils with solutes which are surface active at the air–hydrocarbon surface. Hydrocarbon-based antifoams with low air–oil surface tensions have, in fact, been prepared using long-chain hydrocarbon/polydimethylsiloxane derivatives as surfactants for that surface [7]. In certain circumstances, these antifoams have been found to be more effective than those based solely on the relevant hydrocarbon [7].

Here, we use a similar approach, but we utilize perfluoroalkyl alkanes as surfactants at the hydrocarbon–air surface. Variation of molecular structure and concentration in the hydrocarbon permits variation of air–oil surface tensions in the range 15–30 mN/m [8–12]. The viscosity of the selected hydrocarbon, a liquid paraffin, can be varied up to 10^5 mPa · s by the addition of a hydrocarbon polymer such as poly(isobutene).

II. MATERIALS AND METHODS

A. Synthesis of Perfluoroalkyl Alkanes

Perfluoroalkyl alkanes were synthesized using the method of Rabolt et al. [13]. Perfluoroalkyl iodides were heated with an excess of alkene and a catalytic amount of azoisobutyronitrile under a nitrogen atmosphere for 4 h. The resultant 1-perfluoroalkyl-2-iodoalkanes ranged from liquids to solids depending on chain length and degree of fluorination. The intermediate iodide was reduced by magnesium in methanol according to the method of Hopkins and Suchismita [14]. This proved to be most effective for the shorter-chain-length members of the series. It yielded reaction products which were predominantly alkanes with only a small amount (about 5%) of alkenes. However, the longer-chain-length compounds were less soluble in methanol and it was necessary to heat the reaction mixture. The products of these reactions contained increased (up to about 50%) alkene due to base-mediated elimination. Here, the base was produced by the reaction of the excess magnesium with methanol. All attempts at hydrogenation of the alkene, using a range of catalysts, resulted in failure. It is possible that a large perfluoroalkyl group in close proximity to the alkene function prevents adsorption on the surface of the catalyst.

The presence of alkene contaminants was not considered significant because of domination of adsorption behavior at the air–hydrocarbon surface by the perfluoroalkyl group. A list of compounds prepared for this study is given in Table 1 together with alkene content and some relevant physical properties.

B. Materials

The particles used in antifoams were either ground glass or commercial hydrophobed silica. Ground-glass particles were of size range 8–53 μm and were hydrophobed by treatment with polydimethylsiloxane after the method of Ross

Table 1 Physical Properties of Perfluoroalkyl Alkanes

Compound	Alkene content (wt%)	Physical state at 25°C	Air–oil surface tension (mN/m at 25°C)
$C_6H_{13}C_6F_{13}$	8–10	Liquid	18.6
$C_{10}H_{21}C_6F_{13}$	7.5–8.5	Liquid	—
$C_{12}H_{25}C_6F_{13}$	5–9	Liquid	21
$C_6H_{13}C_8F_{17}$	8	Liquid	18.7
$C_{10}H_{21}C_{10}F_{21}$	37	Solid	—
$C_{12}H_{25}C_{12}F_{25}$	50	Solid	—

and Nishioka [15]. The commercial hydrophobed silica was supplied by Degussa as "Sipernat D17" and "Sipernat D10." D17 has a Brunaeur, Emmett and Teller nitrogen adsorption (BET) surface area of 110 m^2/g and a mean primary particle size of 0.028 μm. D10 has a BET surface area of 90 m^2/g and a mean primary particle size of 0.018 μm.

Liquid paraffin was "Spectrosol" grade supplied by BDH and was passed through an alumina column before use. Poly(isobutene) was supplied by BP and used as received. The grade used for this work was "Hyvis 200," which has a relative density of 0.914 (at 15.5°C) and a viscosity of about 10^6 mPa · s at 25°C.

The samples of polydimethylsiloxane oil of viscosity 50 mPa · s (DC200, Dow Corning) and a silicone antifoam (polydimethylsiloxane/hydrophobed silica, DB100, Dow Corning) were used as received.

Two blends of sodium dodecyl benzene sulfonate were used in this study— a linear sodium dodecyl benzene sulfonate blend (C_{12}LAS) and a technical-grade branched blend (C_{12}bLAS). Two batches of C_{12}LAS were used with chain-length purities of 94.4% and 99% by weight, respectively. The phenyl isomer distribution of both batches was ~23% 2-phenyl, ~19% 3-phenyl, and a total of ~58% 4-, 5-, and 6-phenyl substitution. The technical-grade blend of branched dodecyl benzene sulfonate was supplied by Aldrich and used as received. This material is a blend of different chain lengths with a major proportion of C_{12}. The chain-length distribution was 0.9% by weight C_9, 6.3% C_{10}, 8.7% C_{11}, 64.4% C_{12}, 6.2% C_{13}, 1.9% C_{14}, 1.1% C_{15}, and 0.5% C_{16}, with a mean molecular weight of 346.3. Solutions of both sodium dodecyl benzene sulfonate blends were made at 5 $\times$ 10^{-3} mol/dm^3 in 3.8 mol/dm^3 sodium chloride and used throughout. The air–water surface tensions of these micellar solutions were 30.2 mN/m for C_{12}LAS and 29.05 mN/m for C_{12}bLAS.

Analar sodium chloride (BDH) was used as the electrolyte.

C. Methods

The solubility was assessed at 25°C by first making 50 wt% mixtures of perfluoroalkyl alkanes in liquid paraffin and then diluting these down until clear solutions were obtained. The point at which the solutions just became clear was taken as the solubility limit of perfluoroalkyl alkane in paraffin. With longer-chain materials, which were solid, dissolution was aided by heating. The assessment of solubility was then made after cooling back down to 25°C.

The surface tensions of neat perfluoroalkyl alkanes and their mixtures with liquid paraffin were measured using either Cahn or Beckmann microbalances fitted with a roughened platinum Wilhelmy plate. Viscosities of oils were measured with a Haake rotoviscometer at a shear rate of 10 s^{-1}.

The foam was generated by either cylinder shaking or the Ross–Miles technique. With the former, 0.0025 dm^3 of surfactant solution in a 0.1-dm^3 graduated cylinder was shaken vigorously for 30 times. The volume of air in the foam was then readily measured. Increasing amounts of antifoam may be introduced into the graduated cylinder using either a syringe (precalibrated so that each drop gave a known amount of antifoam) or by weighing it directly on a glass microscope cover slip and adding that to the cylinder. The Ross–Miles technique is described in detail elsewhere [16]. For use with this technique, the antifoam was predispersed in 0.5 dm^3 of the surfactant solution for 2 min at 6000 revolutions per minute using an Ultraturrax high shear mixer.

The particle/oil mixtures were first mixed by hand stirring and then subjected to ultrasonication for 1 min (Dawe soniprobe with intermittent setting on maximum energy with a microtip). The antifoam was then shaken for 5 min. Solutes such as perfluoroalkyl alkanes and polyisobutene were then added and the whole again shaken for 5 min.

III. RESULTS AND DISCUSSION

A. Physicochemical Behavior of Perfluoroalkyl Alkanes in Liquid Paraffin

Perfluorinated alkanes are at best only partially miscible with saturated hydrocarbons, often showing upper critical solution points [17]. This is a consequence, at least in part, of the extremely low polarizability of fluorine giving rise to a weak attractive interaction force with neighboring molecules. The phase separation of a perfluoroalkane permits the remaining hydrocarbon molecules to surround themselves with a greater proportion of similar molecules. This means that the intermolecular potential becomes more negative as a result of the relatively strong attractive interaction force between hydrocarbon molecules.

Weak interaction forces between neighboring perfluoroalkane molecules also mean that the surface tension against air (or vapor) of such materials is low. Mixing with saturated hydrocarbons will, therefore, tend to lower the air–oil surface tension of the latter, as the perfluorocarbon will tend to accumulate at the air–oil surface. However, if this property is to be optimized, the tendency of the perfluoroalkane to phase separate from the hydrocarbon should be mitigated. This may be accomplished by attaching a hydrocarbon chain to the end of the perfluoroalkane. The resulting partially perfluorinated alkane (or perfluoroalkyl alkane) combines an oleophilic alkane ''headgroup'' and an oleophobic perfluoroalkane segment. The balance between solubility and surface activity is determined by the lengths and relative proportions of the two different segments of the molecule.

The solubility, aggregation, and adsorption behaviors of perfluoroalkyl alkanes in pure hydrocarbons have been described in detail elsewhere [8–11]. The solutions of these compounds in hydrocarbons appear to exhibit behavior roughly analogous to that of hydrocarbon-based surfactants in water. Not only do they exhibit adsorption at the hydrocarbon–air surface, but they also appear to aggregate at sufficiently high concentrations and temperatures to form micelle-like entities [8–11]. However, at low temperatures, perfluoroalkyl alkane separates out and micelles are not formed. This behavior leads to the suggestion that a Krafft point exists analogous to that found in aqueous systems [8–11].

Liquid paraffin was selected as the hydrocarbon component for this study. For a given perfluoroalkane chain, solubility in liquid paraffin increased as the alkyl chain length increased. Thus, with a perfluoroalkyl chain length of six carbons and a hydrocarbon chain length of six carbons ($C_6H_{13}C_6F_{13}$), solubility was limited to 16.5 wt% (0.41 mol/dm^3). With the same length of perfluoroalkyl chain and an alkyl chain length of 10 or more carbons ($C_{10}H_{21}C_6F_{13}$ and $C_{12}H_{23}C_6F_{13}$), complete miscibility with liquid paraffin over all compositions was found.

The solubilities of molecules containing perfluoroalkyl and alkyl portions of equal chain length were found to decrease dramatically as the overall chain length increased. Thus, $C_{10}H_{21}C_{10}F_{21}$ was soluble in liquid paraffin up to between 1.5 and 1.75 wt% [$(2.27–2.65) \times 10^2$ mol/dm^3], whereas $C_{12}H_{23}C_{12}F_{23}$ was only soluble up to between 0.25 and 0.5 wt% weight [$(3.17–6.34) \times 10^{-3}$ mol/dm^3]. A plot of log(solubility) against chain length for these compounds is shown in Figure 2. The plot is seen to be approximately linear. If we suppose that changes

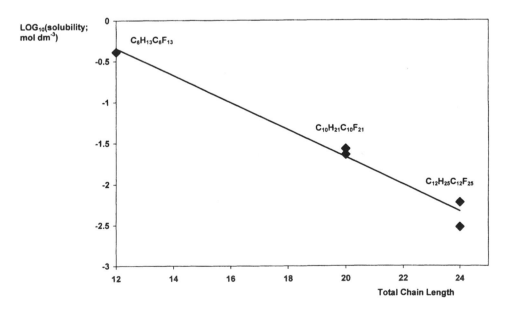

Figure 2 Solubility of symmetrical $C_nH_{2n}C_nF_{2n}$ perfluoroalkyl alkanes in liquid paraffin at 25°C.

in solubility mainly concern the changes in chain length of the perfluoro group, then we can write

$$\frac{d \ln a}{dn} = \frac{\Delta\mu^{\circ}_{(CF2)}}{RT} \tag{3}$$

where n is the chain length of the perfluoroalkyl group, a is the activity of a saturated solution of the perfluoroalkyl alkane in liquid paraffin, R is the gas constant, T is the temperature, and $\Delta\mu^{\circ}_{(CF2)}$ is the standard partial molar free energy of transfer of a CF_2 group from the solvent to the pure perfluoroalkyl alkane. Plots of log(solubility) versus n will be linear if, for these dilute solutions, activity coefficients can be set to unity so that $a \approx s$, where s is the solubility. The gradient of the plot in Figure 2 yields an estimate of $\Delta\mu^{\circ}_{(CF2)}$ of about 1.7 kJ/mol. This estimate of $\Delta\mu^{\circ}_{(CF2)}$ compares with a value for the partial molar free energy of transfer of a CF_2 group from dodecane to pure perfluoroalkane of 1.4 kJ/mol [11]. The similarity of these partial molar free energies is clearly consistent with domination of the solubility behavior in liquid paraffin of the perfluoroalkyl alkanes by the antipathy of the perfluoro group for the hydrocarbon. As implied by Binks et al. [11], this behavior is roughly analogous to the hydrophobic effect in aqueous systems, which has an origin in the antipathy of hydrocarbon chains for water.

The effect of various perfluoroalkyl alkanes on the surface tension of liquid paraffin is shown in Figure 3, in which surface tension is plotted versus log(surfactant concentration). The position of the solubility limit is also indicated on the plots. That the surface tension still declines slightly above this limit presumably reflects the presence of surface-active impurities. From the plots, we see that, in general, the longer the perfluoroalkyl group, the more surface active the compound. Surface pressures $\geq$ 12 mN/m can be achieved with concentrations <3 wt%. Similar findings have been reported elsewhere for solutions of perfluoroalkyl alkanes in both commercial paraffins [12] and pure hydrocarbons [8–11]. In the case of perfluoroalkyl alkanes with alkyl chains of equal length, the surface activity in liquid paraffin increases drastically as the solubility decreases with increasing overall chain length. Here, we should note that although equilibrium was reached rapidly for short-chain-length compounds, it took up to 30 h for solutions of the longer-chain-length compounds to equilibrate. This gradual drift in surface tension with time, over several hours, was probably due to the presence of highly surface-active impurities.

B. Effect of Perfluoroalkyl Alkanes on Hydrocarbon-Based Antifoam Behavior

The antifoams were prepared by mixing hydrophobic particles and liquid paraffin with different proportions of various perfluoroalkyl alkanes. Both ground glass

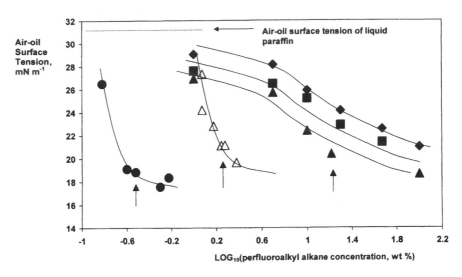

Figure 3 Effect of perfluoroalkyl alkanes on the air–oil surface tension of liquid paraffin at 25°C (equilibrium measurements). ●: $C_{12}H_{25}C_{12}F_{25}$; △: $C_{10}H_{21}C_{10}F_{21}$; ▲: $C_6H_{13}C_6F_{13}$; ■: $C_{10}H_{21}C_6F_{13}$; ◆: $C_{12}H_{25}C_6F_{13}$. Arrows indicate the solubility limits for respectively $C_{12}H_{25}C_{12}F_{25}$, $C_{10}H_{21}C_{10}F_{21}$, and $C_6H_{13}C_6F_{13}$.

and commercial hydrophobed silica particles (Sipernat D17) were used. The antifoams were dispersed in solutions of 5×10^{-3} mol/dm³ sodium dodecyl benzene sulfonate (C_{12}LAS) and 3.8×10^{-2} mol/dm³ sodium chloride. Their effectiveness was assessed by shaking cylinders (30 times) containing the dispersion. Plots of volume of air in the foam against the logarithm of the antifoam concentration are shown in Figures 4 and 5 for ground glass and D17 silica as antifoam promoters, respectively. Air–oil surface tensions of the relevant mixtures of liquid paraffin and perfluoroalkyl alkanes are given in the figures.

It is clear from Figure 4 that the reduction in air–oil surface tension accompanying the addition of perfluoroalkyl alkanes to liquid paraffin produces the expected enhanced antifoam effectiveness. Included in the figure is a comparison with an antifoam prepared using polydimethylsiloxane fluid (of 50 mPa · s viscosity) instead of liquid paraffin/perfluoroalkyl alkane. Clearly, it is possible to prepare antifoams based on mixtures of liquid paraffin and perfluoroalkyl alkane which can yield superior effectiveness to that shown by polydimethylsiloxane-based antifoams with the same particulate promoter.

Similar results were found with hydrophobed silica D17 as antifoam promoter. Thus, a general trend toward enhanced antifoam effectiveness of liquid paraffin with the reduction of air–oil surface tension, upon addition of perfluoro-

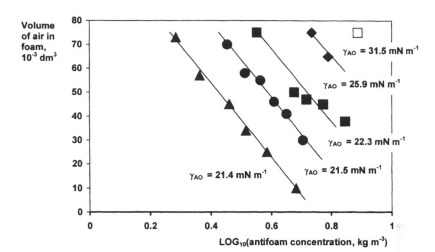

Figure 4 Effect of perfluoroalkyl alkanes on antifoam behavior of hydrophobed glass plus liquid paraffin mixtures (10 wt% glass, 90 wt% oil phase). □: Liquid paraffin alone; ◆: 90% liquid paraffin/10% $C_{12}H_{25}C_6F_{13}$; ■: 90% liquid paraffin/10% $C_6H_{13}C_6F_{13}$; ●: polydimethylsiloxane (of viscosity 50 mPa · s); ▲: 50% liquid paraffin/50% $C_{10}H_{21}C_6F_{13}$. Measurements done by cylinder shaking at ambient temperatures. Solutions: 5×10^{-3} mol/dm³ C_{12}LAS in 3.8×10^{-2} mol/dm³ NaCl.

alkyl alkane, is seen in Figure 5. Here, however, the antifoams were generally an order of magnitude more effective than those prepared using hydrophobed ground glass. This is illustrated clearly in Figure 6. The high relative effectiveness of D17 silica must concern differences in particle hydrophobicity, size, and geometry; and of these, particle size probably dominates. Thus, smaller particle sizes for the silica should mean smaller composite particle/oil droplet sizes. Large particle sizes for the ground glass will mean that particles are only associated with large droplets.

The expected behavior is revealed in the photomicrographs of the relevant antifoam dispersions shown in Figure 7. Here, we see that for ground glass, many smaller droplets do not have particles and a significant number of particles are not associated with droplets at all. Of those droplets which have particles, most reveal only one or two attached particles. With D17 silica, by contrast, all droplets are associated with particles which appear to form a closed packed raft at the oil–water surface. The probability of an oil–particle composite entering a foam film should clearly be higher for D17/liquid paraffin than for ground glass /liquid paraffin. This, in turn, will mean more effective antifoam action. (See Ref. 1 for a detailed account of the relevant mechanistic background.)

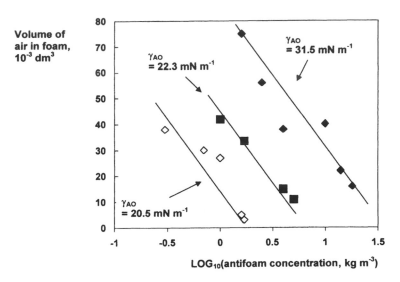

Figure 5 Effect of perfluoroalkyl alkanes on antifoam behavior of hydrophobed silica (D17) plus liquid paraffin mixtures (10 wt% D17, 90 wt% oil phase). ◆: Liquid paraffin alone; ■: 90% liquid paraffin/10% $C_6H_{13}C_6F_{13}$; ◇, ■: 83.5% liquid paraffin/16.5% $C_6H_{13}C_6F_{13}$. Measurements done by cylinder shaking at ambient temperatures. Solutions: 5×10^{-3} mol/dm^3 C_{12}LAS in 3.8×10^{-2} mol/dm^3 NaCl.

Foam generation by the static Ross–Miles technique represents a more reliable and objective method of foam generation than simply shaking graduated cylinders. The Ross–Miles foam heights showing the effect of perfluoroalkyl alkanes on the antifoam behavior of liquid paraffin/D17 mixtures are shown in Figure 8. The antifoam was dispersed as 1 kg/m^3 in 5×10^{-3} mol/dm^3 C_{12}LAS and 3.8×10^{-2} mol/dm^3 sodium chloride using a high shear mixer. Not unexpectedly, we find that both relative foamability and foam stability in the presence of antifoam are diminished when the antifoam contains perfluoroalkyl alkanes.

In general, the initial Ross–Miles foam heights roughly correlate with air–oil surface tensions. The correlation is presented in Figure 9. Here, the air–oil surface tension is plotted against F, which is defined as the ratio of the volume of air in the foam in the presence of antifoam to the volume of air in the foam in the absence of antifoam. It is clear from Figure 9 that lower air–oil surface tensions mean enhanced antifoam effectiveness. However, it is also seen from Figure 9 that antifoams containing $< 1\%$ perfluoroalkyl alkane lie well outside the correlation. It seems probable that the air–oil surface tensions prevailing with droplets of such antifoams may be significantly different from those measured

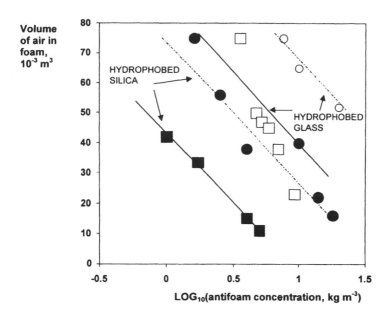

Figure 6 Comparison between hydrophobed ground glass and D17 hydrophobed silica as antifoam promoters for liquid paraffin (10 wt% particles, 90 wt% oil phase). ○: Hydrophobed glass + liquid paraffin alone; □: 10% hydrophobed glass + 81% liquid paraffin + 9% $C_6H_{13}C_6F_{13}$; ●: 10% D17 hydrophobed silica + 90% liquid paraffin; ■: 10% D17 hydrophobed silica + 81% liquid paraffin + 9% $C_6H_{13}C_6F_{13}$. Measurements done by cylinder shaking at ambient temperatures. Solutions: 5×10^{-3} mol/dm³ C_{12}LAS in 3.8×10^{-2} mol/dm³ NaCl.

on the bulk solution. Thus, adsorption of perfluoroalkyl alkane at the air–oil surface of a small oil droplet may cause depletion if the concentration is low.

We can illustrate this effect by a simple example. Consider a droplet of liquid paraffin of radius 1 μm suspended in air and containing 0.3 wt% by weight of $C_{12}H_{25}C_{12}F_{25}$. This concentration represents one of the examples shown in Figure 9, for which the effect on antifoam performance of the perfluoroalkyl alkane is negligible despite measurements of low equilibrium air–oil surface tensions on bulk solutions. If we now suppose that all of this perfluoroalkyl alkane is adsorbed at the air–oil surface of the drop, then we calculate a molecular area of ~1.3 nm². This compares with a molecular area of ~0.26 nm² reported by Binks et al. [8] for the saturated adsorption of $C_{14}H_{29}C_{12}F_{25}$ at dodecane–air surfaces (which approximates the cross-sectional area of an all-trans CF_2 group). It also compares with a rough estimate of the molecular area of 0.3 ± 0.1 nm²

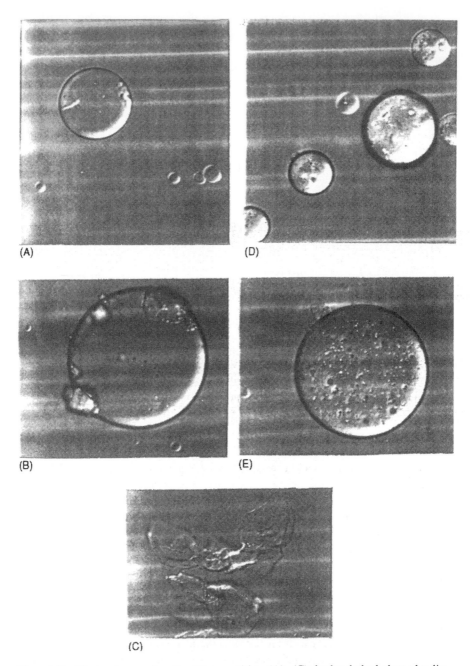

Figure 7 Photomicrographs of antifoam entities. (A)–(C): hydrophobed glass plus liquid paraffin showing free glass particles in (C); (D)–(E): hydrophobed silica (D17) plus liquid paraffin (magnification ×650).

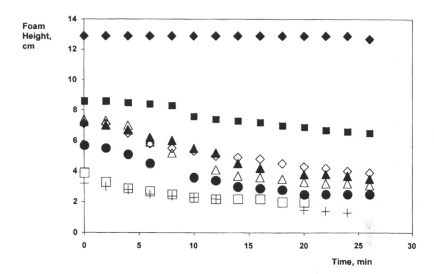

Figure 8 Effect of perfluoroalkyl alkanes on antifoam behavior of hydrophobed silica (D17) plus liquid paraffin mixtures (10 wt% D17, 90 wt% oil phase). Measurements done by the Ross–Miles technique at 25°C. Antifoams predispersed at concentration of 1 kg/m^3. ◆: No antifoam; ■: liquid paraffin alone, γ_{AO} = 31.4 mN/m; ▽: 95% liquid paraffin + 5% $C_{12}H_{25}C_6F_{13}$, γ_{AO} = 28.1 mN/m; ◇: 95% liquid paraffin + 5% $C_6H_{13}C_6F_{13}$, γ_{AO} = 25.7 mN/m; □: 90% liquid paraffin + 10% $C_{10}H_{21}C_6F_{13}$, γ_{AO} = 25.2 mN/m; ●: 98.75% liquid paraffin + 1.25% $C_{10}H_{21}C_{10}F_{21}$, γ_{AO} = 24.8 mN/m; +: 83.5% liquid paraffin + 16.5% $C_6H_{13}C_6F_{13}$, γ_{AO} = 20.4 mN/m; ▲: 99.7% liquid paraffin + 0.3% $C_{12}H_{25}C_{12}F_{25}$, γ_{AO} = 18.8 mN/m. Solution: 5 × 10^{-3} mol/dm^3 C_{12}LAS in 3.8 × 10^{-2} mol/dm^3 NaCl.

obtained from the relevant plot in Figure 3 (using the Gibbs adsorption equation and ignoring activity coefficients). Clearly, the molecular area is significantly higher than would be expected with a close-packed monolayer of perfluoroalkyl alkane. Such high molecular areas suggest low surface pressures. Binks et al. [8], for example, find that for $C_{14}H_{29}C_{12}F_{25}$ at dodecane–air surfaces, molecular areas of ~1.3 nm^2 occur at surface pressures ≪1 mN/m. Depletion will therefore tend to eliminate the effectiveness of low levels of perfluoroalkyl alkane in reducing the air–oil surface tensions of small droplets where the area/volume ratio is large. This is revealed clearly in Figure 10, in which we plot F (for Ross–Miles foamabilities) versus concentration in liquid paraffin of different perfluoroalkyl alkanes. Here, the concentrations were adjusted to compensate for the different surface activities of these compounds in order to yield the same bulk-phase air–oil surface tension. Values of F are seen to decrease with increasing concentration despite the constancy of the air–oil surface tensions.

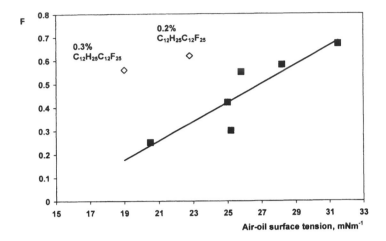

Figure 9 Correlation between air–oil surface tension of perfluoroalkyl alkane/liquid paraffin mixtures and antifoam effectiveness (measured as F = volume of air in foam with antifoam/volume of air in foam without antifoam using the Ross–Miles technique at 25°C with 1 kg/m³ predispersed antifoam consisting of 10 wt% D17, 90 wt% oil phase). Linear least squares fit for all except indicated points gives a correlation coefficient of 0.8. Solution: 5×10^{-3} mol/dm³ C_{12}LAS in 3.8×10^{-2} mol/dm³ NaCl.

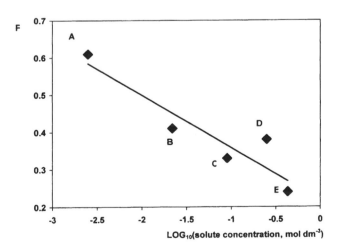

Figure 10 Effect of perfluoroalkyl alkane concentration on antifoam effectiveness. Solutions adjusted to give constant equilibrium air–oil surface tension of 22.5 mN/m. Measured as F = volume of air in foam with antifoam/volume of air in foam without antifoam using the Ross–Miles technique at 25°C with 1 kg/m³ predispersed antifoam consisting of 10 wt% D17 and 90 wt% oil phase by weight A: $C_{12}H_{25}C_{12}F_{25}$; B: $C_{10}H_{21}C_{10}F_{21}$; C: $C_6H_{13}C_8F_{17}$; D: $C_6H_{13}C_6F_{13}$; and E: $C_{10}H_{21}C_6F_{13}$. Solution: 5×10^{-3} mol/dm³ C_{12}LAS in 3.8×10^{-2} mol/dm³ NaCl.

Table 2 Effect of Perfluoroalkyl Alkanes on the Spreading of Liquid Paraffin on the Surface of a Solution of 5×10^{-3} mol/dm^3 C_{12}LAS in 3.8×10^{-2} mol/dm^3 NaCl at 25°C

Solute (wt%)	Air–oil surface tension of liquid paraffin solution (mN/m)	Spreads on C_{12}LAS solution?
None	31.3	No
$C_{12}H_{25}C_6F_{13}$ (5%)	28	No
$C_{10}H_{21}C_6F_{13}$ (1%)	27.6	No
$C_{12}H_{25}C_6F_{13}$ (6.5%)	27.4	Yes (slowly)
$C_{12}H_{25}C_6F_{13}$ (7%)	27.2	Yes (slowly)
$C_{12}H_{25}C_6F_{13}$ (7.5%)	27	Yes (slowly)
$C_{10}H_{21}C_6F_{13}$ (5%)	26.5	Yes
$C_{12}H_{25}C_6F_{13}$ (10%)	26	Yes (rapidly)
$C_6H_{13}C_8F_{17}$ (2%)	25.8	Yes
$C_{10}H_{21}C_{10}F_{21}$ (1.25%)	21	Yes (slowly)
$C_6H_{13}C_6F_{13}$ (16.5%)	20.5	Yes (rapidly)
$C_{12}H_{25}C_{12}F_{25}$ (0.3%)	18.8	Yes (slowly)

As we have seen, the behavior of a droplet of oil at an air–water surface is determined by the spreading coefficient S. It is easy to see from Eq. (1) that if the air–oil surface tension is reduced, this could mean spreading coefficients greater than 0. The possible effect of this on film collapse by a bridging droplet is depicted in Figure 1. As shown by Bergeron et al. [5] and Racz et al. [6], the antifoams which participate in such a process will clearly be subject to droplet breakup during foam generation, leading to a gradual reduction in droplet sizes.

There is both theoretical and experimental evidence to suggest that an optimum droplet size exists for antifoam effectiveness [1]. A decrease in droplet size below that optimum would be expected to produce diminished effectiveness [5]. A process of droplet disproportionation by spreading will therefore ultimately lead to diminished antifoam effectiveness during continuous aeration in the case of an antifoam for which initial values of S are greater than 0 [5]. The reduction of droplet size which is expected to accompany this process means that any advantage for reduction in air–oil surface tension by perfluoroalkyl alkanes may eventually be lost in situations involving continuous aeration for prolonged periods of time.

The effect of perfluoroalkyl alkanes on the spreading behavior of liquid paraffin at the air–water surface of solutions of 5×10^{-3} mol/dm^3 C_{12}LAS in 3.8×10^{-2} mol/dm^3 sodium chloride is presented in Table 2. It is clear that the oil will spread if the air–oil surface tension is sufficiently low (i.e., < 27.6 mN/m). The rate of spreading should increase with increasing S and, therefore, decreasing γ_{AO} [1]. Low spreading rates found for solutions of 1.25% $C_{10}H_{21}C_{10}F_{21}$ and 0.3% $C_{12}H_{25}C_{12}F_{25}$, despite low air–oil surface tensions are probably due to slow transport and/or depletion effects associated with an expanding air–oil surface. Thus, if the oil drop spreads to expand the air–oil surface, then depletion of the perfluoroalkyl alkane can occur. The air–oil surface tension will then rise and the driving force for spreading will decline.

In principle, it should be possible to minimize droplet disproportionation due to $S > 0$ by increasing the viscosity of the hydrocarbon. It seems reasonable to suppose that the role of polydimethylsiloxane viscosity in determining silicone antifoam effectiveness is probably due to such an effect. Thus, an increase in the viscosity of a silicone antifoam (where usually we expect initial values of $S > 0$) produces more effective foam control at the end of prolonged periods of aeration (J. Yorke, unpublished findings).

We may increase the viscosity of a hydrocarbon by the addition of hydrocarbon polymers. Poly(isobutene)s are potentially useful for this purpose. These materials are commercially available, completely miscible with many liquid hydrocarbons, and possess viscosities in the range from 10^2 to 10^6 mPa · s. The viscosities of mixtures of a poly(isobutene) (of viscosity 5×10^5 mPa · s) with liquid paraffin at a shear rate of 10 s^{-1} are shown in Figure 11 by way of an example. The viscosities of commercial silicone antifoams range from 10^3 to 3×10^4/mPa · s. It is clear from Figure 11 that we may produce hydrocarbon-based antifoams with the same range of viscosities if between 20 wt% and 50 wt% of this poly(isobutene) is mixed with liquid paraffin.

The possibility that the performance of perfluoroalkyl alkane/hydrocarbon-based antifoams may be enhanced, over prolonged periods of foam generation, if the viscosity is increased with poly(isobutene) has been explored. Some indication of the effect of prolonged periods of foam generation may be given in experiments where cylinders are repeatedly shaken over long periods. Here, cylinders were shaken 30 times and then left for 30 s. The procedure was repeated nine times. Surfactant solutions were 5×10^{-3} mol/dm^3 branched sodium dodecyl benzene sulfonate (C_{12}bLAS) in 3.8×10^{-2} mol/dm^3 sodium chloride. The results for a variety of antifoams in this C_{12}bLAS solution are given in Figure 12. Both of the hydrocarbon-based antifoams contained the same type and amount of hydrophobed silica (D10) and only differed in surface tension and viscosity. Full compositions are given in Table 3. Comparison was also made with a commercial hydrophobed silica/polydimethylsiloxane antifoam (DB100, Dow Corning) of viscosity ~ 3000 mPa · s.

The differences in the behavior of the two antifoams containing perfluoroalkyl alkane are instructive. Initial antifoam effects correlate with air–oil surface tension. However, the antifoam with the lowest surface tension also has the lowest

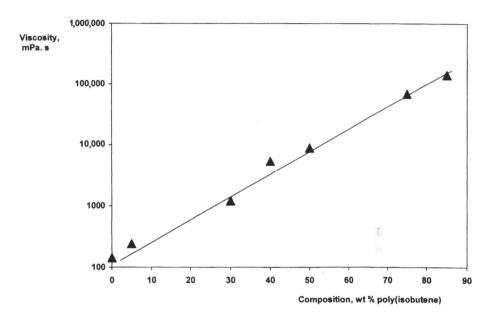

Figure 11 Viscosity of mixtures of poly(isobutene) with liquid paraffin at 25°C (measured at a shear rate of 10 s^{-1}).

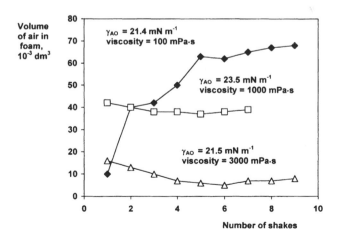

Figure 12 Effect of prolonged aeration on antifoam effectiveness. Measurements made by cylinder shaking at ambient temperature, 1 kg/m^3 antifoam. Antifoam physical properties and compositions are given in Table 3. ◆: antifoam A; □: antifoam B; △: DB100 commercial silicone antifoam. Solution: 5×10^{-3} mol/dm^3 C$_{12}$bLAS in 3.8×10^{-2} mol/dm^3 NaCl.

Table 3 Compositions and Properties of Antifoams Used in Cylinder Shaking Tests to Study Role of Oil Viscosity (Compositions wt%)

Antifoam	Liquid paraffin	Poly(isobutene)	$C_{10}H_{21}C_6F_{13}$	Hydrophobed silica D10	γ_{AO} (mN/m at 25°C)	Viscosity (mPa·s at 25°C)
A	47.5	—	42.5	10	21.5	100
B	48.1	27.0	14.9	10	23.5	1000
DB 100	—	—	—	—	21.5	3000
Commercial silicone antifoam						

viscosity (100 mPa · s). This antifoam exhibits markedly deteriorating performance with prolonged aeration. The foam volume increases by in excess of 600% during the experiment. By contrast, the perfluoroalkyl alkane/hydrocarbon antifoam with the higher air–oil surface tension and higher viscosity (1000 mPa · s) exhibits little deterioration in performance.

The commercial silicone antifoam has about the same air–oil surface tension as one of the perfluoroalkyl alkane/hydrocarbon antifoams. Initial antifoam performance of both antifoams is seen to be similar. However, the relatively high viscosity of the silicone antifoam means no deterioration in performance with continuous aeration under the conditions used here.

We find, then, that low air–oil surface tensions can give rise to a rapidly deteriorating antifoam performance during prolonged aeration, presumably as a result of droplet disproportionation due to spreading. An increase in viscosity of the antifoam oil appears to mitigate this effect.

IV. CONCLUDING REMARKS

We have shown that the addition of perfluoroalkyl alkanes to hydrocarbons can significantly reduce air–oil surface tensions to yield concomitant increases in antifoam effectiveness. However, maintenance of these increases under conditions of continuous foam generation requires increases in viscosity of the antifoam. This may be easily achieved by the use of hydrocarbon polymers such as poly(isobutene)s.

We emphasize that this represents a preliminary study in which a number of fundamental issues have not yet been fully addressed. For example, the effect of changes in nature and size of the antifoam promoter particle have not been systematically studied. Quantitative understanding of the effect of air–oil surface tension and viscosity on the spreading rates and disproportionation of these antifoam mixtures has also not been established. It is desirable that such issues be subject to detailed investigation. This will be necessary if the systems described here are to be used to develop fundamental understanding of the role of oil viscosity and air–oil surface tension in determining aspects of antifoam behavior such as the optimal droplet sizes for maximum effect.

REFERENCES

1. PR Garrett. In: PR Garrett, ed. Defoaming, Theory and Applications. New York: Marcel Dekker, 1992, p. 1.
2. PR Garrett. J Colloid Interf Sci 72:587, 1980.
3. PR Garrett, PR Moore. J Colloid Interf Sci 159:214, 1993.

4. GC Frye, JC Berg. J Colloid Interf Sci 130:54, 1989.
5. V Bergeron, P Cooper, C Fischer, J Giermanska-Kahn, D Langevin, A Pouchelon. Colloids Surfaces A 122:103, 1997.
6. G Racz, K Koczo, DT Wasan. J Colloid Interf Sci 181:124, 1996.
7. RV Kulkarni, ED Goddard, MP Aronson. U.S. Patent 4,514,319 (filed 1983) (to Union Carbide).
8. BP Binks, PDI Fletcher, WFC Sager, RL Thompson. Langmuir 11:977, 1995.
9. BP Binks, PDI Fletcher, RL Thompson. Ber Bunsenges Phys Chem 100:232, 1996.
10. BP Binks, PDI Fletcher, SN Kotsev, RL Thompson. Langmuir 13:6669, 1997.
11. BP Binks, PDI Fletcher, WFC Sager, RL Thompson. J Mol Liquids 72:177, 1997.
12. M Napoli, C Fraccaro, A Scipioni. J Fluorine Chem 51:103, 1991.
13. JF Rabolt, TP Russell, RJ Twieg. Macromolecules 17:2786, 1984.
14. RO Hopkins, L Suchismita. Synthetic Commun 19:1519, 1989.
15. S Ross, G Nishioka. J Colloid Interf Sci 65:216, 1978.
16. M Pauchek, V Veber. Acta Fac Pharm Univ Comenianae 27:27, 1975.
17. JS Rowlinson. Liquids and Liquid Mixtures. London: Butterworths, 1959, p. 171.

11

Application of the Theory of Liquids to Thin Films

Douglas Henderson
Brigham Young University, Provo, Utah

ABSTRACT

By considering the surface, or surfaces, that bound a liquid thin film, as giant particles, it is possible to consider a liquid film as a liquid mixture and, thereby, bring thin liquid films within the context of the theory of liquids. In this paper, some concepts and techniques in the theory of liquids are introduced and some results obtained with these techniques for thin liquid films are reported.

It is a pleasure to recognize the career and contributions of Professor Darsh Wasan, especially when the opportunity arises when he is at such a tender age. It is always gratifying to see a young man succeed. Darsh edits the *Journal of Colloid and Interface Science*, is a senior university executive, and maintains an active research program. Any one of these responsibilities would occupy the full attention of an average scientist. With his amazing energy and highly organized style, he is able to discharge all these responsibilities successfully. He is a human parallel computer. Even with these accomplishments, the attribute of Darsh that I admire most is that he is a considerate gentleman; interactions with him are always a pleasure.

Thin liquid films are a major interest of Darsh. Fortunately, one need not develop special techniques for thin films. The theory of bulk liquids can be adapted by considering the walls or surfaces containing the liquid as giant particles. Hence, a short summary of the theory of liquids is in order.

I. THEORY OF LIQUIDS

The terms "liquid" and "fluid" will be used interchangeably. The basis of any theory is a knowledge of the energy of the particles (usually molecules). It is common to assume *pairwise additivity*:

$$U_N(\mathbf{r}_1, ..., \mathbf{r}_N) = \sum_{i<j} u(r_{ij}) \tag{1}$$

where $r_{ij} = |\mathbf{r}_i - \mathbf{r}_j|$, $\mathbf{r}_i$ being the position of particle i. In Eq. (1), U_N is the total energy of the system and $u(r)$ is the pair interaction potential between a pair of particles.

A common pair interaction is the hard-sphere potential

$$u(r) = \begin{cases} \infty, & r < d \\ 0, & r < d \end{cases} \tag{2}$$

where d is the hard-sphere diameter. For a mixture of hard spheres,

$$d_{ij} = \frac{d_{ii} + d_{jj}}{2} \tag{3}$$

where d_{ii} is the hard-sphere diameter of species i.

There are three common tools in our toolchest for the study of liquids. The first is simulations. The simulation can be based either directly on averaging using Newton's equations of motion or on an average over a random walk. The first method is called the *molecular dynamics* method and the second is called the *Monte Carlo* method. A mindless application of the latter method would lead to an enormous number of improbable configurations, and reliable averages would be achieved exceedingly slowly, or never. It is better to bias the random walk so that the configurations that are generated are fairly probable. The commonest procedure for doing this is to use the *Metropolis algorithm*. Details of this procedure can be found in a review of Barker and Henderson [1].

The second tool in our toolchest is the method of integral equations. These are usually based on the *Ornstein–Zernike* (OZ) equation. Define $g(r_{12}) = h(r_{12}) + 1$ as the probablity of finding molecules 1 and 2 at a separation r_{12}. The function $g(r)$ is called the *radial distribution function* (RDF). The RDF is normalized so that $g(r)$ becomes unity when the two molecules are far apart. The function $h(r)$ is called the *total correlation function*. The OZ equation is obtained by introducing the *direct correlation function* (DCF), $c(r)$. At low densities, where only binary interactions can occur,

$$h(r_{12}) = c(r_{12}) \tag{4}$$

At higher densities, Eq. (4) is extended by adding an indirect contribution to $c(r_{12})$, reflecting the fact that molecule 1 is directly correlated with molecule 3, which also is directly correlated to molecule 2. Because molecule 3 can be anywhere in space,

$$h(r_{12}) = c(r_{12}) + \rho \int c(r_{13})c(r_{23}) \, d\mathbf{r}_3 \tag{5}$$

where $\rho = N/V$ is the density of the particles; V is the volume and N is the number of particles. The integral in Eq. (5) is called a convolution integral and is simpler in k (Fourier)-space. Thus,

$$\tilde{h}(k) = \tilde{c}(k) + \rho[\tilde{c}(k)]^2 \tag{6}$$

or

$$\tilde{h}(k) = \tilde{c}(k)[1 + \rho\tilde{c}(k)] \tag{7}$$

where $\tilde{h}(k)$ and $\tilde{c}(k)$ are the Fourier transforms of $h(r)$ and $c(r)$, respectively. Extending Eq. (7) to the case where more than molecules are involved,

$$\tilde{h}(k) = \tilde{c}(k)\{1 + \rho\tilde{c}(k) + [\rho\tilde{c}(k)]_2 + \cdots\} \tag{8}$$

Summing this expression,

$$\tilde{h}(k) = \frac{\tilde{c}(k)}{1 - \rho\tilde{c}(k)} \tag{9}$$

or, equivalently,

$$\tilde{h}(k) = \tilde{c}(k) + \rho\tilde{h}(k)\tilde{c}(k) \tag{10}$$

In r-space, Eq. (10) becomes

$$h(r_{12}) = c(r_{12}) + \rho \int h(r_{13})c(r_{23}) \, d\mathbf{r}_3 \tag{11}$$

Equation (11) or, equivalently, Eq. (10) is called the OZ equation. The OZ equation is but a definition of the DCF, and by itself, it does not provide a theory. It must be supplemented with a *closure*.

A common closure is the *Percus–Yevick* (PY) closure

$$y(r) = \exp[\beta u(r)]g(r) = 1 + \gamma(r) \tag{12}$$

where $\beta = 1/k_B T$ (T is the temperature and k_B is Boltzmann's constant), $u(r)$ is the pair interaction, and $\gamma(r) = h(r) - c(r)$ is the indirect correlation function, the convolution integral in Eq. (11). Other closures are possible.

Finally, there is perturbation theory. Here, the interaction energy is divided into two terms: a reference energy, $u_0(r)$, and a perturbation energy,

$$u_1(r) = u(r) - u_0(r) \tag{13}$$

In many applications, $u_1(r)$, or its effect on the system, is small. Usually, $u_0(r)$ is the hard-sphere potential.

To a first order, the Helmholtz free energy of the system is

$$\frac{F}{Nk_BT} = \frac{F_0}{Nk_BT} + \frac{1}{2}\beta\rho \int u_1(r)g_0(r) \, d\mathbf{r} \tag{14}$$

where F_0 and $g_0(r)$ are the Helmholtz free energy and RDF of the reference system, respectively. If the reference system is the hard-sphere fluid, the expression due to Carnahan and Starling (CS),

$$\frac{F_0}{Nk_BT} = \eta\frac{4-3\eta}{(1-\eta)^2} \tag{15}$$

where $\eta = \pi\rho d^3/6$ and $\lambda = h/(2\pi mkT)^{1/2}$, is a good approximation. The trivial perfect-gas free energy is not included in Eq. (15). Even when approximating $d = \sigma$, where σ is the value of r such that $u(r) = 0$, and cutting off the potential at 2.5σ, Eq. (14) gives a reasonable description of a fluid of molecules that interact through the Lennard–Jones potential. To some extent, each of these approximations compensates for the errors of the others. Better descriptions are obtained by disgarding these approximations, using the full Lennard–Jones interaction, defining d in a self-consistent manner, and including higher-order terms in β.

Of necessity, the above description is brief. A fuller discussion can be found in the review of Barker and Henderson [1].

II. INHOMOGENEOUS FLUIDS

The above-outlined theory is for bulk or homogeneous fluids, where the density is independent of the position. Near an interface, the fluid is *inhomogeneous* and the density and many other properties vary with position. The atmosphere of the Earth, where the density and pressure vary with altitude, is a familiar example. Thus, we may write, $\rho(\mathbf{r})$. The inhomogeneity in an inhomogeneous fluid may be due to a surface or wall or to an external potential.

The simulation techniques can be applied directly. All that is required is that the external potential be included. For the integral equation approach, the OZ equation may be generalized by placing the density inside the integral to give the OZ2 equation:

$$h(\mathbf{r}_1, \mathbf{r}_2) = c(\mathbf{r}_1, \mathbf{r}_2) + \int \rho(\mathbf{r}_3)h(\mathbf{r}_1, \mathbf{r}_3)c(\mathbf{r}_2, \mathbf{r}_3) \, d\mathbf{r}_3 \tag{16}$$

In addition to a closure, a relation between $\rho(\mathbf{r})$ and $h(\mathbf{r}_1, \mathbf{r}_2)$ and $c(\mathbf{r}_1, \mathbf{r}_2)$ is needed. A force balance equation, such as the Born–Green–Yvon equation, can be used.

A simpler integral equation can be obtained by regarding the bounding surface as a giant particle. Proceeding in this manner, Henderson et al. [2] (HAB) obtained the equation

$$h_w(z) = c_w(z) + 2\pi\rho \int_{-\infty}^{\infty} h_w(z')\, dz' \int_{|z-z'|}^{\infty} r'c(r')\, dr' \tag{17}$$

The function $c(r)$ is the DCF given by Eq. (11). The density profile is given by

$$\rho(z) = \rho[h_w(z) + 1] \tag{18}$$

For simplicity, it has been assumed that the inhomogeneity is planar. Again, a closure is needed.

Density functional (DF) theory can be thought as an analog of perturbation theory. In DF theory, the grand potential is written

$$\Omega = F + \int \rho(\mathbf{r})[v(z) - \mu]\, d\mathbf{r} \tag{19}$$

where F is the Helmholtz free energy, $v(\mathbf{r})$ is the external potential that is the source of the inhomogeneity, and μ is the chemical potential.

We write

$$F = \int \rho(\mathbf{r}) f[\rho(\mathbf{r})]\, d\mathbf{r} \tag{20}$$

where $f[\rho(\mathbf{r})]$ is the free-energy density and is a functional of the density. It is convenient to introduce a local density, $\bar{\rho}(\mathbf{r})$, that is obtained by averaging

$$\bar{\rho}(\mathbf{r}) = \int \rho(\mathbf{r}') W(|\mathbf{r} - \mathbf{r}'|)\, d\mathbf{r} \tag{21}$$

Tarazona et al. [3] have given a simple approximate expression for the weighting function, $W(r)$, that gives good results.

In analogy to Eqs. (14) and (15), but replacing $g_0(r)$ by unity,

$$f[\rho(\mathbf{r})] = k_B T \bar{\eta}(\mathbf{r}) \frac{4 - 3\bar{\eta}(\mathbf{r})}{[1 - \bar{\eta}(\mathbf{r})]^2} + \frac{1}{2} \int \rho(\mathbf{r}') u_1(|\mathbf{r} - \mathbf{r}'|)\, d\mathbf{r} \tag{22}$$

where $\bar{\eta}(\mathbf{r}) = \pi\bar{\rho}(\mathbf{r}) d^3/6$. The equilibrium condition that determines $\rho(\mathbf{r})$ is

$$\frac{\delta\Omega}{\delta\rho(z)} = 0 \tag{23}$$

Further details about density functionals can be found in the article by Evans [4].

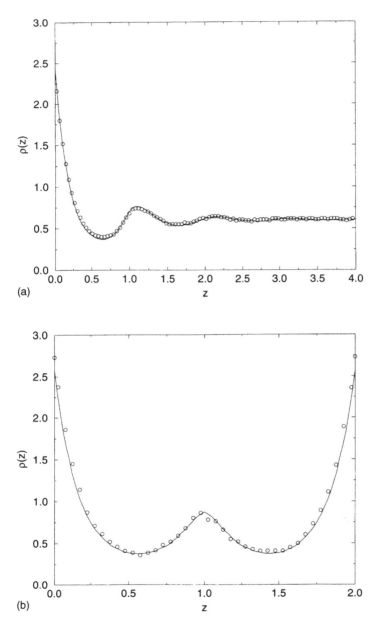

Figure 1 Hard spheres between two hard walls. In (a), the two walls are far enough apart that there is no interaction between the walls. In (b), the walls are close together. The curves give the PY2 results and the points are the simulation results.

III. SOME SELECTED APPLICATIONS

The simplest thin film is a film of hard spheres between two walls. The density profile of such a system is displayed in Figure 1. The walls are far apart in Figure 1a. In this case, there is an inhomogeneous layer near each wall with a region of bulk, or near bulk, fluid in the center. In Figure 1b, the inhomogeneous layers overlap. As the walls are brought closer together, layers of hard spheres are "squeezed" out. The layers seen in Figure 1b account for the oscillations in the force between macroparticles that are seen in the experiment [5]. Hard spheres in a slit pore have been studied by simulations [6], first-order and second-order OZ equations [7], and DF techniques [8].

A thin film of hard spheres between two rough surfaces has been studied recently. There is good agreement between DF theory [9] and simulations [10]. In most applications, the DF that is used is one dimensional. In contrast, the DF for the rough surface geometry is at least two dimensional.

Hard spheres in a wedge is an interesting system because structural changes occur as the wedge widens [11]. This is seen in Figure 2. The hard spheres near the vertex tend to be solidlike, whereas those further from the vertex are fluidlike in structure. This has been seen experimentally [12].

Lennard–Jones molecules near an attractive surface tend to show behavior similar to that shown in Figures 3 and 4. When the gas of Lennard–Jones molecules has a low density or pressure, a monolayer, or near monolayer, is formed (Fig. 3a), whereas multilayer structure is formed at higher pressures (Fig. 3b). The adsorption isotherms are also interesting. At low temperatures, the curves

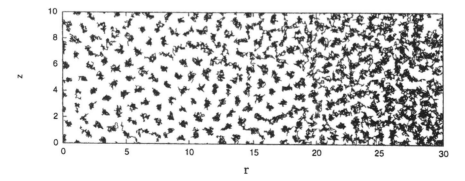

Figure 2 Trajectories, obtained from simulations, of hard spheres in a hard wedge formed by two hard walls that are not parallel. Periodic boundary conditions were applied to the top and bottom surfaces. The sphere trajectories are viewed through one of the hard walls. The hard spheres on the left are near the vertex of the wedge and are solidlike, whereas the hard spheres further from the wedge are fluidlike.

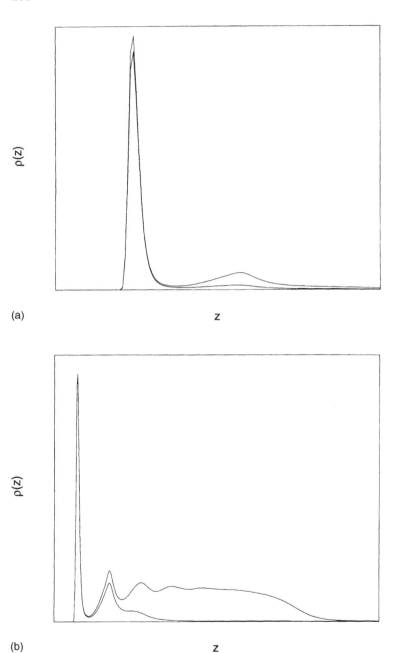

(a)

(b)

Figure 3 Density profiles of a Lennard–Jones gas adsorbed on a surface. The density, or pressure, of the gas is low in (a). The density, or pressure, of the gas is higher in (b). The two density profiles in (b) are for the same density, or pressure. The gas shown in (b) exhibits a first-order prewetting transition.

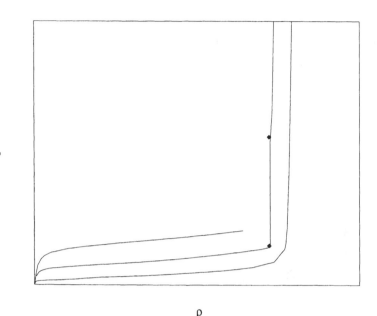

Figure 4 Adsorption isotherms for a Lennard–Jones gas adsorbed on a surface. From top to bottom, the isotherms are for a low temperature where only partial wetting occurs, for a temperature where a first-order prewetting transition occurs, and for a temperature about the surface critical point, where the wetting is continuous.

are continuous and increase to a finite value. In many situations, a "prewetting transition" occurs where two layers of different thicknesses can exist at a single pressure or density. The higher branch of the adsorption isotherm can increase without limit. Above a surface critical temperature, the adsorption isotherm increases continuously and without limit. This behavior is displayed in Figure 4. Wetting transitions are very interesting. A useful introductory account has been given by Evans and Chan [13].

Water in a pore is a system of biological, chemical, and physical interest. Some simulations exist [14] and are given in Figure 5. In Figure 5a, the water molecules are at a low pressure and do not fill the pore. A slight increase in pressure (Fig. 5b) causes them to fill the pore. This behavior might explain why channels in membranes can act like valves, allowing and disallowing the passage of ions or other particles.

Finally in Figure 6, we show the structure obtained from simulations of ions and water molecules [15,16] near a charged electrode. It is to be hoped that such simulations can provide new insights in electrochemistry.

(a)

(b)

Figure 5 Snapshot, obtained from a computer simulation, of water molecules in a pore. In (a), water is at a low pressure and the molecules do not fill the pore. In (b), the pressure is somewhat higher and the water molecules fill the pore.

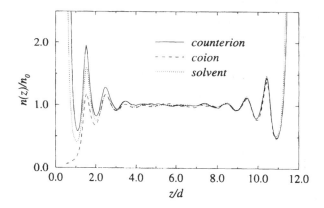

Figure 6 Density profiles of ions and water near a charged surface (left). The surface at the right is uncharged. The two surfaces are far enough apart that there is no interaction between them.

IV. SUMMARY

The theory of liquids can be adapted to provide new insights that have been useful in the study of thin liquid films. These studies indicate that there is a layering near the walls of the pore. Recent experiments support these studies.

ACKNOWLEDGMENTS

This article was written while the author was visiting MCS University in Lublin, Poland. The hospitality of the MCS University Faculty of Chemistry and the Sokolowski family is gratefully acknowledged. Mr. P. Bryk, Dr. S. Sokolowski, and Dr. D. Boda assisted in the preparation of some of the figures. This work was supported, in part, by the National Science Foundation (Grant CHE96-01971) and by the donors of the Petroleum Research Fund, administered by the American Chemical Society (Grant No ACS-PRF 31573-AC9).

REFERENCES

1. JA Barker, D Henderson. Rev Mod Phys 48:587, 1976.
2. D Henderson, FF Abraham, JA Barker. Mol Phys 31:1291, 1976.
3. P Tarazona. Mol Phys 52:81, 1984; Phys Rev A31:2672, 1985; P Tarazona, UMB Marconi, R Evans. Mol Phys 60:573, 1987.

4. R Evans. In: D Henderson, ed. Fundamentals of Inhomogeneous Fluids. New York: Marcel Dekker, 1992, Chap. 3.
5. JN Israelachvili. Intermolecular and Surface Forces. 2nd ed. London: Academic Press, 1992.
6. MS Wertheim, L Blum, D Brakto. In: S-H Chen, R Rajagopalan, eds. Micellar Solutions and Microemulsions. New York, Springer-Verlag, 1990, Chap. 6.
7. D Henderson, S Sokolowski, DT Wasan. J Statist Phys 89:233, 1997.
8. D Henderson, S Sokolowski, DT Wasan. J Phys Chem B102:3009, 1998.
9. D Henderson, S Sokolowski. Phys Rev E57:5539, 1998.
10. M Shoen, S Dietrich. Phys Rev E56:499, 1977.
11. D Boda, K-Y Chan, D Henderson, DT Wasan, AD Nikolov. unpublished work, 1998.
12. P Pieranski, L Strzelecki, B Pansu. Phys Rev Lett 50:900, 1983.
13. R Evans, M Chan. Phys World 9:48 (April 1996).
14. E Spohr, AD Trokhymchuk, D Henderson. J Electroanal Chem 450:281, 1998.
15. D Boda, K-Y Chan, D Henderson. J Chem Phys. 109:7362, 1998.
16. E Spohr. J Electroanal Chem 450:327, 1998.

12

Morphological Pathways of Pattern Evolution and Dewetting in Thin Liquid Films

Ashutosh Sharma and Rahul Konnur
Indian Institute of Technology, Kanpur, India

Rajesh Khanna and Günter Reiter
Institut de Chimie des Surfaces et Interfaces, Mulhouse, France

ABSTRACT

Various stages of evolution of the surface instability and pattern formation are investigated theoretically and experimentally for unstable thin ($\sim$100 nm) fluid films subjected to a long range van der Waals attraction and a shorter range repulsion. The complete three dimensional morphology is resolved based on numerical solutions of the nonlinear 2D thin film equation, and compared with experiments on thin ($<$150 nm) films of polydimethylsiloxane (PDMS) sandwiched between water and bimodal brushes of PDMS chemically grafted on silicon wafers. In the first, linear phase of evolution, initial random nonhomogeneities are quickly reorganized into a small amplitude undulating structure consisting of long "hills" and "valleys." Two distinct pathways of morphological evolution and dewetting are found thereafter depending on the initial film thickness vis-a-vis location of the minimum in the spinodal parameter (intermolecular force per unit volume curve). While dewetting of relatively thick films occurs by the growth of *isolated circular holes*, thinner films dewet by the formation and growth of *droplets*. In the former case, droplets are formed at long times due to the coalescence of holes. Based on the matching of the simulated and experimental patterns, we propose and apply a novel Thin Film Force Microscopy (TFFM) concept, which can determine the unknown intermolecular interactions based on the facile observations of the film pattern on micrometer scales.

I. INTRODUCTION

The problem of stability and spontaneous pattern formation in thin ($\sim$100 nm) fluid films is central to a host of technological applications (e.g., polymeric and metal coatings, foams, emulsions, flotation) and to a diversity of physical and biological thin-film phenomena (e.g., wetting, adhesion, heterogeneous nucleation, colloids, membrane morphology, dry-eye syndrome). Study of thin films is also of fundamental scientific importance in understanding the influence of interfaces on such classical "bulk" phenomena as diffusion, phase separation, and phase change.

After nearly five decades of intense activity in thin films, general principles of the mechanics and thermodynamics of thin films have emerged [1–5]. Like all spinodal processes (e.g., phase separation of incompatible materials), the free surface of an initially uniform thin film is unstable and deforms spontaneously to engender a microstructure when the second derivative of the excess intermolecular free energy (per unit area) with respect to the (local) film thickness is negative, viz. $\partial^2 \Delta G / \partial h^2 < 0$ [1–5]. However, the current theoretical understanding of thin-film patterns and their relationship to the surface properties–intermolecular interactions is very rudimentary, and even misleading, as it is based largely on the linear stability analysis [1–3,6]. Although the one-dimensional (1D) nonlinear simulations [4,7–10] have provided many interesting results, they also cannot provide any clues regarding the full 3D morphology of an unstable thin film. In short, there is as yet no direct way to compare the theory with the thin-film experiments.

This chapter is intended toward a (partial) resolution of the following outstanding questions in the thin-film physics and engineering:

1.   What are the general classes of morphological patterns which can arise spontaneously in a thin ($\sim$100 nm) liquid or liquidlike film on a solid substrate under the influence of antagonistic (attractive and repulsive) short- and long-range surface interactions?
2.   What is the relationship between the morphological structures (patterns of defects, dewetting) and the *form* of intermolecular/surface forces?
3.   Is it possible to address the inverse problem of characterization of the *nanoscale* interactions based directly on the facile observations of the thin-film morphology on *micrometer* scales?

In the context of the first question, it is important to note that Cahn's linear theory of spinodal processes [6] and its extrapolations for the thin films [2] suggested that the surface instability and consequent dewetting could occur *only* in the form of a bicontinuous structure composed of liquid ridges or "hills" and "valleys." However, thin-film experiments show a very rich variety of morphological patterns [11–24], including circular holes. It has been thought [17–19] that the presence of circular holes was necessarily indicative of a "nucleation" induced by the dust particles, preexisting defects, and so forth. Although such a view may be essentially correct for thin films where randomly distributed holes form [18,19], we still do not know why a circular hole (rather than any other shape) is the preferred mode of dewetting over a nonwettable heterogeneous patch which is unlikely to have a circular symmetry in real, ill-defined surfaces.

The last question, which is intimately tied with the first two, has both philosophical and operational significance. Unlike the case of thin fluid films between two *solid* surfaces, where the nanoscale interactions can be directly measured by atomic force microscopy (AFM) or by surface-force apparatus [25], there is as yet no experimental (or theoretical) way to characterize the surface forces in an unstable fluid film sandwiched between a solid and a fluid phase. This issue is of special significance in the characterization of ill-defined, modified, and contaminated solid surfaces and liquids (often polymeric) encountered in actual applications, and even in the best of the controlled experiments.

The main thrust of this chapter is to uncover the variety of morphological patterns which can form spontaneously in an unstable film and to explore the conditions for the selection of a particular pattern. Our recent framework for complete 3D nonlinear simulations [26–28], for the first time, provides a formalism for correlating the film morphology with the interfacial interactions and the film thickness. This should make it possible to directly compare theory and experiments. Among other things, such a formalism will also help address the inverse problem of characterization of surface interactions from the observed morphology.

II. EXPERIMENTS

Experiments were performed with polydimethylsiloxane (PDMS) thin films supported on PDMS brushes chemically grafted on silicon wafers. An additional layer of long PDMS connectors was also grafted (bimodal brushes), which introduced a "soft" longer-range repulsion preventing the rupture of the free PDMS film. A schematic representation of the experimental system is given in Figure 1. Additional details regarding the film and substrate preparation may be found elsewhere [13,21–24].

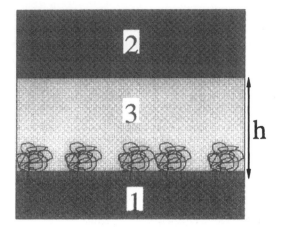

Figure 1 Schematic representation of the experimental system: (1) silicon wafer with grafted short PDMS brush (cloudlike formations); (3) free PDMS film; (2) water. h represents the total thickness of the PDMS phase, including the brush.

For air as the bounding medium, the van der Waals forces for this system are repulsive, leading to perfect wetting and complete film stability. However, an attractive long-range van der Waals force was induced by placing a drop of water on top of the PDMS film, which led to the growth of surface instability. The evolution of the PDMS film was followed at room temperature and in real time by optical microscopy. To investigate the dependence on the film thickness, we performed experiments for total PDMS thicknesses ranging from 30 to 110 nm. Of these total thicknesses, the thickness of the "wet" swollen bimodal brush was about 20 nm.

III. THEORY

A. Excess Intermolecular Interactions/Forces in Thin Films

There are as many (additive) components of the excess free energy as the variety of the underlying intermolecular interactions [1,5,25,29]. Most prominent of these are the following [25,29]: apolar (and universal) van der Waals interactions; polar acid–base and entropic interactions, including the hydration repulsion and hydrophobic attraction (for aqueous media on hydrophilic and hydrophobic surfaces, respectively); electrostatic double-layer interactions (for charged surfaces in ionic solutions); and more complex shorter-range attraction/repulsion engendered by the entropic effects in polymers close to a confining wall or due to chain adsorption/grafting [30,31]. There are four universal classes of the *form* of the free energy for thin films [5,7,32]: type I systems, where both the long- and short-range forces are attractive; type II systems (also related to the Stranski–Krastanov wetting in metallurgy), where a long-range attraction combines with a shorter-range repulsion; type III systems, where all forces are repulsive; and, type IV systems, where a long-range repulsion combines with a shorter-range attraction. Here, we will confine attention to a particular set of type II films [10] which model our experiments on PDMS films [21–24]. Type III systems showing perfect wetting are uninteresting in that their thin films are unconditionally stable. The dynamics and morphology of type I and type IV thin films have been considered elsewhere [26–28,33,34].

The effective Hamaker constant is determined by the individual Hamaker constants (A_{ij}) of the substrate (S), film fluid (f), and the overlying bounding medium (b), viz. $A_s = (\sqrt{A_{ss}} - \sqrt{A_{ff}})(\sqrt{A_{bb}} - \sqrt{A_{ff}})$ [25,29]. The effective Hamaker constant can also be represented in terms of the apolar component of the spreading coefficient, $A_s = -12\pi\, d^2_0\, S_s$ [5,7,8,10]. The apolar (LW) component of the spreading coefficient S_s is defined in terms of the macroscopic apolar components of the surface and interfacial tensions, and d_0 (≈ 0.158 nm) is a cutoff equilibrium distance introduced by the Born repulsion [29].

When the van der Waals component of the substrate surface tension (in air) exceeds that of the film material, the effective Hamaker constant, A, is negative, signifying a net long-range apolar van der Waals *repulsion* which promotes the film stability and wetting. This is almost always the case for aqueous films and for (relatively) low surface-energy polymers (in air) on a majority of substrates (e.g., silicon wafers, glass, mica) [5,7,20,29,34]. The sign of the effective Hamaker constant can, however, be modulated by replacing air by a liquid [e.g., in our experiments, polydimethylsiloxane (PDMS) films on silicon wafers in air are completely wetting ($A < 0$), but replacing air by water (and most other liquids) switches the van der Waals repulsion into a net attraction ($A > 0$), thus destabilizing the film).

The shorter-range, polar, non-van der Waals attraction may be due to the "hydrophobic attraction" for water [25,29,34], as well as to entropic confinement effects some distance away from the surface for polymer films due to adsorption/ grafting at the solid–film interface [30,31]. The latter is responsible for the "autophobic" behavior [13,30,31] of polymer liquids, where a droplet does not completely wet an adsorbed film of the same liquid. Finally, a still shorter-range repulsion leading to spreading close to the solid–polymer interface in polymers can be induced by grafting of long "connector" molecules, which also lead to chain entanglement [21–24]. Molecular parameters (e.g., length, molecular weight, and grafting density of the brush determine the range of the "soft" repulsion). This is the case in our experiments, as discussed later. For the simple nonpolymeric liquids on the other hand, a steep Born repulsion, which arises due to the molecular orbitals overlap, provides the necessary cutoff to prevent the divergence of van der Waals forces at film rupture [28].

Here, we confine our attention to the excess intermolecular interaction free energy composed of a long-range van der Waals attraction combined with a shorter-range repulsion due to the polymer brush. A convenient model of the system is shown in Figure 2, in which a thin film rests on a *coated* substrate [10,14]. The substrate is assumed to be nonwettable, providing a long-range van der Waals attraction, but the thin coating of the substrate (model of the polymer brush) is compeletely wettable, providing a shroter-range repulsion. An analytical representation of the free energy for this system is given by [10]

$$\Delta G = -\left(\frac{A_s}{12\pi h^2}\right)\left[(1 - R)\left(1 + \frac{\delta}{h}\right)^{-2} + R\right] \tag{1}$$

where A_s (>0) is the effective Hamaker constant for the substrate–film–bounding medium system, h is the local film thickness, and δ is the coating thickness (see Fig. 2). $R = A_c/A_s$ and A_c (<0) is the effective Hamaker constant for the coating–film-bounding medium system. Equation (1) demonstrates a long-range attraction ($-A_s/12\pi h^2$) for $h \gg \delta$, which changes into a shorter-range repulsion, ($- A_c/12\pi h^2$), as the film thickness declines locally. The disjoining pressure is defined as $\Pi = -\partial \Delta G/\partial h$, and the spinodal parameter is defined as $\phi_h = \partial^2 \Delta G/\partial h^2$.

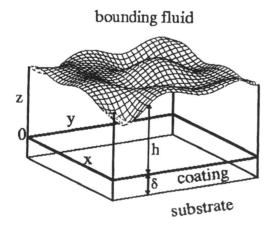

Figure 2 Schematic representation of the model of the surface instability in a thin film on a coated substrate.

Qualitative variations of ΔG and $(\partial^2 \Delta G/\partial h^2 = \phi_h)$ from Eq. (1) are as shown in Figure 3 for type II thin films subjected to a long-range van der Waals attraction combined with a short-range repulsion. The free surface of the film is unstable when the spinodal condition, $(\partial^2 \Delta G/\partial h^2) < 0$, is met. Two sets of parameter values were chosen for simulations (curves labeled 1 and 2 in Fig. 3). The system represented by curves 1 has a higher equilibrium contact angle because the minimum in ΔG is deeper [32]. Qualitatively, these systems model our experiments with thin PDMS films sandwiched between an aqueous medium and silicon wafers grafted with a PDMS "brush," as described in Section II. For illustration, a realistic set of parameters is chosen, but based on a large number of simulations, we have verified that all of the key qualitative morphological features and pathways of dewetting depend on the form of potential, rather than its magnitude. Even a particular analytical representation [e.g., Eq. (1)] is not of consequence, but it is only the *shape* of the potential diagram that matters.

B. Thin-Film Equation

The following nondimensional thin-film equation [26–28], derived from the Navier–Stokes equations, governs the stability and spatiotemporal evolution of a thin film subjected to the excess intermolecular interactions:

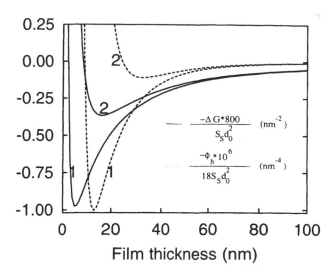

Figure 3 Variations of the free energy per unit area, ΔG, and force per unit volume (spinodal parameter), $\partial^2 \Delta G/\partial h^2$ $(=\Phi_h)$ with film thickness from Eq. (1). $\delta = 20$ nm, $R = -0.01$ and -0.1, for curves labeled 1 and 2, respectively.

$$\frac{\partial H}{\partial T} + \nabla \cdot [H^3 \nabla (\nabla^2 H)] - \nabla \cdot [H^3 \Phi_H \nabla H] = 0 \qquad (2)$$

where $H(X, Y, T)$ is the nondimensional local film thickness scaled with the mean thickness h. The spatial coordinates X and Y in the plane of the substrate are scaled with the characteristic length scale for the van der Waals case [4], $(2\pi\gamma/|A_s|)^{1/2}h^2$. Nondimensional time T is scaled with $(12\pi^2\mu\gamma h^5/A_s^2)$; where γ and μ refer to film surface or interfacial tension and viscosity, respectively. A renormalized real time $t_N = t\,(A_s^2/12\pi^2\mu\gamma) = Th^5$ can also be defined to remove the influence of mean film thickness. Finally, $\Phi_H = (2\pi h^4/|A_s|)(\partial^2 \Delta G/\partial h^2)$ is the nondimensional spinodal parameter. ∇ is the usual two-dimensional gradient operator, $i(\partial/\partial X) + j(\partial/\partial Y)$.

The 1D counterpart of Eq. (2) has been extensively studied cf. Refs. 4 and 7–10. The second term of the thin-film equation denotes the effect of surface tension for a curved surface, which, in a 3D geometry, may be stabilizing (due to "in-plane" curvature as in 2D cases) or destabilizing (due to transverse curvature as in Rayleigh instability of circular cylinders). The third term describes the effect of excess intermolecular interactions, which engender instability by causing flow from thinner to thicker regions in the case of negative "diffusivity" (viz. when the spinodal parameter Φ_H or $\partial^2 \Delta G/\partial H^2$ is negative). The initial growth of instability due to "negative diffusion" is common to all sorts of spinodal decomposition processes. Simulations [7,26] have confirmed the theoretical expectation [5] that the growth of instability eventually saturates leading to a quasi-equilibrium structure when Φ_H becomes positive due to the short-range repulsion.

C. Linear Analysis and Numerical Techniques for Nonlinear Simulations

The linear stability analysis [2,28] of a 3D thin film [Eq. (2)] predicts a dominant nondimensional characteristic length scale of the instability, $\lambda = 4\pi/\sqrt{-\Phi_H}$, which is the diagonal length of a unit square cell of length $L = \lambda/\sqrt{2}$. In order to address the problem of pattern selection, we directly solved the nonlinear thin-film equation numerically over an area of $9L^2$, starting with an initial small-amplitude (≈ 1 Å) random perturbation. A 60×60 grid was usually found sufficient when central differencing in space with half-node interpolation was combined with Gear's algorithm for time marching, which is especially suitable for stiff equations. Details of the numerical methods used are too lengthy to be spelled out here but can be found elsewhere [27,28]. In some cases, automatic grid movement and compression were also used to handle the extreme stiffness, and an ADI (alternate direction implicit) technique was implemented to combine the computational efficiency of explicit integration schemes with the accuracy of implicit schemes.

With this theoretical and experimental background, we can now adress the three fundamental questions posed in Section I regarding the morphology of thin unstable films.

IV. RESULTS AND DISCUSSION

A. Morphological Pathways of Dewetting: Holes, Bicontinuous Patterns, and Droplets

1. Simulations

Based on numerical simulations, we found two completely distinct morphological patterns and their sequence of evolution by which (pseudo) dewetting could occur, depending on the *form* of the potential in the neighborhood of the initial film thickness. In particular, we found that the nonlinear morphological pattern selection depends crucially on the distance between the mean film thickness and the location of the minimum in the $\partial^2 \Delta G / \partial h^2$ versus film thickness curve. Thus, the morphology can be modulated either by changing the potential (by shifting the minimum in the potential to a different thickness) or by changing the film thickness itself (for the same potential). In what follows, we present and contrast the different general pathways of evolution of the surface instability and resulting 3D morphologies.

Figure 4 summarizes the major events in the time evolution of patterns in a relatively thick film (100 nm), which is sufficiently removed from the location of the minimum in the spinodal parameter (Fig. 3, curves 2). The initial random

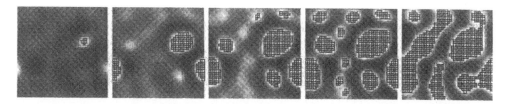

Figure 4   Gray-scale representation of the evolution in a 100-nm-thick film of the system represented by curve 2 of Fig. 3 ($\delta = 20$ nm, $R = -0.1$). The area of each box is $9L^2$. In this, as well as in subsequent figures, the color intensity is proportional to the film thickness, and the hatched zones depict the dewetted areas where the film thickness declines to the equilibrium thickness given by the minimum in the free energy, ΔG (nondimensional equilibrium thickness is 0.17). Pictures 1–5 (from left to right) are at nondimensional time (T) = 60, 72, 82, 90, and 116, and the minimum and maximum nondimensional thicknesses are (0.19, 1.28), (0.17, 1.82), (0.17, 2.31), (0.17, 2.41), and (0.17, 3.37), respectively.

disturbance is first reorganized into a small-amplitude bicontinuous pattern on a length scale close to λ. This stage is reminiscent of the linear concentration field in the spinodal decomposition [2,6]. Long ''hills'' of the structure undergo some fragmentation, whereas the ''valleys'' thin locally to produce largely circular depressions which develop into full-thickness circular holes surrounded by uneven (in height) rims. Contrary to the impression conveyed by the linear theory, not all holes form at the same time. Newer holes continue to appear as the older ones expand. As expected [5,7,26], a true dewetting of the substrate, however, does not occur at the base of these holes (hatched areas), where a nearly flat film of thickness close to the location of the minimum in the free energy (~32 nm) is left behind (pseudodewetting). Hole expansion occurs by the displacement of its (pseudo) contact lines on the equilibrium flat film. The hole axisymmetry is gradually diminished due to the presence of neighboring holes which hinder expansion in certain directions and also due to hole coalescence. Eventually, a repeated coalescence of holes suggests rudiments of a large-scale structure, in which the intervening pools of liquid eventually fragment and slowly transform into increasingly circular droplets (long time results not shown), as shown in the simulation in Figure 5.

Interestingly, a very similar sequence of morphological evolution (bicontinuous pattern → circular holes with rims → hole expansion → hole coalescence) was also shown [28] by simulations for *all thicknesses* in thin films subjected to only a long-range van der Waals *attraction* (type I films). These results indicate a rather general pathway of pattern evolution and dewetting by the formation of isolated circular holes (viz. air-in-liquid dispersion) whenever the initial mean thickness (viz. liquid-phase concentration) is sufficiently high so that the repulsive interactions at the minimum thickness are encountered only after a considerable growth of the instability. The formation of circular holes is, therefore, not *necessarily* indicative of ''nucleation'' by large dust particles, defects, and so

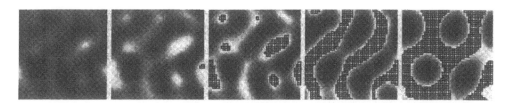

Figure 5 Evolution in a 50-nm-thick film with the same nomenclature and parameters as in Figure 4. Pictures 1–5 (from left to right) are at nondimensional time (T) = 276, 482, 567, 927, and 1519, and the minimum and maximum nondimensional thicknesses are (0.73, 1.22), (0.39, 1.89), (0.36, 2.13), (0.34, 2.48), and (0.34, 3.34), respectively. The nondimensional equilibrium thickness is 0.34.

forth but can *also* occur by a spontaneous growth of surface instability. In particular, simulations show that dewetting by circular holes is encouraged on precisely those heterogeneous patches which are more nonwettable (i.e., display a weak repulsion), so that the spinodal minimum shifts to the smaller thickness and the distance between the spinodal minimum and the mean film thickness is increased.

Figure 5 shows the evolution for a thinner film, which is still to the right of the minimum in the spinodal parameter (Fig. 3, curves 2). The major difference from very thick films is that the depressions lose axisymmetry much before the onset of dewetting and the bicontinuous pattern becomes prominent and persists for longer times. Because depressions encounter repulsion at an earlier stage of their growth, coalescence of neighboring depressions is encouraged before dewetting occurs. Thus, dewetting begins at the base of long valleys rather than at the base of circular depressions. Eventually, dewetting occurs along the entire stretch of long valleys and a bicontinuous pattern of dewetting composed of long liquid threads is established. At this stage, long liquid ridges fragment into droplets due to a difference in the Laplace pressure causing flow from the thinner to thicker regions.

Figure 6 shows the evolution for a 33-nm-thick film which coincides with the minimum in the spinodal parameter. In the first frame of this sequence, the initial depressions have already coalesced rather fully to produce continuous valleys, whereas the initially long liquid ridges have fragmented into droplets at this early stage of evolution. The formation of largely isolated *droplets* surrounded by deep valleys is a prominent feature at the onset of dewetting. Dewetting again occurs gradually all along the valleys (hatched areas). Compared to the earlier picture of dewetting in a 50-nm film, there is now an increasing indication of the dewetted areas surrounding the droplets on all sides. In other words, the liquid (droplet) phase becomes the dispersed phase at an earlier stage of evolution. Both the drop coalescence and fragmentation events continue to occur until isolated

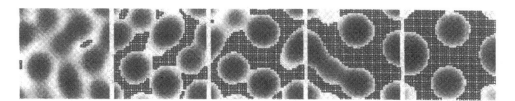

Figure 6   Evolution in a 33-nm-thick film with the same nomenclature and parameters as in Figure 4. Pictures 1–5 (from left to right) are at nondimensional time $(T) = 4198$, 6738, 10787, 15751, and 36211, and the minimum and maximum nondimensional thicknesses are (0.6, 2.06), (0.57, 2.51), (0.56, 2.76), (0.55, 3.07), and (0.54, 4.0), respectively. The nondimensional equilibrium thickness is 0.53.

drops are eventually produced. At very long times (results not shown), ripening of the droplet structure also occurs, whereby the larger droplets grow at the expense of smaller droplets.

An interesting point is that for films thicker than the minimum in the spinodal parameter, dewetting is initiated *before* the formation of isolated mature circlar droplets. There is, however, a drift toward the formation of droplets at increasingly earlier stages of evolution with decreased mean film thickness. For example, contrast the sequence of evolution in Figures 4 and 6. In the former, dewetting occurs by the formation, growth, and coalescence of isolated full-thickness circular holes, and the droplets are produced only at the very last stage of dewetting. In the latter, a prominent droplet phase is visible even at the time of the onset of dewetting.

Finally, in qualitative contrast to the above scenarios of dewetting, Figure 7 depicts a different universal pathway of evolution for unstable films which are thinner than the thickness where the minimum of the spinodal parameter occurs (<32 nm in Fig. 3, curves 2). The initial bicontinuous structure directly produces an array of circular isolated droplets surrounded by circular depressions or valleys (viz. liquid-in-air dispersion). These droplets become increasingly circular due to surface tension and increase in height due to flow from the "valleys" which thin and flatten out, leading to dewetting (hatched areas) *after* the formation of droplets. In this case, the (pseudo) dewetting is accompanied by *maturing of droplets*, rather than by the *expansion of holes*. At long times, ripening of the structure continues by the merger of neighboring droplets due to the Laplace pressure-induced flow from the smaller to larger drops. The same *qualitative* pathway of morphological evolution was seen in a large number of simulations (not shown) for all thicknesses to the left of the minimum in the $\partial^2 \Delta G / \partial h^2$ versus thickness curve, regardless of the numerical values of the parameters which characterize the potential.

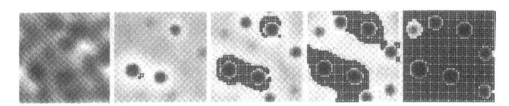

Figure 7 Evolution in a 25-nm-thick film with the same nomenclature and parameters as in Figure 4. Pictures 1–5 (from left to right) are at nondimensional time (T) = 38366, 146421, 211927, 260439, and 413888, and the minimum and maximum nondimensional thicknesses are (0.97, 1.1), (0.78, 3.49), (0.75, 4.28), (0.74, 4.75), and (0.71, 5.22), respectively. The nondimensional equilibrium thickness is 0.69.

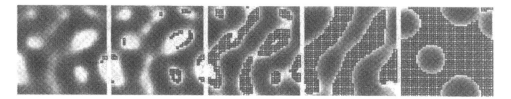

Figure 8 Evolution in a 33-nm-thick film of the system represented by curve 1 of Figure 3 (δ = 20 nm, R = −0.01). Pictures 1–5 (from left to right) are at nondimensional time (T) = 830, 915, 1222, 1636, and 178213, and the minimum and maximum nondimensional thicknesses are (0.2, 2.06), (0.18, 2.35), (0.17, 2.76), (0.17, 3.80), and (0.17, 4.79), respectively. The nondimensional equilibrium thickness is 0.17.

Thus, the morphological pathway of dewetting depends only on the relative location of the film thickness vis-à-vis the location of the minimum in the spinodal parameter. To further illustrate this point, we consider a simulation for a different system represented by curve 1 of Fig. 3, where the minimum occurs at 13 nm rather than at 32 nm as for curves 2. Figure 8 summarizes the evolution for a 33-nm-thick film of this system (curve 1). Qualitatively, the morphological features of the evolution closely resemble those for a 50-nm-thick film of curves 2 (Fig. 5). The evolution is different from the same thickness film of the system represented by curves 2 (simulation in Fig. 6). Thus, as far as the qualitative morphological features are concerned, a 50-nm-thick film of a system displaying spinodal minimum at 32 nm is similar to a 33-nm-thick film of a system with the spinodal minimum at 13 nm. Similarly, Figure 9 shows the evolution of a 25-nm-thick film of the system represented by curve 1 of Figure 3. Even more spectacular differences in the morphology are apparent when we compare this pathway of evolution with a 25-nm-thick film of the system represented by curves

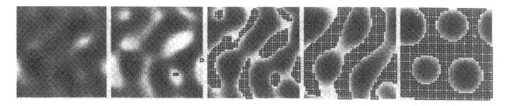

Figure 9 Evolution in a 25-nm-thick film of the system represented by curve 1 of Figure 3 (δ = 20 nm, R = −0.01). Pictures 1–5 (from left to right) are at nondimensional time (T) = 1291, 2523, 3497, 5219, and 172692, and the minimum and maximum nondimensional thicknesses are (0.74, 1.23), (0.27, 2.15), (0.24, 2.64), (0.24, 3.53), and (0.24, 4.32), respectively. The nondimensional equilibrium thickness is 0.24.

2 of Figure 3 (Fig. 7). Thus, the pattern is governed not by the film thickness but by its positioning vis-à-vis the location of the spinodal minimum, which is different for different systems.

One of our main results from the above discussion may be summarized by directly looking at the morphologies of films of different thicknesses at (or near) the onset of dewetting, when the minimum thickness for the first time locally drops to the equilibrium thickness. Beyond this time, (pseudo) dewetting by holes or drops occurs by the retraction of the (pseudo) contact line. Figure 10 shows the morphologies close to the onset of dewetting. For a 100-nm-thick film, dewetting is initiated by a small-diameter steep circular hole (picture 1), and the surrounding film appears more uniform with only rudiments of small amplitude droplets, ridges, and valleys. For a thinner film of 50 nm thickness (picture 2), the morphology changes to wider, deeper, noncircular (but still isolated) depressions and prominent long ridges. At 33 nm (picture 3), depressions have coalesced to form the continuous phase, and the isolated, noncircular droplets become the prominent feature. Finally, at 25 nm (picture 4), and for films to the left of the minimum in the spinodal parameter, completely isolated circular droplets are present at the onset of dewetting. Conceptually, this phenomenon (droplets-in-air → holes-in-liquid) can be thought of as a transient ''morphological phase inversion'' in which liquid film thickness, which is an analog of concentration, is the control parameter.

Figure 11 quantifies the rate at which dewetting of the substrate occurs for films of different thicknesses. Interestingly, after the onset of dewetting, the dewetted area increases as: area $\sim \log(T) \sim \log[A_s^2 t / 12\pi^2 \mu \gamma h^5]$. The slope is al-

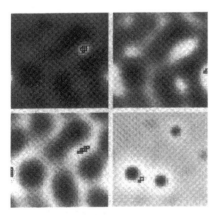

Figure 10 Morphologies at the instant of onset of dewetting in films of different thicknesses of the system represented by curves 2 of Figure 3 ($\delta = 20$ nm, $R = -0.1$). Pictures 1–4 correspond to the mean thicknesses of 100, 50, 33, and 25 nm, respectively.

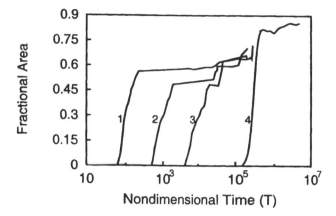

Figure 11 caption area:

Nondimensional Time (T)

Figure 11 Variation of dewetted area with the nondimensional time for films of different thicknesses of the system represented by curves 2 of Figure 3 ($\delta = 20$ nm, $R = -0.1$). Curves 1–4 correspond to 100-, 50-, 33-, and 25-nm-thick films, respectively.

most constant for all films to the right of the spinodal minimum, but higher for a film to the left of it. A sudden decrease in the slope at longer times indicates a quasistable structure, which undergoes ripening, during which the dewetted area increases once again.

2. Experimental Morphologies

Figure 12 for a 85-nm film shows a typical sequence of dewetting by the formation, growth, and coalescence of *circular holes* in our experiments with PDMS

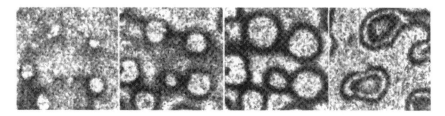

Figure 12 Experimental evolution of the instability in a relatively thick 85-nm PDMS film (viscosity of 1000 cSt, $M_w = 28$ kg/mol) on top of a 20-nm-thick bimodal grafted brush of PDMS. Placing a 1-mm-thick layer of water on top destabilized the film. Pictures 1–4 (from left to right) are taken at 223, 369, 520, and 957, respectively. The size of each micrograph is 200×200 μm^2.

films. Eventually, the intervening liquid fragments and retracts to form isolated droplets, as confirmed also by the simulations. Figure 13 summarizes and contrasts the prominent stages and morphological features in the evolution and dewetting of a relatively thin film (41 nm; series a), a moderately thick film (60 nm; series b) and relatively thick film (85 nm; series c). Clearly, there is increased preference for the formation of isolated droplets (and ridges) at an early stage of evolution as the film thickness is decreased. Dewetting for thicker films occurs by increasingly well-formed depressions and isolated holes. Thus, the simulations reproduce the experimental observations regarding the pattern evolution and morphology quite well.

Although traditionally [2,6,17–19], the mechanism of "spinodal decomposition" has been thought to lead only to bicontinuously undulating pattern (simi-

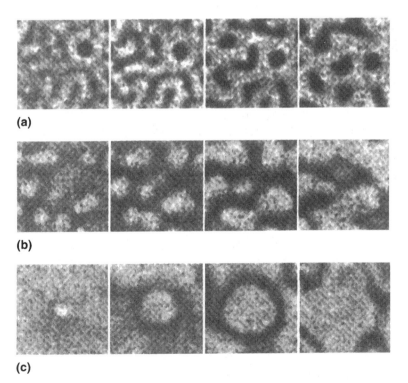

(a)

(b)

(c)

Figure 13 Experimental evolution of the instability in PDMS films. From left to right: (a) 41-nm-thick film at 159, 208, 331, and 515 s; (b) 60-nm-thick film at 64, 122, 330, and 507 s; (c) 85-nm-thick film at 223, 369, 520, and 957 s. Each micrograph represents an area of 50×50 μm^2.

lar to pictures in series a of Fig. 13; and Figs. 7 and 8), both the simulations and experiments clearly show a continuous change in the pattern to produce isolated circular holes as the film thickness increases. Thus, not all holes observed in all thin films are necessarily due to the "nucleation" (e.g., induced by defects and dust particles).

B. Structure–Property (Morphology–Potential) Relationships in Thin Films

1. Which Morphological Features Depend on the *Form* of the Thin-Film Potential?

Based on the above theoretical and experimental results, it now becomes clear that it is the *form* of the potential in the neighborhood of the initial thickness which governs vastly different morphological patterns and their sequence of evolution. For relatively thick films away from the minimum in the spinodal parameter, isolated holes lead to dewetting *before* droplets are formed due to a merger of holes and fragmentation/retraction of the intervening pools of liquid. In contrast, for relatively thin films close to or to the left of the spinodal minimum, a large-amplitude bicontinuous pattern made up of long ridges fragments directly into droplets, which dewet the substrate. For a range of intermediate thicknesses, the initial bicontinuous pattern resolves into a mixture of both drops *and* holes in varying proportions, depending on the distance from the minimum in the potential. An increase in the film thickness increases the relative population of isolated circular depressions of increasing depth and increasingly well-formed rims, until the mature holes are obtained at higher thicknesses. Conceptually, this phenomenon (droplets-in-air → holes-in-liquid) can be thought of as a transient "morphological phase inversion" in which liquid film thickness, which is an analog of concentration, is the control parameter.

An important conclusion is that different morphologies (isolated circular holes, bicontinuous ridges, droplets), and their combinations, can all be produced by the *spinodal decomposition mechanism*, depending on the film thickness vis-à-vis the location of the minimum in the spinodal parameter. Moreover, all the three patterns (holes and drops of varying sizes, bicontinuous) can coexist at a given time, most prominently, on heterogeneous surfaces. As an example, recent experiments of Herminghaus et al. [35] with thin liquid crystal films on silicon show two different coexisting patterns of pseudodewetting: (1) a correlated undulating bicontinuous/droplet structure similar to the one seen in simulations for relatively thin to moderate thickness films and (2) formation of randomly distributed circular holes. Based on our results, we can hypothesize that the circular holes result from a reduction in the soft repulsion leading to a shifting of the spinodal minimum closer to the substrate surface. The heterogeneity of the substrate combined possibly with the loss of molecular ordering close to the substrate can lead to such a scenario. On such randomly distributed patches, dewetting should occur by the formation of holes, because the distance between the spinodal minimum and the film thickness is increased (i.e., the film has become relatively thick via-à-vis location of the spinodal minimum).

2. Which Morphological Features Depend on the Precise *Magnitude* of the Thin-Film Potential?

Whereas the *form* of the potential selects the morphology, it is the precise *magnitude* of $\partial^2 \Delta G / \partial h^2$ which determines the actual (dimensional) characteristic length scale of the pattern, as implied by scalings in Eq. (2) and its linear analysis discussed previously in this chapter. The length scale of the initial bicontinuous pattern was always found to be in agreement with the predictions of the linear analysis. Interestingly, a large number of simulations also showed that about 8–12 holes (or drops) were initially formed on a substrate area of $9L^2$, which confirms the expectation of the linear theory, *before* a significant ripening of drops or hole coalescence occurs. An important conclusion is that although the linear theory can be used as a good approximation for the prediction of length scales (or for the prediction of the potential, $-\Phi_H = 16\pi^2/\lambda^2$ from the observed λ), it fails completely in the prediction of morphological patterns (holes, bicontinuous ridges, drops, etc.).

C. Determination of Intermolecular Interactions Based on Morphology

As discussed in Section I, there is as yet no direct method for the quantification of the potential for unstable thin films. This issue is especially important for the ill-defined, modified (coated, oxidized, contaminated) solid surface and liquids encountered in practice. Even the magnitude of the van der Waals forces in such systems cannot be estimated, the situation with the non-van der Waals forces being considerably more complex.

However, the above discussion on the structure–property relations points to the possibility of determining the interaction potential ($-\Phi_H$ and its dimensional counterpart) by *matching* of the morphological patterns obtained from the theory and experiments. Fortunately, this task is considerably simplified by our finding that the average length scale of the pattern is well predicted by the linear theory, from which an analytical dependence of the length scale on the potential is readily obtained [viz. $-\Phi_H = 16\pi^2/\lambda^2$, where λ is the (nondimensional) length scale]. Thus, if we can determine λ at different thicknesses based on experiments, the variation of Φ_H (and by integration of Φ and ΔG as well) with the film thickness can be readily obtained. We propose the name of thin-film force microscopy

(TFFM) for this technique, which may, in the future, mature as a counterpart of the atomic force microscopy (AFM). The latter can only probe forces between two solid surfaces separated by a thin film (solid–fluid–solid systems), but it cannot be used for unstable thin films in solid–fluid–fluid systems.

Although the present work is largely concerned with the qualitative aspects of the morphological patterns, our preliminary work in the application of the TFFM is briefly illustrated in Figure 14 [21–23] for the sake of completeness. Figure 14 shows the variation of the q vector (inverse of the characteristic length scale) with the film thickness for films of PDMS sandwiched between a bimodal brush and water. The q vector was obtained by taking the Fourier transform of digitized images before the droplets emerged. The same information was also confirmed independently by counting the number of droplets (per unit area). For thicker (>40 nm) films, a dependence of the type $\lambda \propto h^2$ is obtained, thus proving that the instability is caused by the long-range nonretarded van der Waals interactions, because $\lambda \propto [-\Phi_H]^{-1/2}$ and $[-\Phi_H]$ for *nonretarded* van der Waals forces varies as h^{-4}. For thinner films (<30 nm), the exponent in the relation $\lambda \propto h^\alpha$ deviates strongly from its value for thick films (about 2), which clearly shows the influence of the non-van der Waals short-range repulsive forces.

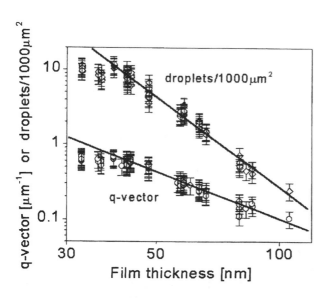

Figure 14   Characteristic q vector and droplets, N (per 1000 μm²) as a function of initial film thickness on logarithmic scales. The characteristic length scale of the pattern is $2\pi/q$. N and q are also related by $N = 1000q^2/4\pi^2$. The straight lines originate from fitting $q \sim h^\alpha$ and $N \sim h^\beta$, for $h > 40$ nm. We obtained $\alpha = -1.97 \pm 0.05$ and $\beta = -3.95 \pm 0.07$, indicating a decay characteristic of the nonretarded van der Waals interaction.

Interestingly, the results show that the van der Waals forces (per unit area), which at 100 nm are considered to be already four to five orders of magnitude weaker than the atmospheric pressure, can cause significant, measurable, macroscopic changes in short times at distances of hundreds of nanometers. A quantitative matching of the linear theory with the experimental q vector for the PDMS films showed [21–23] that the effective Hamaker constant for this system gives a value which is at least about two orders of magnitude larger than the accepted [25] range of the Hamaker constant for *dispersion* forces (10^{-20} J is about the maximum expected for this system). Neither the addition of salt in the overlaying water nor insertion of a thin polystyrene membrane between the PDMS film and water [24] changed the results, showing that electrostatic interactions are not involved. Further details regarding the quantitative aspects may be found elsewhere [21–23]. We close with a conjecture that the ''standard'' assumptions regarding the isotropy and the neglect of permanent dipoles may not be valid for this system. This aspect, and also the fact that the van der Waals interactions even at 100 nm appear to be *nonretarded*, need further serious explorations, as these results have a very far-reaching consequences in the entire area of colloids and interfaces and in their virtually limitless applications in both the physicochemical and biological domains. In this context, it is perhaps also important to recall that there are other ''mysterious'' long-range forces, most notably the hydrophobic attraction in the aqueous media, the existence of which is now universally accepted but the origin and the long-range nature of which continue to be enigmatic from a theoretical viewpoint.

V. SUMMARY

As a concise summary, we conclude by pointing that a framework is now available to correlate both the qualitative and quantitative aspects of morphological patterns in thin films with the excess intermolecular forces and the film thickness. The availability of complete 3D nonlinear simulations also makes it possible, for the first time, to directly compare experiments with the theory. Our (ongoing) work has addressed and at least partially resolved some long-standing problems: What are the property–structure (potential–morphology) relationships in thin films? Can the true morphology be determined by the hitherto extensively used Cahn's linear description of the spinodal processes? Is the formation of circular holes *necessarily* indicative of nucleation? Is it possible to determine the excess nanoscale intermolecular forces in thin films by the facile observations of their structures on microscales?

   We believe that the answers to the above questions will be of significance in the design and interpretation of experiments for uncovering the gamut of thin-film physics and its applications.

ACKNOWLEDGMENTS

The help of A. Casoli, M.-O. David, and P. Auroy in experiments is gratefully acknowledged. One of the authors (A. S.) would like to acknowledge a profound influence on his thinking of the work of Darsh Wasan and his co-workers on the self-organizing structures in thin films. This work was supported by a grant from the Indo-French Centre for the Promotion of Advanced Research/Centre Franco–Indien Pour la Promotion de la Recherché.

REFERENCES

1. BV Derjaguin. Theory of Stability of Thin Films and Colloids. New York: Consultants Bureau/Plenum Press, 1989.
2. A Vrij. Discuss Faraday Soc 42:23, 1966.
3. E Ruckenstein, RK Jain. J Chem Soc Faraday Trans II 70:132, 1974.
4. MB Williams, SH Davis. J Colloid Interf Sci 90:220, 1982.
5. A Sharma. Langmuir 9:861, 1993.
6. JW Cahn. J Chem Phys 42:93, 1965.
7. A Sharma, AT Jameel. J Colloid Interf Sci 161:190, 1993.
8. A Sharma, AT Jameel. J Chem Soc Faraday Trans 90:625, 1994.
9. A Sharma, CS Kishore, S Salaniwal, E Ruckenstein. Phys Fluids A 7:1832, 1995.
10. R Khanna, AT Jameel, A Sharma. Ind Eng Chem Res 35:3108, 1996.
11. G Reiter. Phys Rev Lett 68:75, 1992.
12. G Reiter. Langmuir 9:1344, 1993.
13. G Reiter, P Auroy, L Auvray. Macromolecules 29:2150, 1996.
14. A Sharma, G Reiter. J Colloid Interf Sci 178:383, 1996.
15. W Zhao, MH Rafailovich, J Sokolov, LJ Fetters, R Plano, MK Sanyal, SK Sinha, BB Sauer. Phys Rev Lett 70:1453, 1993.
16. L Sung, A Karim, JF Douglas, CC Han. Phys Rev Lett 76:4368, 1996.
17. R Xie, A Karim, JF Douglas, CC Han, RA Weiss. Phys Rev Lett 81:1251, 1998.
18. J Bischof, D Scherer, S Herminghaus, P Leiderer. Phys Rev Lett 77:1536, 1996.
19. K Jacobs, S Herminghaus, KR Mecke. Langmuir 14:965, 1998.
20. U Thiele, M Mertig, W Pompe. Phys Rev Lett 80:2869, 1998.
21. G Reiter, A Casoli, M-O David, A Sharma, P Auroy. 2nd International Conference on Wetting and Self Organization, Munich, 1998.
22. G Reiter, A Sharma, A Casoli, M-O David, R Khanna, P Auroy. Europhys Lett 46: 512, 1999.
23. G Reiter, A Sharma, A Casoli, M-O David, R Khanna, P Auroy. Langmuir 15:2556, 1999.
24. MO David, G Reiter, T Sitthai, J Schultz. Langmuir 14:5667, 1998.
25. JN Israelachvili. Intermolecular and Surface Forces. London: Academic Press, 1992.
26. A Sharma, R Khanna. Phys Rev Lett 81:3463, 1998.
27. R Khanna, PhD thesis, Indian Institute of Technology, Kanpur, 1998.

28. R Khanna, A Sharma. J Colloid Interf Sci 195:42, 1997.
29. CJ van Oss, MK Chaudhury, RJ Good. Chem Rev 88:927, 1988.
30. KR Shull. J Chem Phys 94:5723, 1991.
31. KR Shull. Faraday Discuss 98:203, 1994.
32. A Sharma. Langmuir 9:3580, 1993.
33. A Sharma, R Khanna. J Chem Phys 110:4929, 1999.
34. A Sharma. Langmuir 14:4915, 1998.
35. S Herminghaus, K Jacobs, K Mecke, J Bischof, A Fery, M Ibn-Elhaj, S Schlagowski. Science 282:916, 1998.

13

Thinning and Rupture of Aqueous Surfactant Films on Silica

S. Simpson, R. K. Jain, V. Raghuraman, and P. Somasundaran
Columbia University, New York, New York

ABSTRACT

The thinning and rupture of thin liquid films on silica is investigated using a modified interferometric technique. Stable films were obtained at higher salt concentrations compared to previous studies, because of an improved cleaning procedure. The free bubble method employed in this investigation approximates flotation more realistically than the captive bubble method. Different rupture mechanisms are observed at different pH for aqueous films of dodecylamine on silica, explaining the pH dependence of quartz flotation using amine. While at low pHs rupture is accompanied by the breaking off of a large drop in the center and subsequent formation of large, irregular droplets; at high pH, ruptured spots grow to large, circular droplets.

I. INTRODUCTION

The rupture of thin liquid films between liquids, gases, and solids plays a governing role in determining the stability of bubbles and droplets in foams, emulsions, froths, and even biological cells [1–5]. The behavior of liquid films on solids is particularly relevant to many processes such as detergency and froth flotation of minerals. In some of these processes such as froth flotation, the film rupture may not involve a complete removal of the liquid on the solid.

Although much work has been done on the rupture of liquid films [6–8] and aqueous films on solids [9,10], very little has been done on the films of surfactant solutions on solid minerals, even though the rupture of these films is the critical step that determines the efficiency of processes such as flotation. In this chapter, we have investigated the role of salt (potassium chloride) and surfactant (dodecylamine) in the thinning and rupture of aqueous films on a mineral (quartz).

II. MATERIALS

The solid substrate was a fused vitreous silica disk (supplied by Oriel Corporation), 12.7 mm in diameter and 6 mm thick, polished flat to within 0.05 wavelength of visible light.

The surfactant used was dodecylamine hydrochloride $(CH_3(CH_2)_{11} NH_2.HCl)$ from Eastman Kodak Company. The pH of the surfactant solutions was adjusted using ACS reagent-grade KOH, supplied by Amend Drug and Chemical Co., and 0.1 N HCl supplied by Fisher Scientific. The pH of all the surfactant solutions was checked before and after the experiment.

The gas was oil-free nitrogen supplied by T.W. Smith Corp., which was passed through ascarite (Arthur H. Thomas Company) to remove CO_2, through distilled water, and then filtered (0.22-μm pore size, 25-mm diameter Milipore MF filter in a stainless-steel holder) to remove dust. The KCl used was ACS reagent grade supplied by Fisher Scientific, which was roasted at 663 K for 3 h prior to use (to remove the organics). The purity of the KCl solutions was checked by the bubble persistence test. The concentrations of the KCl solutions were verified by conductivity measurements. All solutions were made from triply-distilled water (second distillation from potassium permanganate), whose purity was checked by specific conductivity (1.25×10^{-6} mho-s/cm) and surface tension (72.3 mN/m at 295 K).

The silica, and the Teflon cell cover housing the silica plate were cleaned before every measurement using an elaborate stepwise procedure considered to be an improvement over a previously used method [11]. The cell, including the silica plate, was soaked in chromic acid solution for 15 min. The chromic acid was washed away with distilled water. The same procedure was followed with alcoholic KOH (12.5 g KOH, 84 mL 200-proof ethanol and 16 mL distilled water) and with 0.1 M nitric acid (ACS electronic grade, Amend Chemical Co.). As suggested by Aronson and Princen [11], all cleaning solutions were discarded after one use.

The cleaning procedure was checked in the following manner: The cell was filled with either triply-distilled water or salt solution and several film thicknesses were measured. If it did not correspond to the previously measured thickness (evidence of surface nonreproducibility) or if any sign of film instability was observed, the cleaning procedure was repeated.

III. METHODS

The film thickness can be estimated by monitoring capacitance [12] or conductance, or using various techniques such as ellipsometry [6] or interferometry [13]. Interferometry is used in this study because it is the most suitable method for studying the nonequilibrium behavior of films.

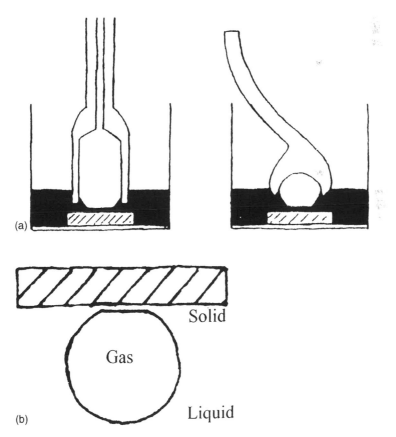

Figure 1 (a) Captive bubble method: gas bubble pressed against solid. (b) Free-bubble method: gas bubble contacts the solid due to buoyancy.

In the past, interferometric work [11,13–16] on the thinning and rupture of aqueous surfactant films was carried out using the captive bubble method. In this method, the gas bubble is gently pressed against the solid substrate, which is submerged in the solution (Fig. 1a). Because the bubble is motionless, kinetic processes such as film thinning and rupture are easily followed.

In the free-bubble method used in this study, the gas bubble is allowed to rise due to buoyancy until it contacts the solid [9] (Fig. 1b). Therefore, this method approximates flotation and other thin-film processes much more realistically than do captive-bubble experiments. However, this method has not been used extensively in the past because of such problems as controlling the motion of the bubble and the effect of mechanical vibrations.

IV. APPARATUS

Figure 2 is a cross-sectional view of the apparatus. Nitrogen gas is released through a capillary tube (0.06 cm outer diameter) at a pressure controlled by a 0.25-mL gas syringe (Fisher Scientific). The capillary holder is equipped with a vertical positioner so that the distance between the capillary tip and the plate bottom and, hence, the velocity of contact can be altered. A cathetometer is used to measure the bubble size. The silica disk is housed in the Teflon cover of the cell, which is also made of silica. The Teflon cover rests on the cell only by gravity, to prevent contamination from sealants. The cover contains openings for the capillary tube and for a suction tube to remove the solution after the measurement.

Cathetometer

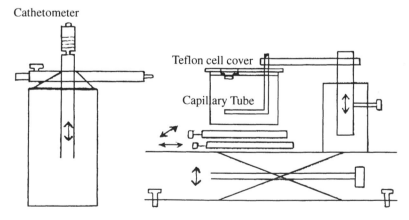

Figure 2 The cell system showing the vertical positioners and cathetometer

The system is viewed from above under monochromatic illumination (a 100-W HBO Hg lamp from Zeiss Corp. fitted with a 546.1-nm Zeiss Corp. or a 435.0-nm Oriel Corp. filter), using a Zeiss Universal microscope. Light reflected by the film may be directed either to the eyepiece or through a phototube for photographic recording, and the light intensity is read from a photometer. The optical arrangement is the same as that used by Aronson and Princen [11]. Various modifications were made to the Zeiss AC stand to damp out random vibrations [17].

V. MEASUREMENTS

A. Thickness

In the microinterferometric method, the relation between the light intensity reflected from the film and the film thickness is expressed by Eq. (1) [18], assuming that the incident beam is perpendicular to the film and that light absorption is negligible within the film:

$$\frac{I}{I_0} = \frac{r_2^2 + r_3^2 + 2r_2r_3 \cos J}{1 + r_2^2 \, r_3^2 + 2r_2r_3 \cos J} \tag{1}$$

where

I is the intensity of the reflected light, I_0 is the intensity of the incident light,

$$r_1 = \frac{n_1 - n_3}{n_1 + n_3}, \qquad r_2 = \frac{n_2 - n_1}{n_2 + n_1}, \qquad r_3 = \frac{n_3 - n_2}{n_3 + n_2} \tag{2}$$

$$J = \frac{4\pi n_2 h}{\lambda}$$

r_1, r_2, and r_3 are the reflectivities of the silica–air, silica–solution, and solution–air interfaces respectively, n denotes the refractive index (values taken from Ref. 19), the subscripts 1, 2, and 3 denote silica, solution, and air, respectively, λ is the wavelength of light used, and h is the film thickness to be determined.

Previous investigators [14,20] used a ratio Δ to account for background illumination:

$$\Delta = \frac{I_f - I_m}{I_M - I_m} \tag{3}$$

where I_f is the recorded intensity for the film, and I_m and I_M are the minimum and maximum recorded intensities of the reflected light, respectively. The advantage of Δ is that random reflection of light from outside sources and from the sides of the gas bubble can be neglected. The equation for film thickness can be derived from substitution of Eq. (3) into Eq. (1) [14,20].

The method just described is suitable only for captive bubbles. Difficulties arise with the higher sensitivity of the free bubbles to vibration due to their increased mobility. A direct measurement of the maximum and minimum intensities with a photometer is not easily achieved due to the small size of the Newton rings [17]. Therefore, a new method better suited to free-bubble experiments for the measurement of film thickness was developed.

This method proposes replacing the maximum and minimum reflected intensities by those of a ruptured film and an infinitely thick film, denoted by I_{rupt} and I_∞, respectively. I_{rupt} is modeled as the intensity of light reflected back from the silica–air interface. I_∞ is modeled as the intensity of light reflected back from the silica–solution interface. In a similar fashion, a ratio Δ_∞ is introduced, so that random reflections from outside sources and the sides of the gas bubble can be neglected:

$$\Delta_\infty = \frac{I_f - I_\infty}{I_{rupt} - I_\infty} \tag{4}$$

As before, substituting Eq. (4) into Eq. (1) and rearranging

$$\cos J = \left\{ \left[(1 - \Delta_\infty)\left(\frac{r_2^2}{1 - r_2^2 r_1^2}\right) + \Delta_\infty\left(\frac{r_4^2}{1 - r_4^2 r_1^2}\right) \right] (1 + r_2^2 r_3^2 - r_1^2 r_2^2 - r_1^2 r_3^2) - r_2^2 - r_3^2 \right\}$$
$$\times \left\{ \left[(1 - \Delta_\infty)\left(\frac{r_2^2}{1 - r_2^2 r_1^2}\right) + \Delta_\infty\left(\frac{r_4^2}{1 - r_4^2 r_1^2}\right) \right] (2r_1^2 r_2 r_3 - 2r_2 r_3) + 2r_2 r_3 \right\}^{-1} \tag{5}$$

from Eqs. (5) and (2), the film thickness h can be calculated.

B. Disjoining Pressure

The disjoining pressure exerted by a stable equilibrium film resists the forces leading to its thinning. It is denoted by Π, and defined as $\Pi = P_f - P_l$, the difference between the film pressure P_f and the bulk liquid pressure P_l. Performing a force balance on the system, as shown in Figure 3,

$$V\Delta\rho g + P_l \pi R^2 - P_f \pi R^2 + 2\pi R\gamma \sin\theta = 0 \tag{6}$$

where, $V\Delta\rho g$ is the buoyancy of the bubble, $P_l \pi R^2$ is the force exerted by the bulk on the bottom cylindrical section of the sphere, $P_f \pi R^2$ is the force exerted by the film acting against the bulk, and $2\pi R \sin\theta$ is the force due to the surface tension of the bubble–liquid interface.

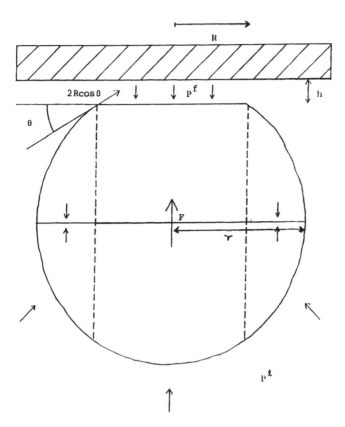

Figure 3 Force balance on a wetting film.

Rearranging and assuming the bubble to be a perfect sphere ($R \ll r$) and $\theta \sim 0$,

$$\Pi = \frac{4gr^3 \Delta\rho}{3R^2} \qquad (7)$$

The capillary pressure P_c is defined as

$$P_c = P_g - P_l$$

where P_g is the gas pressure. For a free bubble,

$$P_c = \frac{2\gamma}{r}$$

where γ is the surface tension of the air–solution interface.

At equilibrium, the film pressure and the gas pressure must be equal (i.e., $P_f = P_g$), hence $\Pi = P_c$. Therefore,

$$\Pi = \frac{2\gamma}{r} \qquad \qquad (8)$$

Combining Eqs. (7) and (8) with known values of ρ and γ for pure water at 295 K (assuming $\Delta\rho \sim \rho$),

$$r^4 = 0.11053 \ R^2 \quad \text{in cgs units. Hence,}$$

$$\Pi = \frac{2\gamma}{(0.11053R^2)^{1/4}} \approx \frac{144.6}{(0.11053R^2)^{1/4}} \qquad (9)$$

Thus, by measuring the film radius, the disjoining pressure can be estimated.

VI. RESULTS AND DISCUSSION

The effect of ionic strength on the thickness of thin films is shown in Figure 4 for a disjoining pressure of 2000 mN/m². At high ionic strengths, electrical double-layer interactions will be suppressed permitting reduction of the film thickness. Read and Kitchener [21] were unable to obtain equilibrium films at elec-

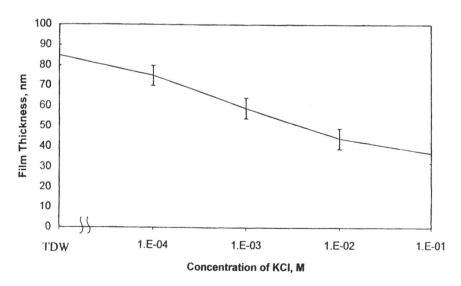

Figure 4   Effect of salt on film thickness at a capillary pressure of 2×10^3 mN/m².

trolyte concentrations greater than 10^{-3} M, due to hydrophobic contamination and dust particles. Equilibrium films for concentrations up to 0.1 M KCl could be obtained in the present case due to the sophisticated cleaning procedure used.

The film thickness–disjoining pressure isotherms obtained experimentally using the free-bubble and captive-bubble techniques are compared in Figure 5a, with that obtained using theoretical predictions for pure water. Read and Kitchener [21] calculated the theoretical values from the tables of Devereux and de Bruyn [22], using the zeta-potentials determined for silica by Jones and Wood [23]. These calculations are based on the Guoy–Chapman diffuse double-layer theory.

It can be seen that the free-bubble method used in the present investigation is in better agreement with theory than the captive-bubble method [11,20]. To show that the cleaning procedure does not account for this difference, data obtained using the cleaning procedure used by previous investigators [11,21] are also plotted. These results show that use of the previous cleaning method reduces the measured thickness, thus increasing the difference between the results given by the two methods. This is expected because the zeta-potential of hydrophobic silica is equivalent to that of clean, hydrophilic silica; thus, the thickness should be the same. The slight decrease in thickness is due to the many hydrophobic spots, which tend to reduce the overall film thickness. The error in the captive-bubble technique could be due to the effect of containing the bubble in a holder and/or the incorrectness of the equation used to compute the disjoining pressure.

The Π–h isotherm for 10^{-2} M KCl solution is plotted in Figure 5b. The theoretical curve is the van der Waals disjoining pressure calculated using values based on optical dispersion data reported by Gregory [24] and reproduced by Read and Kitchener [21]. The good agreement between the measured values and theory confirms the prediction of van der Waals repulsion between silica–water and water–air interfaces. This is because for stable films at high electrolyte concentrations, electrostatic effects are almost completely suppressed.

The study of the effect of amine on the characteristics of thin aqueous films is important because of its widespread use as a collector for quartz flotation. Flotation of quartz using dodecylamine depends on such variables as amine concentration and solution pH, which govern surfactant adsorption and surface tension [25]. Flotation begins at 5×10^{-5} M amine and, at all amine concentrations, is maximum at pH 10.5. Around pH 10.2–10.5, the amine molecule and the ammonium ion exist in equal amounts [26]. It has been hypothesized [27] that they exist as ion-molecule dimers along with monomers in solution at that pH. Flotation is maximum due to the high surface activity of the dimer complex. Complexation decreases the repulsion between the ammonium ions because of shielding by the amine half of the dimer. The adsorption of amine on silica at a concentration of 10^{-5} M has been determined as a function of pH [28].

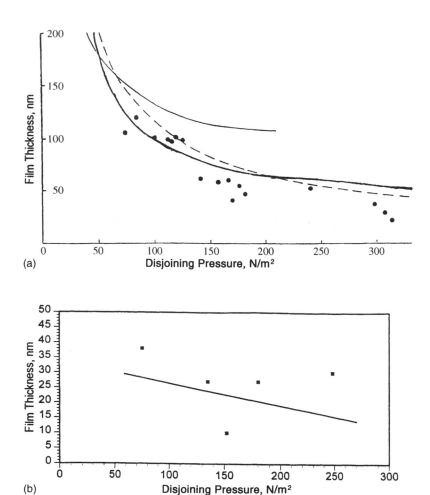

Figure 5 (a) Comparison of h–Π isotherms obtained using the free-bubble and captive-bubble methods. λ = 435.8 nm; 0 M KCl. (●) Results of this study using nitric acid/ethanol cleaning procedure; (---): results of this study using new cleaning procedure. Straight line approximation for h versus log Π: results of Aronson and Princen [11] captive-bubble method (—); theoretical predictions.(—). (b) Comparison of theoretical predictions with experimental data for films of high ionic strength. Salt: 0.01 M KCl; λ = 546.1 nm; ■: points obtained by experiment; (—): theoretical curve (van der Waals disjoining pressure).

Below the point of zero charge (pzc) of silica (pH ~ 2), no adsorption occurred. Adsorption gradually increased until pH ~ 5 as the number of negative sites on silica increased. Adsorption remained constant until it rose to a maximum between pH 10 and 11.5 and then dropped to zero by pH 12. Results obtained for surface tension [26] show that the 4×10^{-4} M amine solution exhibits a minimum at a pH of 10.2 [27].

The attachment (flotation) of bubbles to particles can be attributed to two basic characteristics of the thin film: the film rupture and the advancement of the contact angle. In Figure 6, the flotation of quartz particles [29] is compared with the surface charges [30] and contact angles for ruptured films for 10^{-5} M dodecylamine (DDA) hydrochloride over a range of pH values. Although electrostatic force is not the only force leading to the rupture, it, by far, has the predominant effect. It is clear from Figure 6 that in the high-pH range where flotation is increasing, the electrostatic impetus toward rupture does decrease, whereas the contact angle increases considerably. At low pH, however, it appears that the opposite is true. The contact angle rises only slowly, whereas the surface-charge effect is increasing, although to a lesser extent because of a reduction in silica–water surface charge due to increasing surfactant adsorption. Flotation of quartz with DDA at low pH is attributed to electrostatic effects, although the contact angle must be large enough that the bubble is not dislodged by agitation.

This apparent difference in the predominant force governing attachment at low and high pH leads one to suspect that the rupture characteristics of the thin films may be different at low and high pH. Therefore, the film-rupture mechanisms of amine films on quartz at high and low pH were studied using the free-bubble method.

Before presenting the results, it will be useful to review the following observations made by Scheludko [31,32] for soap films, but equally applicable to aqueous films on silica:

1. Thick films will always have a convexity in the center; this is known as a dimple.
2. Upon thinning, the convexity decreases and, at a small thickness, the film appears to be plane parallel.
3. Upon further thinning, either a stable film will form or the film will rupture when it reaches a critical thickness. After rupture, either the gas will spread on the surface or a very thin film will form. This film is white and corresponds to black films formed by soap films [7].

At very low concentrations of DDA, stable films are formed at all pH values. When 10^{-8} M DDA is reached, the film continuously thins, droplets appearing on its surface (Fig. 7). After 2 days, the film appears as a ruptured film (Fig. 8), actual rupture never having occurred.

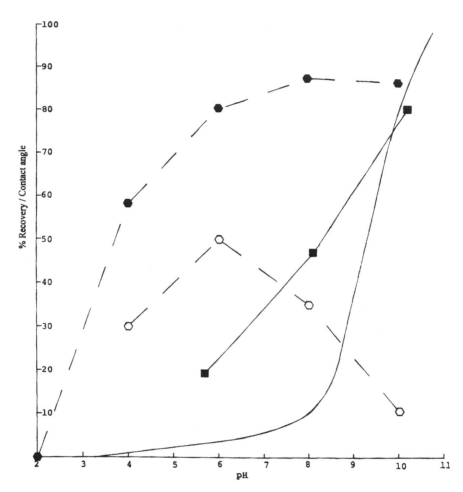

Figure 6 Quartz flotation compared with surface charges and contact angle as a function of pH (10^{-5} M DDA). (—): percentage recovery [29]; ●: $-\psi_z$ silica–water interface [30]; ○: $+\psi_z$ air–water interface [30]; ■: contact angle θ.

As the concentration is increased, actual rupture occurs. At low/natural pH (~5.8), this manifests in what we term *peripheral rupture* (Fig. 9a), in which an area on the periphery of the film ruptures and instantaneously begins to expand around the edge of the film. When the two ends meet (in Fig. 9a, this has just occurred at the top), a central drop is broken off. As the edge expands to the final contact angle, this large drop migrates to the center of the ruptured film, followed by the formation and growth of large, irregular droplets (Fig. 9b), in

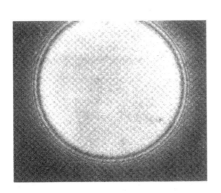
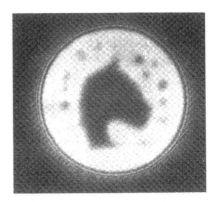

Figure 7 Film thinning and formation of droplets (1×10^{-8} M DDA, pH $\sim$ 5.6).

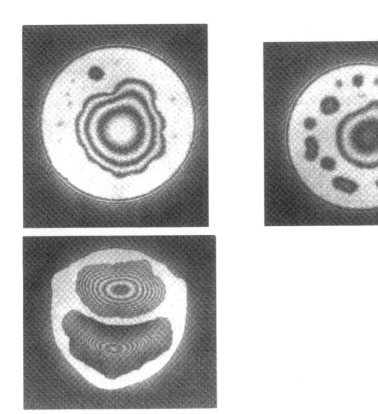

Figure 8 Irregular film (after 2 days) (1×10^{-8} M DDA, pH $\sim$ 5.6).

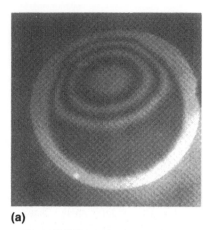

(a)

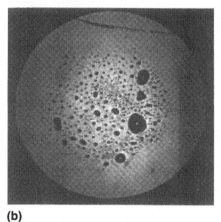

(b)

Figure 9 (a) Peripheral rupture; note the central drop breaking off. (b) Formation and growth of large, irregular droplets (1×10^{-6} M DDA, pH $\sim$ 5.6).

agreement with the observations of Schulze [33] and Aronson and Princen [11].

The characteristics of rupture are different at high pH ($\sim$10.5), as shown in Figure 10a. This type of rupture, which we call *spot rupture*, begins with the formation of ruptured spots at the periphery which grow, meet each other, breaking off a central drop, and, finally, expand as a single edge to the final contact angle. As at low pH, droplets form and grow but are circular in shape (Fig. 10b).

The differences in rupture characteristics can be explained on the basis

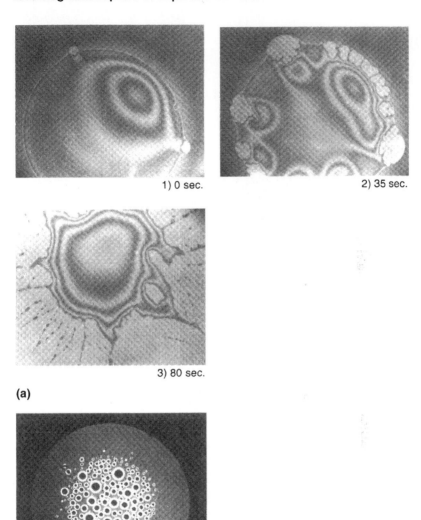

1) 0 sec. 2) 35 sec.

3) 80 sec.

(a)

(b)

Figure 10 (a) Spot rupture (1×10^{-6} M DDA, pH $\sim$ 10.5) at time (1) 0 s, (2) 35 s, and (3) 80 s. (b): Formation and growth of circular droplets (1×10^{-6} M DDA, pH $\sim$ 10.5).

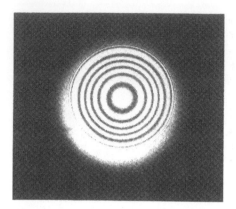

Figure 11   Spontaneous dimple formation (elapsed time $\sim$ 10 min) ($1 \times 10^{-2}\, M$ DDA, pH $\sim$ 5.6).

of the previous observations on flotation. At low pH, the predominant force is electrostatic, and once one spot is ruptured, because its edges are closer to the silica surface, it expands rapidly because the two interfaces are strongly attracted. At high pH, however, this attraction is not as strong and the first hole formed does not expand over the whole film. Instead, many spots have time to form, their expansion and the final rupture being due to the advancement in contact angle.

If the concentration of the cationic surfactant is increased [above the critical micelle concentration (cmc)], stable films are once again formed (at natural pH). This is attributed to the formation of a bilayer adsorbed at the silica–solution interface. Figure 11 shows a stable film formed in a $2 \times 10^{-2}\, M$ DDA solution at pH 5.6 with spontaneous dimple formation. In earlier studies [11] with another cationic surfactant, stable films were observed at concentrations above the cmc.

VII. SUMMARY

In this study, we were able to obtain stable films at saturation concentrations of the electrolyte because of an improved cleaning procedure. As predicted by the electrical double-layer theory, the film thickness decreased as the ionic strength increased.

The free-bubble method is seen to be in better agreement with the theoretical predictions than the captive-bubble method. The free-bubble method, although less convenient, has no unknown effects that exist when the gas is contained in a holder and, in addition, has a clearly defined and easily calculable disjoining pressure. Moreover, the free-bubble technique is a more realistic approximation of processes like froth flotation.

The results obtained for films formed at high ionic strength confirm the prediction of van der Waals repulsion between the silica–water and air–water interfaces.

The observed rupture characteristics of thin dodecylamine films on silica are different at low and high pH, when the concentration is below the cmc. At low pH, the rupture occurs due to a large central drop breaking off, followed by the formation of large, irregular droplets. At high pH, on the other hand, many ruptured spots form, gradually growing into larger, but circular droplets This difference in mechanism explains the marked pH dependence of quartz flotation using an amine.

ACKNOWLEDGMENT

The authors acknowledge the support of the National Science Foundation.

REFERENCES

1. RD Kirkpatrick, MJ Lockett. Chem Eng Sci 29:2363, 1974.
2. RS Allan, GE Charles, SG Mason. J Colloid Sci 16:150, 1961.
3. GE Charles, SG Mason. J Colloid Sci 15:236, 1960.
4. GDM MacKay, SG Mason. Can J Chem Eng 203, (October 1963).
5. TD Hodgson, JC Lee. J Colloid Interf Sci 30:94, 1969; 30:429, 1969.
6. BV Derjaguin, ZM Zorin. Proc. 2nd Int. Congr., Surface Activity, 1957, Vol. 2, p. 145.
7. KJ Mysels, K Shinoda, S Frankel. Soap Films. London: Pergamon Press, 1959.
8. HM Princen, JTG Overbeek, SG Mason. J Colloid Sci 24:125, 1967.
9. BV Derjaguin, MM Kussakov. Acta Physicochim URSS 10(1):25, 1939; 10(2):153, 1939.
10. BV Derjaguin, L Landau. Zh Eksper Teor Fiz 11:802, 1941.
11. MP Aronson, HM Princen. Colloid Polym Sci 256:140, 1978.
12. H Sonntag. Proc. III Int. Congr. Surface Activity, 1960, Vol. 2, p. 610.
13. A Scheludko, D Exerowa. Dokl Akad Nauk SSSR 127:149, 1959.
14. TD Blake. J Chem Soc Faraday Trans I 71:192, 1975.
15. E Manev. MS thesis, University of Sofia, 1961.
16. E Manev, RJ Pugh. Society of Mineral Engineers Annual Meeting, 1992.
17. S Simpson, MS thesis, Columbia University, New York, 1979.
18. A Vasicek. Optics of Thin Films. Amsterdam: North-Holland, 1960 English edition, New York: Interscience, 1960.

19. H Landolt. Physikalish-chemish Tabellen. Berlin: Springer-Verlag, 1883.
20. TD Blake, JA Kitchner. J Chem Soc Faraday Trans I 68:1435, 1972.
21. AD Read, JA Kitchner. J Colloid Interf Sci 30(3):391, 1969.
22. OF Devereux, PL De Bruyn. Interaction of Plane Parallel Double Layers. Cambridge, MA: MIT Press, 1963.
23. G Jones, LA Wood. J Chem Phys 13:106, 1948.
24. J Gregory. PhD thesis, University of London, 1964.
25. A Bleier, ED Goddard, RD Kulkarni. Structural Effects of Amine Collection Flotation, AM Gaudin Memorial, Vol. 1, Soc Min Eng/AIME, MC Fuerstenau, ed., 117, 1976.
26. P Somasundaran. Int J Mineral Proc 3:35, 1976.
27. K Ananthpadmanabhan, P Somasundaran, TW Healy. Soc Min Eng AIME Trans 266:2003, 1978.
28. I Petrov, I Panaitov, D Panaitov. Fac. Chim 69(2):37, 1974.
29. R Chew. Undergraduate Research Project, Columbia University, 1979.
30. HJ Schulze. Colloid Polym Sci 252:730, 1975.
31. A Scheludko. Adv Colloid Interf Sci 1:391, 1967.
32. A Scheludko. Colloid Z 155:39, 1957.
33. HJ Schulze. Proc. Int. Conf. Colloid Surf. Sci. (Budapest), 1975, Vol. 1, pp. 179–186.

14

Newton Black Films: From Simple Surfactants to Proteins

J. J. Benattar, F. Millet, M. Nedyalkov,* and D. Sentenac
Commissariat à l'Energie Atomique, Saclay, France

ABSTRACT

Black films of surfactants are simple free standing bilayer systems involving most of the basic physical interactions existing in more complex structures such as biological membranes. In this paper we report three different investigations of black films by means of x-ray reflectivity. In the first part we will show, using a combination of disjoining pressure measurement and x-rays, that it is possible to show the role of hydration forces in surfactant black films. In the second part, we will discuss the behavior of free-standing vertical films of long associating polymers in order to separate the effects produced by the confined geometry of a film from those due to the polymer physics. In the last part we will report the first results of the confinement of a single layer of proteins in a surfactant Newton Black Films (NBF).

I. INTRODUCTION

Black films [common (CBF) or Newton (NBF)] are the final stages of thinning of soap films due to the draining of water in the absence of evaporation [1,2].

---

* *Current affiliation*: University of Sofia, Sofia, Bulgaria.

They are generally formed from solutions of usual surfactants; they are called black because they do not reflect the natural light as a result of their smaller thickness compared to the visible wavelengths. Depending on the salt concentration, two types of black films can be observed. For the CBF, a balance between van der Waals attraction and the double-layer repulsion forces determines the equilibrium thickness [3,4]. The CBFs are known to be thicker (because they contain an aqueous core) than the NBFs, which involve more subtle short-range forces. Free amphiphilic black films were generally studied by class ical techniques [optical reflectance, infrared (IR), absorption, electrical, and contact-angle measurements] [5–10]. We have shown that x-ray reflectivity is the most powerful technique because it is sensitive to electron density gradients, which are strong at the two air–film interfaces. A few years ago, we demonstrated that the NBF was thinner than usually expected and well organized [11]. It consists of two opposite walls of molecules, without any liquid water between them. The aqueous core is reduced to an ultimate hydration layer of the polar heads and the roughness is limited to the capillary waves. Such features are general even for very different surfactants [12]. The NBF is thus a very good model system for the direct study of interfacial phenomena such as interactions between surfactant bilayers. For example, we have investigated black films of amphiphilic diblock copolymers (polyelectrolytes) [13]. More recently we have finalized a method for the investigation of black films, which combines the force measurement to the structure determination of the films by x-ray reflectivity, which we have employed for the study of hydration effects.

II. X-RAY REFLECTIVITY AND THIN FILMS

Let us recall that an x-ray reflectivity experiment consists of the measurement of the ratio $R(\theta) = I(\theta)/I_0$, where I_0 is the intensity of the incident beam and $I(\theta)$ that of the beam reflected by a surface at an angle θ at the specular position. The analysis of the experimental reflectivity curve is made using an optical method [14,15] in which the system is considered as a succession of homogeneous slabs characterized by three parameters: the thickness, the density, and the interfacial roughness. The main advantage of free-standing films arises from their high electron density gradients at the interfaces. The reflectivity profiles display very strong "Kiessig fringes" that originate from the interference of the beams reflected on each side, which enable accurate structure determinations and give the overall film thickness. The present reflectivity experiments were performed using a high-resolution reflectometer for vertical surfaces (OptiX from Micro-controle). A copper tube is used as an x-ray source ($\lambda = 1.5405$ Å) and a small vertical slit (100 μm) ensures a low divergence (0.15 mrad). A horizontal slit (1.25 mm) limits the height of the illuminated area of the film.

A. Structure and Interactions in Black Films: A New Method Combining X-Ray Reflectivity and Force Determination

Recently, we have carried out new experiments on black films, combining the determination of the film structure using x-ray reflectivity with the force determination by the measurement of the applied pressure on the film (Fig. 1). It is well known that the forces involved in membranes and black films are essentially van der Waals forces, electrostatic forces (DLVO theory), and short-range microscopic forces. These forces (e.g., steric forces and hydration forces) act when the thickness of the film becomes very small (<10 nm). For the study of NBFs, the experiments using classical optical methods are not appropriate because the medium is not continuous but "molecular," and x-rays are better suited than light. We have built a new cell based on the principle of the porous plate method and adapted it for our specific x-ray reflectivity equipment. The pressure applied on the film is varied by means of a syringe connected to the cell, whereas the porous plate is connected to the capillary tube at the atmospheric pressure (Fig. 2). The pressure in the cell is $P_0 + \Delta P$ and ΔP is measured by means of differential

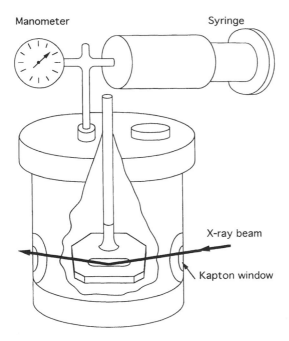

Figure 1 Schematic of the setup installed at the center of the reflectometer. The films are formed in an airtight cell to maintain a saturated vapor atmosphere.

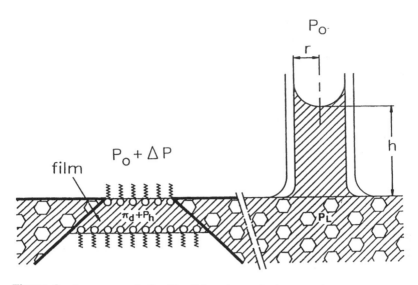

Figure 2   A macroscopic flat film (15 × 4 mm) is drawn at the extreme upper surface of the hole of the porous plate initially filled with the solution. The disjoining pressure of the film is measured by the application of an additional pressure in the cell. The film structure is determined by x-rays.

manometers. For a film at equilibrium, the disjoining pressure Π_d [16] is given by

$$\Pi_d = \Delta P + \frac{2\gamma}{r - \Delta\rho g h}$$

where r and h are defined in Figure 2, γ is the surface tension, g is the gravity acceleration, and $\Delta\rho$ the density difference between the solution and the air.

B. A First Application: Evidence of Hydration Effects Black Films with Electrolytes

Using our new experimental cell, we have investigated the interactions within free, suspended black films drawn from aqueous solutions of anionic Aerosol-OT (AOT) with two added electrolytes, LiCl and CsCl [17,18]. We have accurately measured the long-range force law within a CBF of AOT with the two electrolytes. During the disjoining pressure measurements, the pressure was increased by steps of 500 Pa until rupture (the NBF has not been reached).

The isotherms are plotted with respect to the separation between the molecular planes (Fig. 3). One can note that the slope above the 0.05 M electrolyte tends to bend at the lowest pressure and this is the signature of the attractive dispersion forces. At the 0.05 M electrolyte, the isotherms for LiCl and CsCl are quasi-identical. At the 0.1 M electrolyte, an additional repulsion arises in the LiCl isotherm and a similar gap is present at the 0.2 M electrolyte. The shape of the isotherms with Cs ions presents a rather good agreement with the mean field electrostatic theory, whereas the isotherms obtained with Li cannot be taken into account. In conclusion, the observed gap between isotherms could be due to the decreased Debye screening length of the LiCl systems by reducing the net charge of the Li counterion, induced by a local screening of the surrounding water molecules.

C. Free-Standing Films of Long Associating Polymers

Emulsions of oil in polar media are known to be thermodynamically unstable systems. The more efficient stabilizers seem to be hydrophobically modified poly-

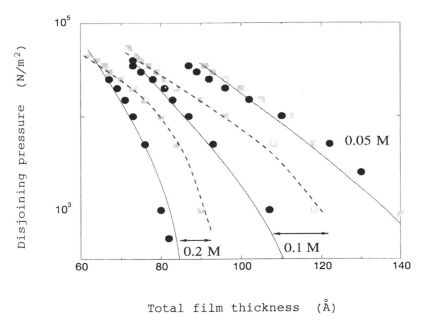

Figure 3 Π_d isotherms of free-standing films of AOT (2.5 mM), with (from right to left) concentrations of 0.05, 0.1 and 0.2 M CsCl (circles) and LiCl (squares); fits are made using the Lifshitz theory for dispersion forces and the Poisson–Boltzmann equation for electrostatic forces. The dashed line is a guide for the reader's eyes.

electrolytes because they combine the long-range electrostatic repulsion of ionic surfactants and the steric hindrance of polymers [19]. The motivation of this work was to study the behavior of free-standing vertical films of such long associating polymers in order to separate the effects produced by the confined geometry of the film from those due to the polymer physics. The associating polymers were the hydrophobically modified poly(acrylic acid) sodium salt (HMPAANa) labeled as follows: $(M_w, 10^{-3})$ τCn, where τ is the degree of grafting (mol%) and Cn is the number of carbon atoms in the alkyl chain.

We found that our polymers could form very stable free-standing films whose equilibrium thickness is independent of the degree of grafting and of the salt concentration; the only relevant parameters were the polymer concentration and the molecular weight. We also found a concentration threshold at the critical aggregate concentration (cac): Below this threshold, the thickness of the films was independent of the macroscopic properties of solutions; the film is composed solely of two monomolecular polymer walls, each of them being adsorbed at one side of the film. Above this threshold, nonadsorbed polymer molecules are trapped in the film and are linked to the adsorbed molecules via the hydrophobic aggregates; the film thickness increases and the film behaves like a physical gel. The polymer concentration inside the film is possibly slightly higher than that in the solution. In Figure 4, we report some examples of the thickness dependence with the concentration, and in the inset of Figure 4, we show that the thickness

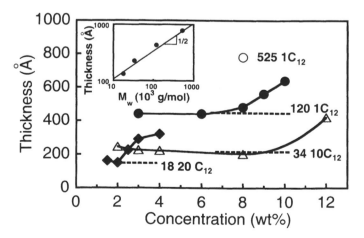

Figure 4   Concentration dependence of the film equilibrium thickness for various molecular weights of the polymer backbone: O: 525 $1C_{12}$; ●: 120 $1C_{12}$; Δ: 34 $10C_{12}$; ◆: 18 $20C_{12}$. Inset: Dependence of the film thickness on the polyelectrolyte weight-average molecular weight M_w. The thickness scales as $M_w^{1/2}$.

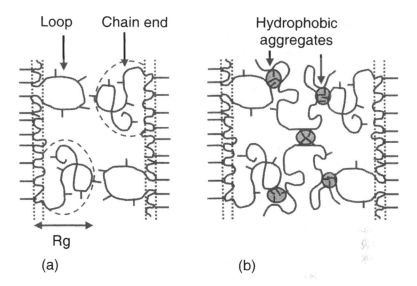

Figure 5 Schematic of the structure of an amphiphilic associating polyelectrolyte film. (a) $C < C_t$: the radius of gyration determines the film thickness. (b) $C > C_t$: the adsorbed molecules are connected to the bulk via the hydrophobic aggregates. The film thickness increases rapidly with concentration. A transition from a bimolecular film to a nonpourable physical gel occurs.

of the films increases roughly as the square root of the average molar weight (i.e., the radius of gyration) and versus the polymer concentration. Our results are in good qualitative agreement with theoretical and experimental works, which conclude that the thickness of the film of an amphiphilic polymer is controlled by the radius of gyration. In Figure 5, we show the film structure below and above the threshold concentration.

III. THE TWO DIMENSIONAL CONFINEMENT OF BIOLOGICAL MOLECULES

The last step of our experimental processes is obviously devoted to the use of an NBF to investigate the interfacial properties of biological systems. A few studies have reported attempts to form black films of proteins. Nevertheless, where microscopic CBF and NBF have been obtained, their structures were either bilayers of denatured proteins, more complex multilayer films, or thick films [20–25]. We have managed to realize an alternatively stabilized free-standing film, confining a single protein layer in a surfactant bilayer.

The nonionic surfactant $C_{12}E_6$ (hexaethylene glycol monododecyl ether) was chosen for its ability to form a stable NBF. The respective surfactant and protein ranges of possible concentrations were found empirically after testing the film stability. We found the minimum concentration of the surfactant which provides stable films only when mixed with BSA (C_{C12E6} = 0.075 mg/mL ~ 2 cmc). This concentration, used for all the experiments, allows the formation of stable mixed films. This is the first experimental proof of the presence of BSA within the NBF. The surfactant concentration for which large stable films can be obtained without BSA is C_{C12E6} = 0.5 mg/mL ~ 15 cmc. The pH of the solutions is stable roughly for 1 h after the preparation and was found to be 7.4 ± 0.1, above that of the BSA isoelectric point (4.8) [24]. At this pH, the BSA is known to be globular with an ellipsoidal shape of dimensions 4.16 nm and 14.09 nm.

The first reflectivity experiments were carried out on a film drawn from a solution at C_{BSA} = 0.5 mg/L. The reflectivity curve, recorded immediately after film formation, differs surprisingly little from that of the pure surfactant. At this stage, a first conclusion can be drawn: The protein interacts with the surfactant to stabilize the film, but its concentration within the NBF remains very low. After this first observation, we increased the C_{BSA} to 1 mg/mL, keeping C_{C12E6} constant. The reflectivity profile indicated only a slight increase of the thickness (~1 nm). Our next experiment carried out at C_{BSA} = 2 mg/mL was crucial because we observed a remarkable time-dependent "swelling" of the film, characterized by a continuous shift of the "Kiessig fringes." After a few hours, a stable reflectivity profile was reached. In order to understand this phenomenon and to be more quantitative, another series of experiments was carried out at a higher concentration: C_{BSA} = 4 mg/mL (Fig. 6). The experimental results are consistent with previous experiments because a swelling was again observed. We call the difference between the overall thickness of the NBF and that of the pure surfactant (6.3 nm) extra thickness; it represents the matter swelling the NBF. The lowest value (0.3 nm) is observed just after the drainage (t = 0) and corresponds to the initial thickness of a pure $C_{12}E_6$–NBF with a very small amount of protein. After 45 h, there is a plateau, which indicates the end of thickness evolution. The system then reaches an equilibrium state in which the overall thickness is 10.3 ± 0.2 nm, corresponding to a 4-nm final extra thickness. This extra thickness remains constant (30 h) until the film bursts. We attribute the swelling of the film to the protein insertion and not to water because the increase of the overall film thickness is much smaller than that resulting from a transition to a CBF due to the formation of an aqueous core (in general, >10 nm). Thus, the swelling cannot be accounted for by the adsorption of water. The film remains an NBF whose overall thickness is smaller than that characteristic of a CBF. The swelling is therefore due to the sole insertion of protein. The last crucial problem is to locate the protein with respect to the surfactant and to interpret the maximum extra thickness value (~4 nm). This value may correspond either to roughly twice the

REFLECTIVITY

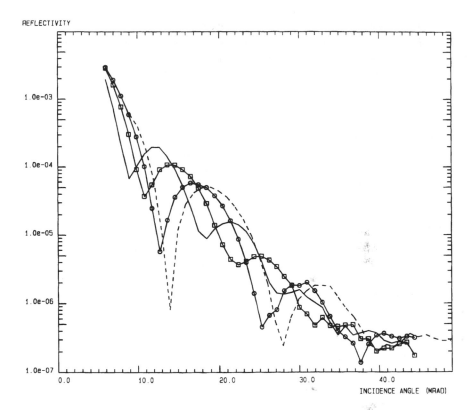

Figure 6 Set of experimental reflectivity profiles exhibiting well-defined interference "Kiessig fringes" indicative of a homogeneous film. Here, a time-function shift of these fringes toward smaller angles is observed; it evidences the swelling of the initial film due to the protein insertion. The different profiles correspond to recordings carried out on a pure $C_{12}E_6$ Newton black film (dashed line) and on a BSA–$C_{12}E_6$ NBF (C_{C12E6} = 0.075 mg/mL and C_{BSA} = 4 mg/mL), after 1 h (circles), after 10 h (squares), and after 18 h (solid line). The central core thicknesses between the surfactant walls are respectively 0, 0.3, 1.5, and 3 nm.

size of an unfolded molecule or to the width of a native molecule (4.1 nm) situated in the central core of the NBF. To obtain a definitive answer, we first formed a stable pure $C_{12}E_6$–NBF and then injected the solution of pure BSA with a syringe. To make this experiment possible, we increased C_{C12E6} to 0.5 mg/mL and, subsequently, we also increased C_{BSA} to 6.6 mg/mL, thus reducing the film stability. We made a series of reflectivity profiles at regular time intervals and we again observed the swelling process. This is a direct experimental proof of the protein insertion within the NBF. It is thus clear that at equilibrium the new system is a "sandwich NBF" whose central core is a single layer of protein molecules. An important question concerns the state (unfolded or denatured) of the protein itself. As a result, the extra thickness can be attributed to the central core formed of a close-packed single layer of BSA molecules, probably in a native state. In the NBF, the protection against the denaturation could result from the "2D encapsulation" by the surfactant walls. Our observations may be explained by a very simple model [26] by comparing the BSA chemical potential in the solution, at the Langmuir (air–water) interface, and in the NBF. We have made the first NBF containing a single and close-packed layer of proteins. We found a time-dependent insertion of protein after the NBF formation, which we explained by the difference in chemical potential between the NBF and the solution after the drawing. In summary, we believe that the use of NBFs for the study of specific interactions in membranes is very promising. The NBFs associated with the x-ray reflectivity technique should provide a new tool for structural biology.

REFERENCES

1. R Hooke. Communication to the Royal Society, 1672.
2. I Newton. Optiks Book II, Part I, London: Smith and Watford, 1704.
3. J Israelachvili. Intermolecular and Surface Forces. San Diego: Academic Press, 1985.
4. EJW Verwey, JThG Overbeeck. Theory of the Stability of Lyophobic Colloids. Amsterdam: Elsevier, 1949.
5. MN Jones, KJ Mysels, PC Sholten. Trans. Faraday Soc. 62:1336, 1966.
6. S Clunie, JF Goodman, BT Ingram. In: E Matijevic, ed. Surface and Colloid Science. New York: Interscience, 1971, Vol. 3, p. 168.
7. J Mysels. J Phys Chem 68:3441, 1964.
8. A Scheludko, D Exerowa. Kolloid Z 165:148, 1956.
9. JM Corkill, JF Goodman, DR Haisman, SP Harold. Trans Faraday Soc 57:821, 1961.
10. A De Feijter, A Vrij. J Colloid Interf Sci 70:456, 1979.
11. O Bélorgey, JJ Benattar. Phys Rev Lett 66:313, 1991.
12. JJ Benattar, A Schalchli, O Belorgey. J Phys I (France) 2:955, 1992.
13. P Guenoun, A Schlachli, D Sentenac, JW Mays, JJ Benattar. Phys Rev Lett 74:3628, 1995.
14. M Born, E Wolf. Principles of Optics. 6th ed. London: Pergamon Press, 1984, pp. 51–60.
15. F Abeles. Ann Phys (Paris) 5:596, 1950.
16. T Kolarov, R Cohen, D Exerowa. Colloids Surfaces 42:49, 1989 and references therein.
17. D Sentenac, JJ Benattar. Phys Rev Lett 81:160, 1998.
18. JJ Benattar, A Schalchli, D Sentenac, F Rieutord. Prog Colloid Polym Sci 105:113–117, 1997.

19. P Perrin, F Lafuma. J Colloid Interf Sci 197:317, 1998; F Millet, JJ Benattar, P Perrin. Phys Rev E 60:2045, 1999.
20. PR Musselwhite, JA Kitchener. J Colloid Interf Sci 24:80–83, 1967.
21. D Platikanov, GP Yampol'skaya, N Rangelova, ZhK Ahgarska, LE Bobrova, VN Izmaila, Kolloid Zh 43:177, 1981.
22. DC Clark, M Coke, R Mackie, C Pinder, DR Wilson. J Colloid Interf Sci 138:207–219, 1990.
23. GK Marinova, TD Gurkov, OD Velev, IB Ivanov, B Campbell, RP Borwankar. Colloids Surfaces 23–24:155–167, 1997.
24. Th Peters. Adv Protein Chem. 37:161, 1985.
25. DE Graham, MC Phillips. J Colloid Interf Sci 70:403–439, 1979.
26. JJ Benattar, M Nedyalkov, J Prost, A Tiss, R Verger, C Guilbert. Phys Rev Lett 82:5097, 1999.

15

Self-Assembly of Amphiphilic Molecules in Foam Films

Elena Mileva and Dotchi Russeva Exerowa
Institute of Physical Chemistry, Bulgarian Academy of Sciences, Sofia, Bulgaria

ABSTRACT

The structural reorganizations of smaller self-assembled aggregates (premicelles) in foam films are studied. The micellar entities form at lower amphiphile concentrations below the bulk cmc-value. A general theoretical scheme is proposed that connects the bulk theory of self-assembly with the specific interactions in the thin films. This scheme is used to suggest a model mechanism for the explanation of the following experimental observations: (1) at low surfactant concentration the surface tension/concentration curves ($\Delta\sigma/C_s$) display a well-defined plateau sector; (2) in microscopic foam films obtained from a solution whose concentration is in the plateau region, the so-called "black dots" are observed. These are miniature nonspreading spots. The foam films that pass through this stage are unstable and break up quickly. The appearance of "black dots" is explained with the existence and the structural reorganization of the self-assembled structures inside the film and on its interfaces.

I. INTRODUCTION

The thin liquid films, and foam films in particular, have been for many years an actively investigated research topic in the physical chemistry of disperse systems (e.g., Refs. 1–3). They have specific properties that make them useful models for the study of molecular forces (e.g., Refs. 1 and 4–8). Particularly interesting is the interplay of long-range and short-range interactions that occur during the formation and the evolution of foam films. Recently, new theoretical and experimental results have been accumulated that distinguish the black films as very significant for the proper understanding of the nature of these interactions (e.g., Refs. 1, 2, and 9).

Investigations of microscopic thin liquid films (the film radius is in the range of *10–500* μm) allow the direct determination of notable parameters of the films. Of particular importance is the possibility of working with very low surfactant concentrations. Thus, this offers a valuable opportunity to register the start of the self-assembly of amphiphilic molecules and the consecutive structural changes in the film in the course of the drainage process.

The self-assembly is one of the oldest topics in interface chemistry. In recent years, the investigations of solutions of amphiphilic substances have proved to be an intensively developing domain of the physical chemistry of interfaces [10–13]. The onset and the growth of miscellaneous self-assembled structures are a result of the delicate balance of the forces that act inside the micellar aggregates, on the one hand, and those acting between the micelles, on the other hand. Two major peculiarities should be particularly outlined: (1) The structural entities in the amphiphilic systems do not maintain permanent identities but are continuously exchanging fragments with one another. (2) Every variation in the system's conditions—composition of the solution (surfactant and salt concentrations, presence of additives), temperature, external fields, proximity of interfaces—influences the intra-aggregate and the interaggregate interactions. Thus, the overall distribution of the amphiphilic structures is constantly affected.

These general self-assembling tendencies should clearly manifest themselves in thin liquid films formed from bulk amphiphile solutions containing self-assembled structures. This was first mentioned in our recent paper [14]. During the formation and the drainage of films, the immediate vicinity of each surfactant molecule and aggregate becomes entirely different from the respective homogeneous bulk case. Therefore, the microscopic foam films are a very suitable model system for the study of the self-assembling phenomenon in amphiphilic solutions. The well-developed methods of thin-film-evolution registration allow the determination of the onset and the reorganization of micellar species in amphiphilic systems. Thus, during the film-drainage process, it is possible to follow the structural changes in confined amphiphilic systems, exposed to the additional field of the disjoining pressure.

The aim of this chapter is to summarize the existing experimental data on foam films that are, in our opinion, indicative of the presence and the reorganization of self-assembled structures. On this basis, a general scheme is proposed that attempts to connect the theory of the bulk self-assembly with the particular characteristics of the thin liquid films.

II. EXPERIMENTAL

The microinterferometric method operates with the measuring cell of Scheludko and Exerowa [1–2,15–18]. It is schematically shown in Figure 1.

The microscopic film of radius 100–500 μm is formed in the middle of a biconcave drop, situated in a glass tube of diameter 0.2–0.6 mm, by withdrawing the liquid from it. Studies of foam films of solutions of several low-molecular-weight amphiphilic substances (sodium hexyl sulfate, sodium octyl sulfate, sodium undecyl sulfate, sodium tetradecyl sulfate, and sodium dodecyl sulfate) are performed for different electrolyte concentrations.

When the concentration of the stabilizing surfactant at the interfaces of the film is near the close-packing value of the adsorption layers [Γ_∞ (the maximum

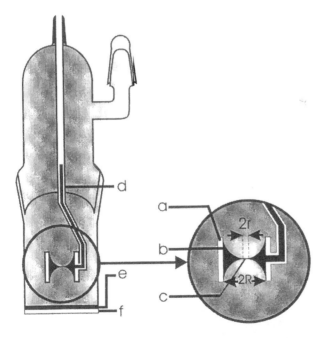

Figure 1 Schematic of the Scheludko–Exerowa measuring cell for the study of microscopic foam films. (a) Glass tube film holder; (b) biconcave drop; (c) microscopic foam film; (d) glass capillary; (e) surfactant solution; (f) optically flat glass.

surfactant concentration) and critical micelle concentration (cmc)] and the added electrolyte quantity is higher than a value $C_{el,cr}$ so as to surmount the first barrier in the disjoining pressure isotherm, black spots appear [1]. Such spots usually expand and form stable black films (e.g., Refs. 1–6, 8, and 9). These black formations have been well studied both with regard to their structure and stability. It is also established that at low amphiphile concentrations, miniature nonspreading spots with radii of about 5×10^{-4} cm appear. We have given them the name "black dots" (Fig. 2).

The dots never develop into ordinary black spots or black films; these foam films are unstable and break up quickly. The black dots have never been paid special attention although it was perceived they mark, in a sense, the beginning of structural reorganizations inside the films [1,9]. As a rule, they are persistently observed when the salt concentration exceeds $C_{el,cr}$ but the surfactant concentration is one order of magnitude lower than the cmc and Γ_∞.

It is well known that there is a close relationship between the black foam film formation and the properties of the surfactant adsorption layers [1,8,9]. Thus, it is very informative to examine the surface tension isotherms $\Delta\sigma(C_s)$ which characterize the decrease in the equilibrium values of the surface tension σ at the solution–air interface of the initial surfactant system from which the films were formed. The respective curves are shown in Figures 3 and 4. They were obtained for a broad range of amphiphile concentrations. The data in Figure 3

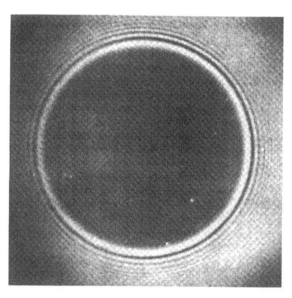

Figure 2   A photograph showing a black dot of radius 5×10^{-4} cm.

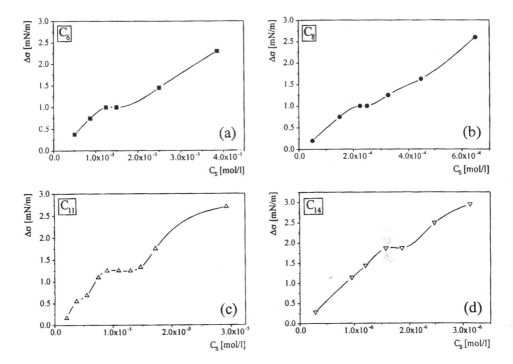

Figure 3 Dependencies of the surface pressure $\Delta\sigma = \sigma_0 - \sigma_b$ on the concentration of sodium alkylsulfates for $C_{el} = 0.1$ mol/l (KCl) and $t = 20°C$: (a) hexyl sulfate; (b) octyl sulfate; (c) undecyl sulfate; (d) tetradecyl sulfate.

were obtained by the conventional Du Nouy technique (the precision was ± 0.1 mN/m) [8]. The electrolyte concentration is 0.1 M/L KCl at a temperature of 20°C.

More precise results (see Fig. 4) were obtained via the spherotensiometric techniques (accuracy ± 0.01 mN/m) [19,20]. Here the temperature is 22°C and the added salt is NaCl.

The following observations are of particular importance. Each of the isotherms in Figures 3 and 4 contains a plateau sector at low amphiphile concentrations. It is below the bulk cmc value and the close-packing concentrations at which the usual black spots and black films appear. The width of the plateau depends on the type of the surfactant and on the concentration of added electrolyte. The corresponding investigations of the foam films from solutions with concentrations in the plateau regions have shown that it is here that the black dots persistently appear (Fig. 5). This indicates that some structural reorganization in the foam film is to be expected in these concentration limits. The crude estima-

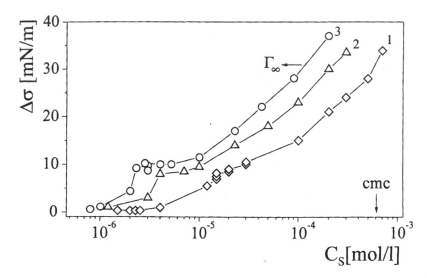

Figure 4 Surface tension isotherms of sodium dodecyl sulfate solutions at different electrolyte concentrations (NaCl) and temperature t = 22°C: (1) C_{el} = 0.1 mol/l; (2) C_{el} = 0.35 mol/l; (3) C_{el} = 0.5 mol/l.

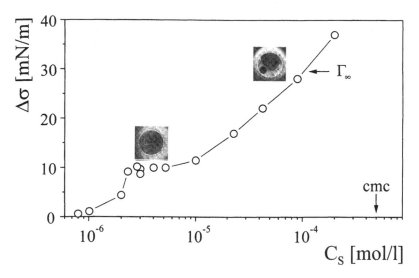

Figure 5 Surface tension isotherms of sodium dodecyl sulfate solutions at electrolyte concentration (NaCl) C_{el} = 0.5 mol/l and t = 22°C.

tions of the adsorbed quantity on the interfaces based on the Gibbs adsorption equation reveal that the average surface coverage is less than the values of the dense monolayer.

Thus, in what follows, we will focus our attention on the explanation of the possible onset and reorganization of these amphiphilic structures.

III. THEORETICAL BACKGROUND

There have been several attempts at an explanation of the above-mentioned observations by concentrating on the state of the adsorption layers. The presence of horizontal plateau portions in the surface pressure–surfactant concentration curves at low amphiphile concentrations as well as the persistent appearance of black dots in the resultant foam films are obviously interconnected. This stimulated us to seek a better explanation of these experimental results. We found it in the interconnections of the specific properties of the foam films with the state of the adsorption layers on their interfaces and the existence of self-assembled structures.

The starting point of the analysis is the notion that at surfactant concentrations lower than the cmc, self-assembled structures (premicelles) appear. This idea came out in the early studies of the bulk amphiphilic systems [21–24]. Later, the classical cmc viewpoint (i.e., the presumption that only micellar aggregates containing a considerable number of monomers can form at certain threshold concentration) almost completely shielded the premicellar concept. In recent years, that notion has been revived. This is based on the accumulating experimental evidences on solid–liquid interfaces. It has been established that under particular conditions, smaller aggregates (premicelles, hemimicelles, admicelles, etc.) are formed in the vicinity and on the interfaces [25–28]. This was observed at surfactant concentrations lower than the bulk cmc values. Several numerical experiments also show that the general principles of statistical mechanics do predict the formation of bulk premicellar aggregates of various sizes and forms [28,29]. The diphilic character of the surfactant molecules results in the respective aggregation on the interfaces as well [30–32]. Inevitably, this phenomenon should affect the adsorption properties of the surfaces.

Recently, we have proposed the hypothesis that the onset of black formations in thin liquid films is related to both the bulk and interface amphiphilic structures [14]. Here, we shall summarize the major stages of the analysis and try to use it in the new interpretation of the experimental facts that are described in Section II.

It is greatly convenient to view the foam film as a three-layered model: the film bulk phase and two-dimensional (2D) surface phases (air–liquid interfaces). It is assumed that the self-assembly is possible in all the layers and that they are in thermal, mechanical, and chemical equilibrium.

A. Bulk-Film Self-Assembly

The characteristic quantities are denoted with the superscript b. The condition for chemical equilibrium states that

$$\mu_n^b = n\mu_1^b, \qquad \mu_n^b = \left(\frac{\partial G^b}{\partial N_n^b}\right)_{\{N_n'\},p,T}, \qquad \mu_1^b = \left(\frac{\partial G^b}{\partial N_1^b}\right)_{\{N_n'\},p,T} \tag{1}$$

where G^b is the total free energy of the solution, N_n^b are the numbers of the respective n-mers, and the number distribution of the other premicellar aggregates is denoted by $\{N_n'\}$. G^b is presented as a sum of four terms [33–35]:

$$G^b = G_f^b + G_m^b + G_{\text{int}}^b + G_h^b = G_0^b + G_h^b \tag{2}$$

Here G_0^b accounts for the self-assembling factors of the bulk-solution micellization when the role of the interfaces is negligible, whereas G_h^b accounts for the particular properties of the thin liquid film. G_f^b models the change in the free energy when solvent molecules are added to pure solvent and when aggregates, consisting of i monomers, are placed in the pure solvent [14,33]:

$$G_f^b = N_w^b \mu_w^{0,b} + \sum_i N_i^b \overline{\mu}_i^{0,b} \tag{3}$$

The aggregate formation is primarily governed by short-range intermonomer interactions and it is assumed that this term does not depend on the film thickness.

The free-energy portion, G_m^b, is the mixing energy of the aggregates, the monomers, and the water molecules:

$$G_m^b = kT\left(N_w^b \ln X_w^b + \sum_i N_i^b \ln X_i^b\right) \tag{4a}$$

$$X_w^b = \frac{N_w^b}{N_w^b + \sum_i iN_i^b}, \qquad X_n^b = \frac{N_n^b}{N_w^b + \sum_i iN_i^b} \tag{4b}$$

The term G_{int}^b accounts for the interactions among the various amphiphilic species. The respective Gibbs energy portions are formally presented as follows [14,33,34]:

$$G_{\text{int}}^b = \frac{1}{2}\sum_i N_i^b \sum_j N_j^b \frac{\mu_{ij}^b}{V_f}, \qquad V_f = A_f h = N_w^b v_w + \sum_i iN_i^b v_a \tag{5}$$

In Eq. (5), μ_{ij}^b stands for the bulk interaction potential between an i-mer and a j-mer. We assume that the interaggregate interactions do not explicitly depend on the film thickness. This is acceptable for smaller aggregates but not for very long-range interaction forces. The notations v_w, and v_a are used for the effective volumes of a water molecule and an amphiphile molecule, respectively.

The consecutive statistical mechanical modeling of the last term of G_h^b for a multicomponent system is an elaborate task. Therefore, we present a coarse scheme for the self-assembling phenomenon in the foam films, and only a general expression is used here. It connects the model with either experimentally obtained or model-calculated disjoining pressure isotherms [36]:

$$G_h^b = 2\sigma_f(h)A_f, \qquad 2\sigma_f(h) = 2\sigma_b + \int_h^\infty \Pi(h')dh' \tag{6a}$$

$$A_f = \sum_i N_i^s a_i^s + N_w^s a_w^s \tag{6b}$$

This free-energy portion introduces the particular interactions in the foam films via the dependence of the disjoining pressure on the film thickness. The notation σ_b stands for the measurable surface tension of the surfactant solution from which the thin liquid film is formed; it is assumed to be constant in the following, in conformity with the experiment [8,19,20]. A_f is the area of the film surface; N_i^s, and N_w^s are the numbers of the i-mers and of the water molecules of the surface layer, respectively; a_i^s and a_w^s are the effective areas assigned to every surface aggregate and water molecule, respectively, that form the adsorption layer. Here, we have assumed that every aggregate can take only one position on the interface.

Thus, the size-distribution curve of the micellar species for the bulk of the film is obtained as

$$X_n^b = X_n^{0,b} \exp\left(\frac{\sum_j (n\mu_{1j}^b - u_{nj}^b)}{kTV_f} \right)$$

$$\times \exp\left[\frac{2}{kT}\{2\sigma_f(h) - \sigma_b + \Pi(h)h\}\left(n\frac{\partial A_f}{\partial N_1^b} - \frac{\partial A_f}{\partial N_n^b} \right) \right] \tag{7a}$$

$$X_n^{0,b} = (X_1^b)^n \exp\left(\frac{\mu_n^{0,b} - n\mu_1^{0,b}}{kT} \right), \qquad \mu_n^{0,b} = \overline{\mu}_n^{0,b} + kT, \qquad \mu_1^{0,b} = \overline{\mu}_1^{0,b} + kT \tag{7b}$$

B. Aggregative Adsorption

All quantities characterizing the interfaces are denoted with a superscript s. The size-distribution curve of the 2D amphiphilic structures is obtained in analogy to the bulk micellization following the scheme of Eqs. (1–6). The result is

$$
X_n^s = X_n^{0,s} \exp\left(\frac{\sum_j (nu_{1j}^s - u_{nj}^s)N_j^s}{kTA_f}\right) \exp\left(-\frac{\Delta a_n^s}{2kTA_f^2}\sum_i N_i^s \sum_j N_j^s u_{ij}^s\right)
$$

$$
\times \exp\left(-\frac{\Delta a_n^s}{kT}\sigma_f(h)\right) \exp\left[\frac{v_a \Pi(h)}{kT}\left(n\frac{\partial N_1^b}{\partial N_1^s} - \frac{\partial N_n^b}{\partial N_n^s}\right)\right] \qquad (8a)
$$

$$
\times \exp\left[-\frac{\Delta a_n^s}{kT}\int_\infty^h \left(\frac{\partial \Pi}{\partial h'}\right)h'dh'\right]
$$

$$
X_n^{0,s} = (X_1^s)^n \exp\left(-\frac{\mu_n^{0,s} - n\mu_1^{0,s}}{kT}\right),
$$

$$
X_n^s = \frac{N_n^s}{N_w^s + \displaystyle\sum_j iN_j^s}, \qquad \Delta a_n^s = na_1^s - a_n^s \qquad (8b)
$$

In the above X_n^s is the surface mole fraction of the interface n-mers. $\mu_1^{0,s}$ and $\mu_n^{0,s}$ are the standard chemical potentials respectively of a monomer and an n-aggregate in the 2D layers. They carry the information about the intrinsic surface self-assembling properties of the amphiphile on the interface of a bulk solution before a thin film is formed from it. In analogy to Eqs. (7), the surface tension σ_f of the film interface and the disjoining pressure both enter the size-distribution curve. However, they become important only if $\Delta a_n^s \neq 0$ and $n(\partial N_1^s/\partial N_1^s) - \partial N_n^b/\partial N_n^s \neq 0$. Provided that larger premicellar aggregates prevail, $\Delta a_n^s > 0$ and the disjoining pressure plays a serious role.

C. Bulk Film–Interface Rearrangement

The periphery of the foam film is connected with the bulk-solution phase from which it is obtained via the so-called film meniscus. The total balance of the film–interface amphiphilic structures should depend on the film–meniscus equilibrium. As a first step, it is assumed that the film drains in such a way that the total mole fraction of the added amphiphile remains constant. Hence,

$$
X = X^b + X^s = \sum_i iX_i^b + 2\sum_i iX_i^s = \text{const} = X_1^b + 2X_1^s
$$

$$+ \sum_i iX_i^{0,b} \exp\left(\frac{\sum_j (iu_{ij}^b - u_{ij}^b)N_j^b}{kTV_f}\right)$$

$$\times \exp\left[\frac{2}{kT}[2\sigma_f(h) - \sigma_b + \Pi(h)h]\left(i\frac{\partial A_f}{\partial N_1^b} - \frac{\partial A_f}{\partial N_i^b}\right)\right]$$

$$+ 2\sum_i iX_i^{0,s} \exp\left(\frac{\sum_j (iu_{ij}^s - u_{ij}^s)N_j^s}{kTA_f}\right) \exp\left[\frac{v_a\Pi(h)}{kT}\left(i\frac{\partial N_1^b}{\partial N_1^s} - \frac{\partial N_i^b}{\partial N_i^s}\right)\right] \tag{9}$$

$$\times \exp\left(-\frac{\Delta a_i^s}{kT}\sigma_f(h)\right) \exp\left[-\frac{\Delta a_i^s}{kT}\int_\infty^h \left(\frac{\partial \Pi}{\partial h'}\right)h' \, dh'\right]$$

$$\times \exp\left(-\frac{\Delta a_i^s}{2kTA_f^2}\sum_k N_k^s \sum_j N_j^s u_{kj}^s\right)$$

If the bulk of the film is thicker (i.e., we have a thin liquid layer), the disjoining pressure may be neglected. Let A_f, the surface area of the film interfaces, also remain constant and independent of the bulk amphiphile concentration. Then, Eq. (9) transforms into

$$X = X_1^b + 2X_1^s + \sum_i iX_i^{0,b}\exp\left(\frac{\sum_j (iu_{ij}^b - u_{ij}^b)N_j^b}{kTV_f}\right)$$

$$+ 2\sum_i iX_i^{0,s} \exp\left(\frac{\sum_j (iu_{ij}^s - u_{ij}^s)N_j^s}{kTA_f}\right) \tag{10}$$

$$\times \exp\left(-\frac{\Delta a_i^s \sigma_b}{kT}\right) \exp\left(-\frac{\Delta a_i^s}{2kTA_f^2}\sum_k N_k^s \sum_j N_j^s u_{kj}^s\right)$$

When smaller entities are predominant in the bulk of the film, in thinner layers their mole fraction will increase. As long as the surface mole fractions remain insensitive to film thickness, the fraction of the monomers both in the bulk of the film and on the interfaces is decreased. Thus, the rearrangement of the amphiphilic species upon thinning in the absence of the disjoining pressure results in additional premicellization. So, there is an increase of the number of the smaller aggregates in the bulk.

If the foam films are so thin that the disjoining pressure becomes operative, its effect on the bulk structure reorganization may be observed only if $i(\partial A_f/\partial N_1^b) - \partial A_f/\partial N_i^b \neq 0$. One should remember that usually $\Delta a_i^s \geq 0$. In the case of higher electrolyte concentrations, the electrostatic component of the disjoining pressure is effectively depressed [1,2]. Hence, for negative disjoining pressures [$\Pi_{vw}(h) \sim (-K/h^3)$, K is a constant], the quantity of the respective amphiphilic structures decreases for thinner films both in the bulk and on the interfaces. However, as is easily seen from Eq. (9), this effect is, to a certain degree, depressed in the bulk, because of the increasing influence of the interaggregate interaction term. Thus, for thinner films, the mole fraction of the amphiphilic structures is decreased and the size-distribution curve is shifted toward the monomers. Therefore, the reorganization of the amphiphilic entities results in an overall disintegration of both the surface and the bulk aggregates, thus increasing the mole fractions of the respective monomers upon film thinning.

IV. DISCUSSION

The general scheme presented in the previous section outlines the overall tendencies of the reorganization of existing (pre)micellar aggregates in thicker liquid layers [$\Pi(h) = 0$] and in thin liquid films [$\Pi(h) \neq 0$]—foam films in particular. They are formed from amphiphilic solutions containing self-assembled structures. This scheme allows a new explanation and interpretation of the above-stated experimental results.

Let us start with the bulk amphiphilic solution (Figs. 3 and 4) at low surfactant concentration (from the plateau sector of the $\Delta\sigma/C_s$ curves). When the electrolyte quantity is high, favorable conditions for the formation of premicellar aggregates are created, both in the bulk solution and at the interfaces. A variant of expression (10) may serve for qualitative analysis of this case:

$$X = X_1^b + X_1^s \sum_i iX_i^{0,b} \exp\left(\frac{\sum_j (iu_{1j}^b - u_{ij}^b)N_j^b}{kTV_f}\right)$$

$$+ \sum_i iX_i^{0,s} \exp\left(\frac{\sum_j (iu_{1j}^s - u_{ij}^s)N_j^s}{kTA_f}\right)$$

$$\times \exp\left(-\frac{\Delta a_i^s \sigma_b}{kT}\right) \exp\left(-\frac{\Delta a_i^s}{2kTA_f^2} \sum_k N_k^s \sum_j N_j^s u_{kj}^s\right)$$

(10')

The existence of a plateau region in the $\Delta\sigma/C_s$ curves is due to the following: Upon an increase of the total amphiphile amount (X), the additional monomers go to form smaller aggregates in the bulk—(pre)micelles. The surface coverage is practically not affected. Thus, the surface tension of the interface remains unchanged. Therefore, at a low surfactant concentration a plateau region is formed. Its location and width depend on the electrolyte concentration and on the type of the amphiphilic molecules. As observed experimentally (Fig. 3), the shorter surfactant chains result in a broader plateau region. One possible explanation is that smaller surfactant chains facilitate the onset of smaller oligomers in the bulk. For a given amphiphile (Fig. 4), the higher electrolyte concentration also causes a widening of the plateau sector. Here, the self-assembly is obviously aided by the increased screening of the headgroups of the amphiphile molecule.

When thin layers $[\Pi(h) = 0]$ are formed from the plateau concentration-range solutions, the process may be regarded as a gradual approach of two interfaces at constant overall amphiphile concentration $(X = \text{const})$. Before the onset of the disjoining pressure in the confined volume, the tendency for bulk micellization is somewhat enhanced for thinner layers and higher electrolyte concentrations [see Eq. (10)]. This effect is connected only with the interaggregate interaction portion of the bulk-distribution curve. The polydispersity of the bulk is decreased because the tendency for smaller aggregates formation continues. The existing structures on the interfaces are not seriously altered with respect to the pure bulk solution case.

In foam films $[\Pi(h) \neq 0]$, the action of the disjoining pressure is very important [Eq. (9)]. At a high electrolyte concentration, the leading component is the van der Waals constituent Π_{vw}. Its action results in enhanced destruction of the existing self-assembled structures both in the bulk and at the film interfaces. Thus, the number of the free monomers increases. As long as the conditions for the bulk aggregation become less favorable by the "force field" of the disjoining pressure, it is reasonable to expect that the newly released monomers would prefer to go to the air–liquid interface, thus ensuring better coverage on the interfaces.

When the electrolyte concentration exceeds the critical value and the film drains to about 30–40 nm, it becomes unstable and fluctuations in its thickness are observed [2]. In the thinner areas, the existing aggregates are additionally decomposed and, thus, release additional monomer quantities. This creates conditions for the outbreak of narrower quasiequilibrium regions in the film. On the interfaces of these thinner areas in the film, the monomer molecules form adsorption layers which are approximately as packed as is the conventional stable black spots and films. These thinnings are registered as "black dots" (Fig. 2). The dots appear when the electrolyte quantity is high enough so that an effective suppression of the electrostatic component of the disjoining pressure is achieved. Thus, at least in principle, it becomes possible for black formations to be observed.

The total amphiphile quantity, however, remains insufficient and cannot ensure the dense coverage of the interfaces, as in the case with the usual black spots and films. Therefore, films with black dots should be very unstable and break up immediately as observed experimentally.

It should be recognized that the quantitative verification of the theoretical scheme presented here demands a specially designed experiment aimed at the registration and the visualization of the self-assembled structures in foam films. The theoretical analysis in itself remains at the level of just outlining the prevailing tendencies. The first next step should be a detailed modeling of the standard chemical potentials of the aggregates.

V. CONCLUSIONS

A new possibility for the onset of self-assembly in thin liquid films, foam films in particular, is presented in this chapter. It is shown that the appearance of black formations in foam films (denoted as ''black dots'' in our case) is connected with the presence and reorganization of amphiphilic structures in the initial bulk solution. We have been successful in showing that the foam film is a detector of the self-assembled aggregates. Upon the formation of films, the conditions in the anisodiametric bulk and on the interfaces alter dramatically. Therefore, the self-assembly in foam films should be analyzed with due account both of the specific interactions in the confined volume of the film and with the proper modeling of the self-assembly conditions that appear as a result of thin-film formation. The particular ''force field'' of the disjoining pressure gives unique possibilities for the registration of the gradual formation and destruction of the micellar aggregates existing in the film. In our opinion, the well-developed Scheludko–Exerowa thin-film technique is particularly suitable for the identification and registration of (pre)micellar entities in the vicinity of and at fluid interfaces.

REFERENCES

1. D Exerowa, P Kruglyakov. Foam and Foam films. Amsterdam: Elsevier, 1998.
2. A Scheludko. Adv Colloid Interf Sci 1:391, 1967.
3. B Deryaguin, N Churaev, V Muller. Surface Forces, New York: Consultants Bureau, 1987.
4. D Exerowa, A Nikolov, M Zacharieva. J Colloid Interf Sci 81:419, 1980.
5. D Exerowa, T Kolarov, Khr Khristov. Colloids Surfaces 22:171, 1987.
6. T Kolarov, R Cohen, D Exerowa. Colloids Surfaces 42:49, 1989.
7. J de Feijter. In: IB Ivanov, ed., Thin Liquid Films. New York: Marcel Dekker, 1988.
8. D Exerowa, A Scheludko. Bull Inst Phys Chem 3:79, 1963.

9. D Exerowa, A Nikolov, M Zacharieva. J Colloid Interf Sci 81:419–429, 1981.
10. Current Opinion Colloid Interf Sci 1:327,367, 1996.
11. Current Opinion Colloid Interf Sci 2:279, 1997.
12. W Gelbart, A Ben-Shaul. J Phys Chem 100:13,169, 1996.
13. S-H Chen, JS Huang, P Tartaglia, eds. Structure and Dynamics of Strongly Interacting Colloids and Supramolecular Aggregates in Solution, NATO ASI Series, Ser. C: Mathematical and Physical Sciences Vol. 369. Dordrecht: Kluwer Academic, 1992.
14. E Mileva, D Exerowa. Colloids Surfaces A 149:207–216, 1999.
15. A Scheludko, D Exerowa. Commun Dept Chem Bulg Acad Sci 7:123, 1959.
16. D Exerowa. Commun Dept Chem Bulg Acad Sci 11:739, 1978.
17. D Exerowa, M Zacharieva, R Cohen, and D Platikanov. Colloid Polym Sci 257: 1089, 1979.
18. D Exerowa, D Kashchiev, D Platikanov. Adv Colloid Interf Sci 40:201, 1992.
19. A Nikolov, G Martynov, D Exerowa, W Kaishev. J Colloid Interf Sci 81:116, 1981.
20. D Exerowa, A Nikolov. In: KL Mittal, ed. Surfactants in Solution. New York: Plenum Press, 1984, Vol. 4, p. 1313.
21. J McBain. Kolloid Z 12:256, 1913.
22. P Ekwall. Kolloid Z. 80:77, 1937.
23. J Stauff. Kolloid Z. 96:246, 1941.
24. J McBain, W Dye, S Jonson. J Am Chem Soc 61:321, 1939.
25. S Niu, K Gopidas, N Turro, G Gabor. Langmuir 8:1271, 1992.
26. B Jonsson, P Wängnerud, B Jönsson. Langmuir 10:3542, 1994; F Tiberg, B Jönsson. Langmuir 10:3714, 1994.
27. S Manne, J Cleveland, H Glaub, G Stucky, P Hansma. Langmuir 10:4409, 1994.
28. M Vold. Langmuir 8:1082, 1992.
29. B Smit, P Hibers, K Esselink. In: S-H Chen, JS Huang, P Targalia, eds. Structure and Dynamics of Strongly Interacting Colloids and Supramolecular Aggregates in Solution. NATO ASI Series, Ser. C: Mathematical and Physical Sciences Vol. 369. Dordrecht: Kluwer Academic, 1992, p. 519.
30. T Gu, B Zhu, H Rupprecht. Prog Colloid Polym Sci 88:74, 1992.
31. J Israelachvili. Langmuir 10:3774, 1994.
32. V Fainerman, R Miller. Langmuir 12:6011, 1996.
33. D Blankschtein, G Benedek, G Thurston. J Chem Phys 85:7268, 1986.
34. P Missel, N Mazer, G Benedek, C Young, M Carey. J Phys Chem 84:1044, 1980.
35. E Mileva. J Colloid Interf Sci 178:10, 1996.
36. D Kashchiev. Surface Sci 220:428, 1989.

16

Formation of 2D Structures of Micrometer-Sized Latex Particles Inside Thinning Foam Films

Krassimir P. Velikov
Van't Hoff Laboratory for Physical and Colloid Chemistry, Debye Institute, Utrecht University, Utrecht, The Netherlands

Orlin D. Velev
University of Delaware, Newark, Delaware

ABSTRACT

Thinning foam films containing micrometer-size polystyrene latex microspheres were studied by interferometric microscopy. The microsphere dynamics depends on the type of film stabilizer, the particle concentration and hydrophobicity, and rate of film formation. No entrapment of particles between the surfaces was possible in films stabilized by an anionic surfactant. In films containing protein, a limited number of particles were caught inside the film area due to the decreased mobility of the interfaces. Extraordinary long-ranged (>100 μm) capillary attraction leads to two dimensional (2D) particle aggregation and formation of metastable structures. When the microspheres were partially hydrophobized by the presence of cationic surfactants, the film opening and thinning was sterically inhibited by a layer of particles simultaneously adsorbed onto the two interfaces. The particles within this layer show excellent 2D hexagonal ordering. The 2D crystallization process can be controlled by changing the pressure inside the film meniscus. The formation of the structures in the film is governed both by the particle-particle and particle-meniscus capillary interactions.

I. INTRODUCTION

The free liquid films formed between the bubbles of foams or the droplets of emulsions quite often contain solid particles, either from their original composition or as impurities. Examples of similar systems containing solid particles inside thin films are the so-called Pickering emulsions [1–4], some food emulsions containing protein aggregates [5], ice creams, as well as latex and metallic dyes and coatings. A phenomenon of particular interest and importance caused by micelles inside foam films is the stratification [6–10]. It has been shown that the stratification is a major factor of stabilization of liquid films in emulsions and foams and that this phenomenon can also be observed with nanometer-sized latex particles and proteins [11–13].

Another recently emerging area where thin liquid films containing particles have been the subject of increased interest is the assembly of ordered two-dimensional (2D) and multilayered materials. Such 2D structures can possibly find use in high-technology applications like data storage, microelectronics, and optical devices [14,15]. Two types of supported thin liquid films have been used as assembly sites for latex particles [16–18] and globular proteins [15,19]. The first type are wetting films onto solid substrates, where the 2D crystallization has been carried out by liquid evaporation and convective drag of the particles toward the drying area [16,17]. Wetting aqueous films onto fluorinated oil [18] or liquid mercury [15] have also been successfully used as 2D crystallization sites. In this latter case, capillary forces attracting the particles between the mobile film interfaces enhance the assembly process. However, one case that had not been investigated in detail is the use of films with two free air–liquid interfaces (foam films) as crystallization sites for micrometer-sized latex particles.

The present study was aimed to explore, by direct observation of the dynamics and structure, formation of micrometer-sized latex particles confined between the surfaces of thinning foam films. We investigated the role of the surfactant type and concentration, the particle hydrophobicity, and the rate of film formation.

II. MATERIALS AND METHODS

The latex suspensions contained vol 10% of polystyrene microspheres (Dynospheres from IDC, USA) 7 μm in diameter. Due to the presence of sulfate groups on the surface, the polystyrene latex particles are charged negatively with a zeta-potential $\approx$ −70 mV. Three ionic surface-active substances (obtained from Sigma) were alternatively added to the aqueous phase to stabilize the aqueous films against rupture: sodium dodecyl sulfate (SDS), dodecyl trimethyl ammonium bromide (DTAB), and hexadecyl trimethyl ammonium bromide (HTAB).

Bovine serum albumin (BSA) obtained from Sigma (Cat. No. A-3803) was also used as the stabilizer. The solutions for the experiments were prepared with de-ionized water from a Millipore Milli-Q RO system (Organex grade).

The experiments were carried out by a Scheludko–Exerowa-type model cell for microinterferometric observation of liquid films [20]. The foam thin liquid films (TLF) were formed by sucking liquid out of a biconcave meniscus held in a capillary of inner radius of 3.20 ± 0.05 mm. To prevent evaporation and rupture of the film, the cell was installed in a closed chamber saturated with water vapor.

The particle dynamics and crystal evolution were followed by microinterferometric observation in reflected monochromatic illumination [21] and recorded with a video camera. Selected frames from the record were processed on a personal computer equipped with a TARGA 16/32+ video digitizer board to reconstruct the film-thickness profile.

III. RESULTS

It turned out that the major factor affecting the particle dynamics and the resulting structures was the type of surface-active film stabilizer used; therefore, the results will be presented according to the type of surfactant used. The initial concentrations of the surfactants were slightly above the critical micelle concentration (cmc), where the adsorption layers are complete but no complex interference from the micelles due to depletion or stratification can be expected.

A. Anionic Surfactant: SDS

The experiments were performed with solutions containing 0.016 M SDS. Because of its negatively ionized head, SDS should repel the negatively charged particles away from the surface. This repulsion, together with the capillary forces and the low interfacial viscosity, favors the expulsion of the particles out of the film area and into the surrounding meniscus. All latex microspheres were immediately pushed out of the film area during the first few seconds of film thinning and before finishing the thinning stage of film evolution (Fig. 1). No entrapment of the microspheres whatsoever could be achieved by varying the rate of film formation and the local particle concentration.

B. Protein: BSA

We chose to work with BSA concentration of 0.1 wt%, which should ensure a complete saturation of the monolayer [22]. The pH of the obtained solutions (≈6.4) was above the isoelectric point of BSA (≈4.7) [23] and the net charge

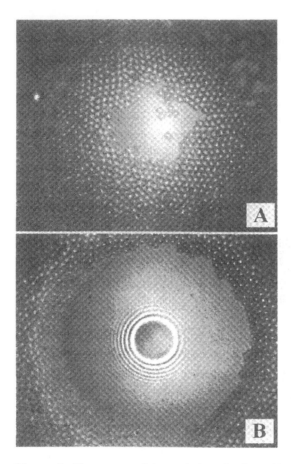

Figure 1 Two stages of the evolution of a foam film stabilized by anionic surfactant
(SDS) and containing latex particles. When the film thickness becomes lower than the
particle diameter, the latex microspheres are expelled out of the film area. The frames in
(A) and (B) are separated by a time period of ~2 s.

of the molecules is negative; therefore, the particles should again be repelled
from the surfaces. However, due to the high interfacial viscosity of the BSA
adsorption layer (up to 2000 surf. poises [24]), the film surfaces become immobilized and the motion of the liquid inside is strongly impaired [25], favoring the
entrapment of the microspheres within the film and their compression between
the surfaces. This was indeed observed experimentally and a number of different
cases could be distinguished, depending on the local concentration of sedimented
particles between the surfaces of the film.

When the number of particles between the surfaces was small, only a few particles could be entrapped inside the film by quickly forming it in less than 1 s. The entrapped particles were typically collected inside the thicker dimple [25] in the film center. They tended to aggregate in 2D groups, attracting each other from distances that could reach up to 100 μm (Fig. 2). This aggregation is obviously governed by the so-called lateral capillary forces [26–29]. The small ordered 2D aggregates survived for a short time and were "spit" out at a later stage of the dimple evolution. On many occasions, this stage was not

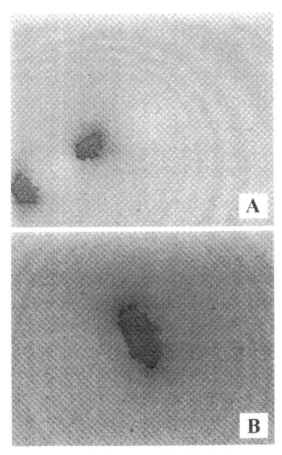

Figure 2 Example of the action of the long-ranged attractive capillary forces between particles entrapped inside a foam film with BSA. The two separate 2D latex aggregates (A) are attracted and merge into a larger one (B). The time elapsed between the two frames shown is 1 s. Note the ordering of the particles within the aggregates.

reached because the entrapped microspheres caused rupture of the otherwise stable film.

At intermediate latex concentrations, more and more particles were caught between the surfaces. At a certain degree of coverage ($\sim$50% by area), the appearance of the particle aggregates changed from single clusters to "bridges" and 2D foamlike formations (Fig. 3). These clusters are similar to 2D structures observed and reported earlier for particle dynamics in films above the surface of fluorinated oil [17]. The cluster formation on fluorinated oil is a result of the capillary forces too. In almost all cases, the particulate bridges and "2D foam" were eventually broken and pushed out by the thinner films formed in the areas void of particles between the bridges. This, however, strongly slows down the film evolution, and unlike the case of SDS, the particle islands and structures show a clearly visible rigidity sometimes greatly changing the shape of the otherwise circular film meniscus. At a higher initial coverage of the lower surface with latex particles, the thinning of the film is more retarded and its shape deviates further from equilibrium.

C. Cationic Surfactants: DTAB and HTAB

The surfactant concentrations used were 0.0012, 0.006, and 0.03 M for DTAB and 0.001 M for HTAB, all of which ensured excellent stability of the foam films. The surfactant concentration was varied and its effect was found to be significant for the behavior of the latexes. Cationic surfactants cause a major shift of the properties of the polystyrene microspheres by their hydrophobization. This phenomenon utilized by us earlier [18,30] results from the coupling of the negative

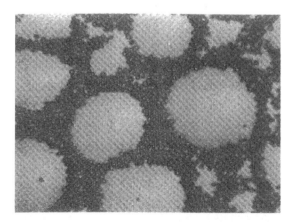

Figure 3 Metastable 2D foamlike particle structures formed inside a BSA-stabilized film.

charges on the beads surfaces with the positively charged surfactant molecules. The hydrophobized microspheres tend to adsorb onto the water–air interfaces because of increased hydrophobic attraction. The effect of the cationic surfactant on the film–particle dynamics was most conveniently observed at the intermediate concentration of DTAB ($0.006\ M$), when the partially hydrophobized microspheres adhered strongly to the film interfaces, but no aggregation in the bulk suspension occurred. The basic stages of film evolution are presented in Figure 4. When the film was formed for the first time, the particles were pushed out into the surrounding meniscus and the film area was void of particles. During the evolution of this first film, however, the particles in the ring closest to the film periphery came into contact with the upper surface and adhered to it, linking together the lower and the upper surfaces (Fig. 4A). When the surfaces were withdrawn to close the film, these particles were sucked in the middle (Figs. 4B and 4C). If the applied reverse pressure was not high enough to detach the upper surface from the layer of particles, the beads that had bound the surfaces together remained within the area of the next opened film (Fig. 4D). During the closure of the second film, these microspheres were joined by a new group of particles that had adhered to the surfaces at the periphery. Thus, in a consecutive manner, each subsequent opening and closing of the film added new particles to the ones already entrapped between the interfaces.

Most interesting is the fact that the multistep procedures described lead to latex microsphere organization into high-quality 2D arrays with almost no lattice defects. The size of the single-crystalline domains obtained could reach up to $3 \times 10^4\ \mu m^2$. Thus, it was demonstrated that the method allowed controllable formation of large 2D crystals from latex particles modified with different types of cationic surfactant. The increase of the surfactant concentration leads to lower quality of the particle arrays inside the film, due to some degree of lateral and bulk coagulations of the latex beads. In this case, the formed 2D arrays did not possess the excellent ordering present in the lower surfactant concentrations but showed a higher degree of rigidity and resistance to further manipulation of the film.

The observed formation of ordered 2D layers bears some resemblance to the models proposed for the interpretation of the phenomenon of stratification [6–13]. In order to find out whether we can control the formation of ordered multilayers of particles, a few experiments at high particle concentrations and different film stabilizers were conducted. Eventually, we were able to momentarily observe formation of a bilayer structure and a transition to a monolayer in the case of films with high particle concentrations and stabilized by a mixture of 0.1 wt% BSA and $8 \times 10^{-4}\ M$ HTAB (where the protein immobilizes the surfaces and the surfactant hydrophobizes the microspheres). Although the multilayers existed only for less than 1 sec, we were able to eventually observe that both the bilayer and the monolayer were ordered in a hexagonal lattice and there was a

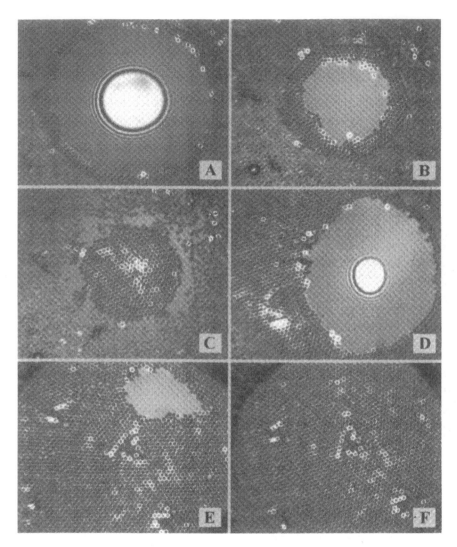

Figure 4 Consecutive frames of the formation of a 2D ordered array from latex particles inside a foam film stabilized by 0.006 M DTAB: (A) During the formation of the first film, the particles are pushed out into the surrounding meniscus and some adsorb to both surfaces; (B) when the film is closed, the adsorbed particles are dragged toward the center; (C) the ''zipped'' particles reorganize into a 2D crystal; (D) opening of a second film outside the 2D array; (E) closing of the fourth film; (F) the whole film area is occupied by a large 2D crystalline array.

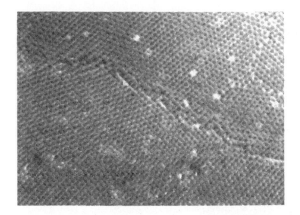

Figure 5   A momentary coexistence of a metastable particle bilayer and monolayer during the thinning of a film with high concentration of particles. The transition region is seen between the top left side and bottom right side of the frame. The upper region is the monolayer.

sharp transition boundary between them (Fig. 5). Unlike stratification however, the formed bilayer was highly metastable and the transition to a monolayer was very quick and uncontrollable.

IV.   DISCUSSION

A.   Colloidal Chemistry Aspects

Our data reveal the three major parameters controlling the particle dynamics in the film: the particle–surface interactions, the capillary forces, and the interfacial viscoelasticity. The capillary interactions play a major role in the 2D crystallization observed, unlike the systems studied earlier [16–18], where the assembly was executed mainly by the convective flux caused by water evaporation. To estimate the energy of capillary attraction between spherical latex particles confined within a free TLF, we slightly modified the theory of Kralchevsky and coworkers [26–29] for capillary interactions between equal spheres immersed in a film on a solid substrate. We assume that the total capillary interaction energy in foam films is two times the corresponding energy between particles on a solid substrate with one liquid interface. This assumption is invalid in the presence of disjoining pressure inside the film, but it should hold well in our case, where the films are thicker than a few micrometers. The parameters used were interfacial tension $\sigma = 54$ mN/m (determined independently by tensiometry) and contact angle $\alpha = 41°$ as measured by others [31]. The results for the capillary interaction energy normalized by the thermal energy k_bT for films of different thicknesses are shown in Figure 6. In the films studied, the surfaces meet the latex particles without formation of a curved meniscus at a thickness of ≈ 5 μm, and no interactions are present at this thickness. Capillary interaction could occur both above and below this thickness, but the menisci around the particles have different curvatures (see the top and bottom insets in Fig. 6). Notably, both of these interactions are attractive and it is seen from the figure that the interaction energy can increase to many orders of magnitude above k_bT for distances as large as 100

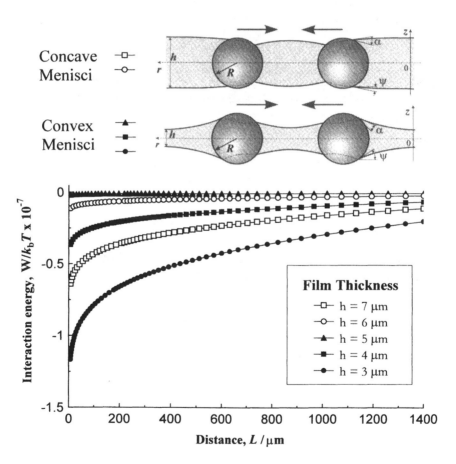

Figure 6 Calculated capillary interaction energy acting between two microspheres of radius 3.5 μm for different values of the film thickness h and for the cases of concave and convex menisci. The interactions are always attractive and can explain the 2D aggregation, but not the difference between the stable and unstable formations.

times the particle diameters. Such highly attractive interactions should cause a quick approach and 2D ordering of the spheres. This is in agreement with the experimental data and explains the strong long-ranged attraction and aggregation observed in films with BSA or cationic surfactants.

An important stability factor to be considered is the capillary interaction between the particles and the curved menisci surrounding the film. The particle meniscus–cell meniscus interaction can be attractive or repulsive depending on the sign of $\sin \psi_1 \sin \psi_2$ (where ψ_1 and ψ_2 denote the slope angles at the cell and particle menisci, respectively). The cell is composed of hydrophilic glass and the film meniscus is always "convex" (i.e., $\sin \psi_1$ is always negative). The meniscus around the particles is convex ($\sin \psi_2 < 0$) for hydrophilic particles and concave ($\sin \psi_2 > 0$) for hydrophobic particles attached to the surfaces of a thicker film. According to the theory [32], when $\sin \psi_1 \sin \psi_2 > 0$, protruding particles will be attracted to the film periphery. This explains why the 2D particle aggregates are expelled out in thinning SDS or protein-stabilized films (Fig. 7A). As observed, in this case only metastable structures may be formed in films with high interfacial viscoelasticity. On the contrary, the interaction will be repulsive ($\sin \psi_1 \sin \psi_2 < 0$) and the particles will be pushed toward the film center in the case of thick films with concave menisci around partially hydrophobic particles (Fig. 7B). This is exactly the case observed with thickening DTAB or HTAB films, where the particles adhered between the surfaces were repelled from the film periphery and compressed in the center. In summary, although the particle–particle capillary attraction always enhances the 2D crystallization, intrinsically stable structures are obtained only in films with repulsive capillary forces between the particles and the meniscus.

Besides the crystallization aspects discussed in Section IV.B, the "zipping" mechanism of structure formation may be relevant to stabilization and flocculation of Pickering emulsions. It has been known for a long time [1] that the stabilization of emulsions by solid particles is possible only when the particles are partially hydrophobic, as is the case with our latexes in the presence of HTAB. Our work shows that this is the only system in which stable particulate structures were formed inside the film. We, therefore, believe that organized particle layers, similar to the ones observed with hydrophobized particles, may be formed between the droplets of Pickering emulsions, playing a major role in their rheological behavior and stability against coalescence.

B. Colloidal Crystallization Aspects

The experiments with cationic surfactants have suggested a new method for controlled growth of 2D ordered arrays via multiple compression of the three-phase contact line in free foam films. By using this method, we believe we are the first to obtain single-crystalline arrays larger than 1000 particles/domain from parti-

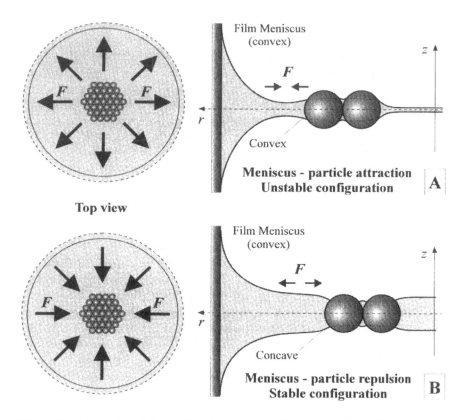

Figure 7 Schematics of the capillary interactions acting between the particle arrays and the meniscus around the film. The arrays are in a stable configuration only when the particles are partially hydrophobic, as is the case in the presence of cationic surfactant.

cles larger than a few micrometers. The formation of such high-quality arrays from large latex microspheres by alternative methods on a solid substrate is close to impossible due to the defects caused by particles adhering to the solid substrate. Our method appears technologically more complicated than the ones for 2D assembly reported earlier [16,19], but it has the potential advantage of creating single-crystalline arrays of larger size. Its practical application for the fabrication of 2D arrays requires the exploration and development of procedures that allow the extraction of the assembled layers as separate solid-state materials. One procedure that can be used to achieve this is the ultrafast freezing of the foam films by, for example, quick plunging in liquefied gasses [33]. Once the film with the ordered particles inside is frozen, it can be detached from the cell and placed on an appropriate substrate and the water can be removed by melting or sublimation.

V. CONCLUDING REMARKS

The investigations reported here outline three major cases of dynamics of micrometer-sized latex particles inside free foam films: unstable configurations in films with anionic surfactant, formation of metastable structures in the presence of protein, and formation of stable layers in films containing cationic surfactants. We expect that these three cases are relevant and can be observed in a wide variety of particle-stabilized foams and emulsions. In addition, a new method for controlled growth of 2D crystals by meniscus pulsation inside thin foam films has been suggested.

ACKNOWLEDGMENTS

This material was presented at the Special Symposium honoring the 60th birthday of Professor Darsh Wasan. The authors would like to acknowledge Professor Wasan's advise, motivation, and support to many young scientists in this area of research. The authors also gratefully acknowledge the support of Professor F. Durst and the financial support of this collaborative research by the "Volkswagen-Stiftung." K.P.V. is grateful for a Ph.D. fellowship from FOM, a part of Netherlands Organization for Scientific Research (NWO).

REFERENCES

1. Th F Tadros, B Vincent. In: P Becher, ed. Encyclopedia of Emulsion Technology. New York: Marcel Dekker, 1983, Chap. 3.
2. DE Tambe, MM Sharma. Adv Colloid Interf Sci 52:1, 1994.
3. ND Denkov, IB Ivanov, PA Kralchevsky, DT Wasan. J Colloid Interf Sci 150:589, 1992.
4. R Aveyard, BP Binks, PDI Fletcher, TG Peck, CE Rutherford. Adv Colloid Interf Sci 48:93, 1994.
5. OD Velev, AD Nikolov, ND Denkov, G Doxastakis, V Kiosseoglu, G Stalidis. Food Hydrocolloids 7:55, 1993.
6. AD Nikolov, DT Wasan, PA Kralchevsky, IB Ivanov. In: N Ilse, I Sogami, eds. Ordering and Organization in Ionic Solutions. Teaneck, NJ: World Scientific, 1988, p. 302.
7. AD Nikolov, DT Wasan, PA Kralchevsky, IB Ivanov. J Colloid Interf Sci 133:1, 13, 1989.
8. AD Nikolov, DT Wasan, ND Denkov, PA Kralchevsky, IB Ivanov. Prog Colloid Polym Sci 82:87, 1990.
9. V Bergeron, CJ Radke. Langmuir 8:3020, 1992.
10. AD Nikolov, DT Wasan. Colloids Surfaces 128:243, 1997.

11. DT Wasan, AD Nikolov, PA Kralchevsky. Colloids Surfaces A 67:139, 1992.
12. K Koczo, AD Nikolov, DT Wasan, RP Borwankar, A Gonsalves. J Colloid Interface Sci 178:694, 1996.
13. G Sethu, AD Nikolov, DT Wasan. Proc. 1998 Annual AIChE Meeting, 1998.
14. S Hayashi, T Kumamoto, T Suzuki, T Hirai. J Colloid Interf Sci 144:538, 1991.
15. K Nagayama. Nanobiology 1:25, 1992.
16. ND Denkov, OD Velev, PA Kralchevsky, IB Ivanov, H Yoshimura, K Nagayama. Nature 361:26, 1993.
17. ND Denkov, OD Velev, PA Kralchevsky, IB Ivanov, H Yoshimura, K Nagayama. Langmuir 8:3183, 1992.
18. GS Lazarov, ND Denkov, OD Velev, PA Kralchevsky, K Nagayama. J Chem Soc Faraday Trans 90:2077, 1994.
19. K Nagayama. Colloids Surfaces A 109:363, 1996.
20. A Scheludko. Adv Colloid Interf Sci 1:391, 1967; H Sonntag, K Strenge. Coagulation and Stability of Disperse Systems. New York: Halsted Press, 1972; D Exerowa, D Kashchiev, D Platikanov. Adv Colloid Interf Sci 30:429, 1992.
21. OD Velev, TD Gurkov, Sv K Chakarova, BI Dimitrova, IB Ivanov, RP Borwankar. Colloids Surfaces A 83:43, 1994.
22. DE Graham and MC Phillips. J Colloid Interf Sci 70:3, 1979.
23. Th Peters. Adv Protein Chem 29:205, 1969.
24. JV Boyd, JR Mitchell, L Irons, PR Musselwhite, P Sherman. J Colloid Interf Sci 45:478, 1973.
25. IB Ivanov. Pure Appl Chem 52:1241, 1980.
26. PA Kralchevsky, VN Paunov, IB Ivanov, K Nagayama. J Colloid Interf Sci 151:79, 1992.
27. PA Kralchevsky, VN Paunov, ND Denkov, IB Ivanov, K Nagayama. J Colloid Interf Sci 155:420, 1993.
28. PA Kralchevsky, K Nagayama. Langmuir 10:23, 1994.
29. VN Paunov. Langmuir 14:5088, 1998.
30. OD Velev, K Furusawa, K Nagayama. Langmuir 12:2374, 2385, 1996.
31. A Hadjiiski, R Dimova, ND Denkov, IB Ivanov, R Borwankar. Langmuir 12:6665, 1996.
32. PA Kralchevsky, ND Denkov, VN Paunov, OD Velev, IB Ivanov, H Yoshimura, K Nagayama. J Phys Condensed Matter 6:A395, 1994.
33. ND Denkov, H Yoshimura, K Nagayama. Phys Rev Lett 13:2354, 1996; ND Denkov, H Yoshimura, K Nagayama. Ultramicroscopy 65:147, 1996.

17

Adsorption and Exchange of Whey Proteins onto Spread Lipid Monolayers

Michel Cornec, Daechul Cho, and Ganesan Narsimhan
Purdue University, West Lafayette, Indiana

ABSTRACT

The dynamics of surface pressure (π) and of surface concentration (Γ) of ^{14}C radiolabeled whey proteins (Bovine Serum Albumin (BSA) and β-lactoglobulin) adsorbed onto spread monolayers of lipid (egg yolk lecithin or 1-monopalmitoyl-rac glycerol) at the air–water interface was measured. The adsorption of BSA onto spread lecithin monolayers of 1.07 and 0.64 nm^2/molecule was enhanced at short times, indicating synergism due to possible dissolution of BSA molecules into loosely or moderately packed lecithin layers. Both surface pressure and surface concentration were increased in the presence of the spread lecithin monolayer. However, at longer times, lecithin was found to be dominant in the monolayer. The rate of adsorption of ^{14}C labeled β-lactoglobulin was enhanced onto spread monolayer of C-16 monoglyceride of 0.19 and 0.27 nm^2/molecule at short times. However, the amounts of protein adsorbed after 10 h were lower with the values being 0.8 and 1.2 mg/m^2 respectively. Spreading of lecithin or monoglyceride monolayer of a close packed area after 2$^1/_2$ h onto air–water interface with adsorbed BSA or β-lactoglobulin led to complete displacement of the protein from the interface, possibly because of the surface pressure and exclusion effects.

I. INTRODUCTION

Many food products such as salad dressing, mayonnaise, and ice cream are foams and emulsions. Such dispersions are thermodynamically unstable. However, their stability may be enhanced by the adsorption of amphiphilic molecules to the surface of the dispersed phase [1,2]. These molecules generally fall in two classes: macromolecules, such as proteins, and low-molecular-weight surfactants, such as monoglycerides and phospholipids. Surfactants adsorb and orient at fluid–fluid interfaces and thus reduce the interfacial tension between the phases. Proteins form a condensed viscoelastic film of highly self-interacting molecules at the interface, which resists local deformation [3,4]. In contrast, surfactants form a fluid adsorbed layer in which adsorbed molecules can diffuse laterally toward regions of high surface tension conferring stability via the Marangoni effect [5]. Individually, the viscoelastic and Marangoni mechanisms are very effective at stabilizing foams and emulsions but are mutually incompatible. Low-molecular-weight surfactants, because of their much lower molecular weight, pack more efficiently at the interface and thus reduce the interfacial tension to a greater extent than proteins [6]. On the other hand, intermolecular interactions between molecules of a low-molecular-weight surfactant are much weaker and development of high mechanical strength is not possible [7]. For these reasons, both classes of molecules are used in food systems. This may cause a problem, as competition between the two mechanisms arises, leading to instability of the system [8–26].

Despite a considerable amount of work (reviewed in Refs. 27–29), little is known, in a quantitative sense, about the molecular mechanisms of protein–surfactant interactions at the surface. Surface properties (structure, stability, and mechanical and dynamic properties) of protein–lipid complexes depend on the mechanisms of protein–lipid interactions at the surface. Reported mechanisms for protein–surfactant interactions involved either electrostatic interactions between ionic surfactant headgroup and the charged macromolecules [30–37] and/or hydrophobic interactions [38,39]. Electrostatic interaction were studied by Cornell and co-workers [30,40]. Using the film balance technique, they investigated the adsorption of various whey proteins from solution onto a spread monolayer of phospholipids. They observed that the amount of protein bound to the lipid was more important when both molecules carried opposite net charges and this amount was decreased with an increase of the ionic strength of the solution. They also found that penetration of protein into a mixed monolayer of phospholipids could occur at a higher surface pressure than the equilibrium surface pressure obtained for the adsorption of the proteins and they concluded that the formation of pure protein patches was unlikely and that portions of the protein were intercalated in the lipid monolayer. On the other hand, Du and co-workers [39] showed that the penetration of an enzyme, the glucooxydase, into a glycolipid monolayer was easier for the longer aliphatic chain of the glycolipid. Their results suggest that the hydrophobic interaction is the predominant force in the interactions. The formation of protein–lipid complexes may result in a structural change of the former component [41,42], which can affect the structure of the adsorbed layer as well as the film formation capacity and, thus, the formation and stability of food emulsions and foams. In addition, the physical state of the lipid phase has been shown to be important for film formation. It was observed that protein adsorbed to a greater extent in expanded lipid monolayers than in condensed ones [43]. In addition, it was observed that the lipid which formed condensed films at the air–water interface did not form homogeneous films with proteins. On the other hand, those exhibiting liquid-expanded monolayer behavior tended to produce well-mixed monolayers with proteins [44].

Most of the previous work on mixed lipid–protein monolayers has focused on the interactions between protein and polar phospholipids. Less attention has been directed toward mixed monolayers formed with protein and monoglycerides [45,46] despite the fact that they are abundantly used in food systems [47,48], with the major uses in baked goods as cake emulsifiers, crumb softeners, and dough strengtheners [48]. They are also commonly used as stabilizers for margarine and diet spreads and in icing and whipped toppings. They help increase the water absorption in an icing and improve the aeration properties, and in extruded products such as cereals, pastas, and pet foods, they provide some lubrication during extrusion processing. In addition, they are used as dispersing agents in coffee whiteners and as a crystal modifier in confectionery or compound coatings.

In this study, a mixed system of β-lactoglobulin and 1-monopalmitoyl-rac glycerol was studied. Changes in surface concentration of the proteins were probed by using ^{14}C-radiolabeled proteins. Results were compared with those obtained with another system composed of [^{14}C] BSA and egg-yolk lecithin.

II. EXPERIMENT

A. Materials and Apparatus

Bovine serum albumin (BSA), β-lactoglobulin, 1 monopalmitoyl-rac glycerol (C16:0) and egg-yolk lecithin (with L-α headgroup) were purchased from Sigma Chemical Inc. (St. Louis, MO). Reagent-grade n-hexane and ethanol were purchased from Aldrich Chemical Co. Monoglyceride solutions were made in hexane–ethanol (v/v = 9:1) and were used within 2 days. Isotopes of [^{14}C]formaldehyde (37.3%) and [^{14}C]sodium stearate were purchased from Sigma Chemical Inc. Sodium cyanoborohydride (NaCNBH$_3$) was purchased from Aldrich Chemical Inc. All the experiments were carried out at pH 7.4 using a 10 mM commercial sodium phosphate buffer containing 0.9% NaCl. Throughout the experiments, ultrapure deionized water was used.

A Langmuir minitrough (with dimensions of $330 \times 75 \times 6.5$ mm) from KSV Instruments (Helsinki, Finland) was used for both surface pressure and surface concentration measurements. A gas proportional detector (Ludlum model 120, with a 2×2-in. Mylar window) with a digital scaler/counter (Ludlum model 520) was used for detecting radioactivity, in counts per minute (cpm), from the adsorbed monolayer at the air–water interface. Radioactivity was measured under P-10 gas (10% methane in argon) flowing at 55 mL/min through the detector chamber.

B. Methods

1. Radiolabeling of Proteins

β-Lactoglobulin or BSA (30 mg) were dissolved in 0.05 M phosphate buffer (pH 7) and mixed with 0.1 M sodium cyanoborohydride and [^{14}C]formaldehyde (102 μCi) and allowed to react for 2 h at room temperature [49]. After the reaction, the mixture was dialyzed against a 0.05 M phosphate buffer for 30 h at 4°C. for complete removal of unreacted species. BSA and β-lactoglobulin were found to have respectively 3.1 and 1 amide group labeled per molecule on the average (2.14 and 2.52 μCi/mg of protein, respectively), as analyzed with a scintillation counter (Tri-carb 4000, from Packard Instruments). Because the degree of modification due to radiolabeling is small, it was assumed that the surface properties of the protein were not significantly affected. Comparison of the spread monolayer isotherm of native and radiolabeled bovine serum albumin as reported by Cho et al. [50] indicated no significant differences in the surface activity between the two because of radiolabeling. In addition, Graham and Phillips [51] showed that radiolabeling did not affect the surface activity of different proteins by comparing the adsorption data at air–water interfaces using a radiotracer and ellipsometry.

2. Surface Pressure–Molecular Area Isotherms

The Langmuir trough was first filled with phosphate buffer. Then, the surface was cleaned by sweeping it with the Teflon barrier and any surface-active contaminants were removed by suction (aspiration) of the interface. The lipid (lecithin or monoglyceride) was spread over the clean air–water interface by applying the solution dropwise from a Hamilton syringe. Protein spread monolayers were prepared using the Trurnit's monolayer spreading method [52]. Aliquots of 50 mL of a 0.0247 wt% protein solution were dripped from the top of a glass rod (5 mm in diameter and 5 cm long) positioned across the air–water interface. The solution was spread uniformly on the interface. As detailed previously, there was a negligible loss of protein to the bulk due to desorption [53]. Consequently, it was assumed that all the proteins spread adsorbed at the air–water interface.

For mixed monolayers, first a protein (BSA or β-lactoglobulin) monolayer was formed with Trurnit's method and allowed to rest for 10 min. Then, various aliquots of lipid solution in hexane–ethanol were spread at several spots on the surface. The monolayers were allowed to equilibrate for another 10 min. Then, the surface area was compressed by moving the minitrough Teflon barrier at a constant speed of 4 cm^2/min and the surface pressure was continuously recorded. Immediately after the end of the compression stage, the area was expanded at the same rate.

3. Adsorption from Solution

The adsorption of [^{14}C] proteins onto spread lipid monolayers was studied as described previously [50]. In brief, the trough was filled with the buffer solution without any protein (surface tension = γ_0) and the surface was carefully aspirated to remove any surface impurities before the surface pressure was adjusted to zero ($\pi = \gamma_0 - \gamma$). Then, the protein solution was gently poured into the trough. After that, a lipid monolayer was spread using a Hamilton syringe and the surface pressure and surface concentration (via radioactivity measurements) were monitored up to 10 h. In another set of experiments, the lipid was spread on top of adsorbed protein monolayers prepared by allowing the proteins to adsorb for 150 min.

4. Calibration of Ludlum Gas Proportional Detector

The counts per minute were converted to surface concentrations by calibrating the Ludlum gas proportional detector with radioactive samples of known surface or bulk concentrations, as described previously [50]. The bulk radioactivity calibration procedure of Hunter et al. [49] was employed for background correction using ^{14}CH$_3$COONa. [^{14}C] Proteins (BSA or β-lactoglobulin), rather than [^{14}C]stearic acid, was used for calibration of the surface radioactivity, because the use of [^{14}C]stearic acid tends to underestimate the surface concentration due to its much smaller molecular size compared to the proteins [50,54]. A total of 175 mL of 0.01 M phosphate buffer (pH 7, containing 0.9% NaCl) was poured into the Langmuir minitrough and a spread monolayer of [^{14}C] protein was formed using Trurnit's method The detector was placed at a distance of 3 mm above the air–water interface and the steady-state counts per minute was measured. The area of the air–water interface was compressed in stages to provide several surface radioactivities, which were used, for the detector calibration.

III. RESULTS

A. Surface Isotherms

The surface pressure–molecular area (π–A) isotherms of lecithin, BSA, 1-mono-palmitoyl-rac glycerol, and β-lactoglobulin are shown in Figs. 1a–1d, respec-

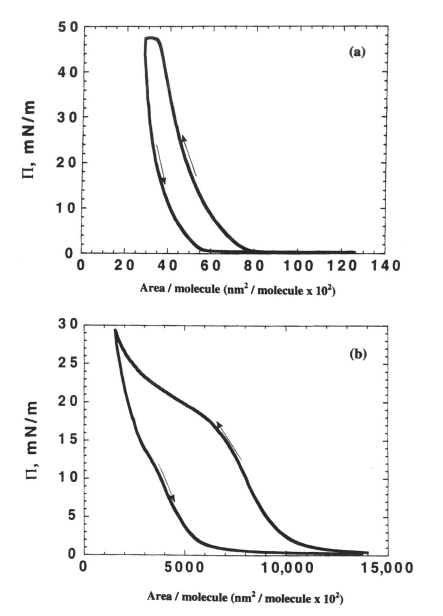

Figure 1 π–$\overline{A}$ (area per molecule) isotherm of (a) egg lecithin, (b) BSA, (c) 1 monopal-mitoyl-rac glycerol (C16: 0), and (d) β-lactoglobulin. (Fig. 1a and 1b reprinted with permission from Langmuir, 13:4710, 1997. Copyright 1997, American Chemical Society.)

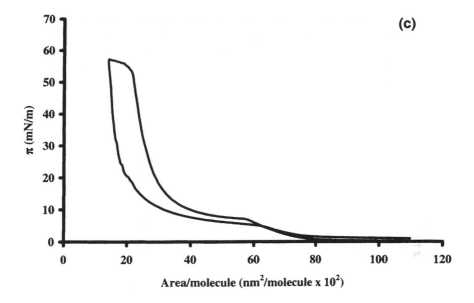

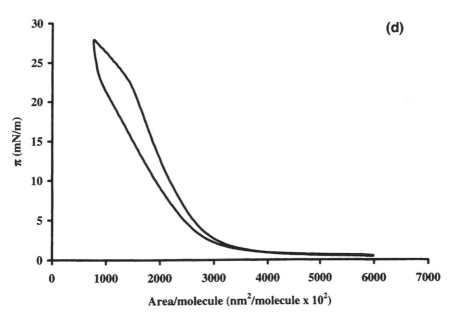

tively. The BSA monolayers were gaslike at an area greater than 100 nm²/molecule, beyond which an apparent phase transition was noted to liquidlike with much lower monolayer compressibility. Another pronounced transition, probably to a more condensed monolayer phase, seems to occur for $\bar{A}$ ranging from 70 to 30 nm²/molecule, $\pi \approx 18$–30 mN/m. Upon compression, molecules of BSA are forced to adopt a close-packed form with only solvated polar amino acids submerged in the subphase. At a higher surface pressure (> 25 mN/m), irreversible desorption of the protein to the bulk phase or irreversible surface coagulation may occur [55,56]. The value of $\bar{A} = 15$ nm²/molecule at the maximum surface pressure (30 mN/m) is close to the theoretical minimum area, $A_{min} = 12.50$ nm²/molecule (cross-sectional area of prolate spheroid of dimensions $4 \times 4 \times 14$ nm) where the protein molecules are expected to be close packed and oriented with their major axis normal to the surface. The same trend was observed in the case of β-lactoglobulin. The β-lactoglobulin monolayer behaved gaslike for $\bar{A}$ greater than 40 nm²/molecule and liquidlike for $\bar{A}$ between 14 and 27 nm²/molecule. At $\pi \approx 22$ mN/m, an inflection point is observed in the isotherm. Above this π, whole protein molecules or segments of the protein are probably squeezed down into the subphase. However, the fact that the compressed monolayer was able to recover its initial state upon decompression and that no shift in the molecular area did occur upon subsequent compression ruled out massive loss of polypeptide molecule to the subphase. However, the observed hysteresis suggests that the recovery is slow. As observed for BSA, the value of $\bar{A}$ at maximum compression (7.63 nm²/molecule at $\pi = 27$ mN/m) was close to the theoretical minimum area ($\bar{A} = 10$ nm²/molecule for a sphere of dimensions 3.58×3.58 nm).

Lecithin monolayers showed small surface pressures for $\bar{A}$ ranging from 1.26 to ~ 0.80 nm²/molecule. Upon further compression, π increased quickly until it reached a plateau at $\bar{A} = 0.36$ nm²/molecule and $\pi_{max} = 41$ mN/m. The isotherm suggests that the lecithin monolayer was gaslike in the first range and liquidlike in the second. A monolayer collapse was observed at π_{max}. The monolayer also exhibited hysteresis, with the molecular areas being smaller at the same π for the expansion cycle. The isotherm of the C16 monoglyceride monolayer showed a gradual increase in the surface pressure until a plateau occurred at $\bar{A}$ between 0.40 and 0.60 nm²/molecule. The plateau reveals a phase transition from a liquid-expanded (LE) state to a liquid-condensed (LC) state. Gehlert et al. [57], using Brewster angle microscopy (BAM), observed, in the region corresponding to the beginning of the plateau some condensed-phase domains surrounded by a homogeneous fluid phase of low density. As the molecular area was decreased, the domains grew in area at the expense of the fluid phase. Upon further compression, the domains started to overlap, filling the gaps in the condensed phase, which was accompanied by an increase in the surface pressure. Above a surface pressure of 25 mN/m, the domains were compressed to close packing without a visible gap. Hysteresis was also observed for C16 monoglyceride monolayers, with $\bar{A}$ being smaller at a certain π during the expansion than the compression cycle.

The compression cycle of the π–$\bar{A}$ isotherms of the mixed layers formed with BSA and lecithin were found to appear similar to that of the BSA monolayer. However, the specific area at which each monolayer transition occurred depended on the composition, and the condensed phase was observed for a wider range of molecular areas for a lower-molecular, weight ratio of lecithin to BSA ($R_{L/B}$) [50]. Whereas lecithin-rich monolayers exhibited smoother phase transitions, BSA-rich monolayers exhibited sharper transitions with smaller slopes. Expansion of the mixed monolayer led to a strong hysteresis behavior (Fig. 2), with average molar areas being smaller for the expansion cycle than for the compression cycle. The surface pressure reached at maximum compression was closer

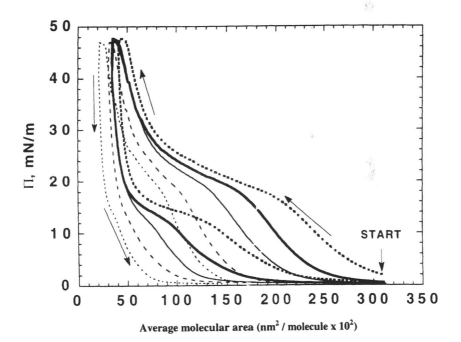

Figure 2 π–$\bar{A}$ isotherms of mixed monolayers for varied mass ratio of lecithin to BSA: (---) $R_{L/B} = 0.5$; (—) $R_{L/B} = 0.67$; (——) $R_{L/B} = 1.0$; (- - -) $R_{L/B} = 1.5$; (···) $R_{L/B} = 2.0$ The average molecular area of the mixture, $\bar{A} = A/(n_L + n_{BSA})$, where A is the total area and n_L and n_{BSA} are the total number of molecules of lecithin and BSA, respectively. (Reprinted with permission from *Langmuir*, 13:4710, 1997. Copyright 1997, American Chemical Society.)

to that of the lecithin monolayer than that of the BSA monolayer. This further indicated that lecithin seemed to be the main contributor to the surface pressure. This was more pronounced at the close-packed limit, at which BSA might have been expelled from the surface layer. Mixed-monolayer isotherms are shown in Figure 3 for weight ratios of 1-monopalmitoyl-rac glycerol to β-lactoglobulin ($R_{C16/\beta\text{-lg}}$) from 3 to 0.33. Analysis of the shape of the isotherms obtained shows the effect of the composition on the surface pressure and on the area during compression. A LE–LC transition is visible in all isotherms at a π of ∼7–8 mN/m. The shift of the transition to higher areas when the proportion of β-lactoglobulin is increased points out the presence of the two constituents at the interface. Nevertheless, the conservation of the LE–LC transition is evidence of the great influence of the monoglyceride on the properties of the mixed monolayer during compression [58,59]. At higher surface pressures, the mixed-monolayer isotherms exhibited another transition, which was also present in the isotherm of the pure

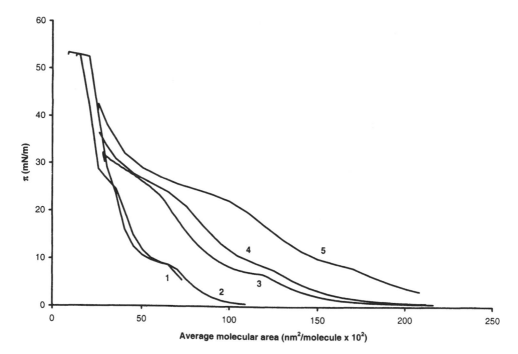

Figure 3 π-$\overline{A}$ isotherms of mixed monolayers for varied mass ratio of 1-monopalmi-toyl-rac–glycerol to β-lactoglobulin; curve 1: $R_{C16/Blg}$ = 3; curve 2: $R_{C16/Blg}$ = 2; curve 3: $R_{C16/Blg}$ = 1; curve 4: $R_{C16/Blg}$ = 0.5, curve 5: $R_{C16/Blg}$ = 0.33. The average molecular area of the mixture, $\overline{A}$ = $A/(n_L + n_{Blg})$, where A is the total area and n_L and n_{Blg} are the total number of molecules of 1-monopalmitoyl-rac–glycerol and β-lactoglobulin, respectively.

β-lactoglobulin. C16–monoglyceride-rich monolayers exhibited smoother phase transitions than β-lactoglobulin-rich monolayers. At the lowest $R_{C16/\beta\text{-lg}}$, the transition was reduced to a kink, which may correspond to the expulsion of the protein molecules from the mixed film resulting in the reduction of the compressibility of the film to a level similar to that of pure monoglyceride. As the film was further compressed, the isotherms of the mixed monolayer asymptotically approached that of the monoglyceride monolayer. At a high enough surface pressure, the isotherms for different compositions of the mixed monolayer coincided, which indicated that the β-lactoglobulin molecule was eventually completely squeezed out from the 1-monopalmitoyl-rac glycerol monolayer. However, the collapse surface pressure was observed to be lower for mixed monolayers than for pure monoglycerides. Hysteresis was observed for the C16 monoglyceride and β-lactoglobulin mixed monolayers (results not shown), with the average molar areas being smaller during the compression cycle.

B. Adsorption and Exchange for Protein–Lipid Surfactant Mixtures

Interactions of proteins and spread lipid monolayers were investigated by spreading a lipid monolayer on an adsorbing protein solution. First, control experiments were performed in order to validate the methodology used in this study. Spreading of solvent only (hexane–ethanol mixture without any lipid) on an adsorbing protein solution resulted in a small overshoot in π but no significant change in Γ was noted. It was concluded that the spreading solvent did not significantly affect protein adsorption. The second control experiment consisted in spreading 50 µg of lipid in solvent on a clean interface. A sudden increase in π with a small overshoot was observed. Because π was observed to remain constant after spreading, it was concluded that the spread lipid did not experience any desorption. Therefore, no attempt was made to monitor separately the lipid surface concentration, which was assumed to be constant.

1. BSA–Lecithin Mixtures

Immediately following spreading of lecithin (50 mg, 0.64 nm²/molecule) on a nearly clean surface of [¹⁴C] BSA solution (Fig. 4), the surface pressure jumped to 15 mN/m within 1 s. Then, π increased gradually until it reached a plateau at 21 mN/m. A rapid increase in the surface concentration of [¹⁴C] BSA (to 0.4 and 0.1 mg/m² for a BSA bulk concentration of 1.25 and 2.5×10^{-4} wt%, respectively) was also observed within 1 min following the spreading of the lipid monolayer. No further increase in Γ was observed. This initial rapid increase in Γ was not due to any interaction between the BSA molecule and the solvent used for spreading the lipid monolayer, because spreading of the solvent alone was ob-

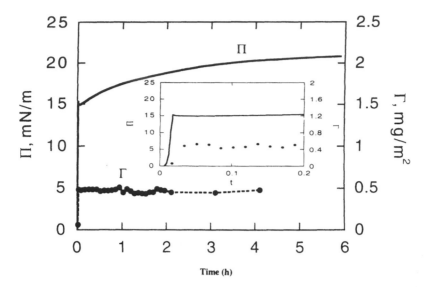

Figure 4   Time-dependent changes of π (surface pressure) and Γ (surface concentration) when a solution of [^{14}C] BSA (c_b = 1.25 ppm) was poured in the trough, followed by immediate spreading of lecithin (50 μg or 0.64 nm^2/molecule). Inset: Time-expanded π(t) plot. (Reprinted with permission from *Langmuir*, 13:4710, 1997. Copyright 1997, American Chemical Society.)

served not to significantly affect the evolution of both π and Γ with time. Therefore, it was concluded that the presence of the lecithin monolayer at the interface enhanced the adsorption of the BSA in a short time. Indeed, the coefficient of diffusion of the protein was estimated to be 0.25 mg/m^2/min or, about 11 times faster than that when BSA adsorbed on a clean interface. It is to be noted that lecithin and BSA increased the surface pressure synergistically, at least for short times, because the sum of the surface pressures for lecithin and BSA alone was less than that of the mixture for short times and gradually approached the mixture curve for longer times (Fig. 5).

To test how an adsorbed BSA monolayer would respond to spreading of a lecithin monolayer, the spreading was delayed by 2 h, when the surface monolayer was nearly at steady state. It was observed that the effect of the spread lecithin monolayer on the protein adsorption depended on the amount of lipid spread. When 30 μg (1.07 nm^2/molecule) of lecithin was spread, BSA adsorption was enhanced. If this amount was increased to 50 μg (0.64 nm^2/molecule), the protein surface concentration dropped slowly from 1.2 to 0.76 mg/m^2. On the other hand, π surged up to 22 mN/m (Fig. 6). Spreading of more lecithin (80

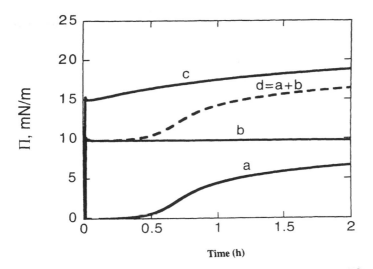

Figure 5 Evolution of π for lecithin monolayer, pure BSA solution, and BSA solution with a spread lecithin monolayer. In the case of the lecithin monolayer, time refers to the time after spreading. In the case of the BSA solution, time refers to the time elapsed after aspiration of the interface. Curves a and b denote the π(t) of BSA and lecithin, respectively. Curve c is for the mixture. Curve d is the sum of curves a and b. (Reprinted with permission from Langmuir, 13:4710, 1997. Copyright 1997, American Chemical Society.)

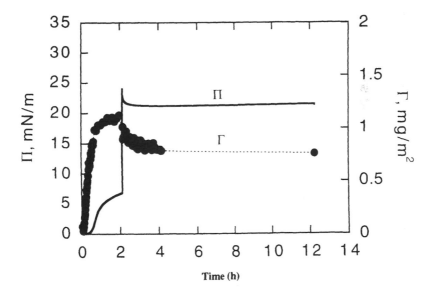

Figure 6 Time-dependent changes of π and Γ when 50 μg (0.64 nm²/molecule) of lecithin were spread on the surface of an adsorbing solution of [¹⁴C] BSA (c_b = 1.25 ppm) after 120 min of adsorption. (Reprinted with permission from Langmuir, 13:4710, 1997. Copyright 1997, American Chemical Society)

μg, 0.39 nm²/molecule) caused a faster and nearly complete desorption of BSA from the interface with a corresponding increase in π from 5 to 45 mN/m. In summary, if spreading small amounts of lecithin seemed to enhance BSA adsorption, large amounts induced displacement of the protein.

2. β-Lactoglobulin–1-Monopalmitoyl-rac Glycerol Mixtures

Spreading 12.35 μg (1.10 nm²/molecule) of 1-monopalmitoyl-rac glycerol on a nearly clean interface was observed to slightly enhance the adsorption of β-lactoglobulin (Γ = 1.82 mg/m² instead of 1.63 mg/m² when no lipid monolayer was at the interface). As observed previously, the initial rate of adsorption was increased. Increasing the amount of C16 monoglyceride spread at the surface caused the steady-state surface concentration of β-lactoglobulin to decrease. Indeed, Γ was 1.19 and 0.86 mg/m² respectively when 24 μg (0.275 nm²/molecule) and 73 μg (0.19 nm²/molecule) were spread (Fig. 7). However, it is to be noted

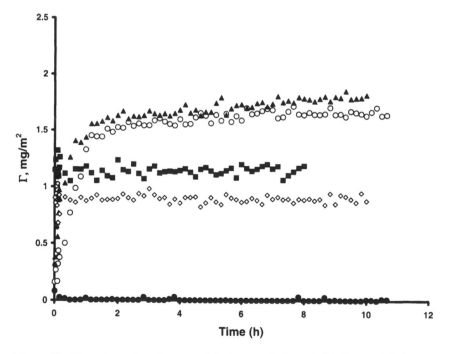

Figure 7 Time-dependent changes of Γ when a solution of [¹⁴C]β-lactoglobulin (c_b = 1 ppm) was poured in the trough, followed by immediate spreading of 1-monopalmitoyl-rac glycerol (C16:0): ○: no monoglyceride spread; ▲: 12.35 μg (1.10 nm²/molecule), ■: 24.5 μg (0.275 nm²/molecule), ◇: 71 μg (0.19 nm²/molecule). ●: 113 μg (0.12 nm²/molecule)

that the initial rate of adsorption was faster as the spread amount of monoglyceride was increased. Protein was excluded from the interface when a close-packed monolayer (0.12 nm²/molecule) of C16 monoglyceride was present at the interface.

The response to an adsorbed β-lactoglobulin monolayer to the spreading of a monoglyceride monolayer was also tested by spreading different amounts of monoglycerides on an adsorbing protein solution. Results are shown in Figure 8. As observed previously for BSA–lecithin mixtures, the effect of the spread monoglyceride monolayer on the adsorption of β-lactoglobulin depended on the amount of monoglyceride spread. Protein adsorption was found to be enhanced when 24 μg (0.27 nm²/molecule) was spread at the surface. If this amount was increased to 73 μg (0.19 nm²/molecule), the amount of protein adsorbed was first increased, followed by a sudden desorption, and Γ reached a steady-state value of 1.40 mg/m². β-Lactoglobulin was found to be displaced from the interface by spreading 113 μg (0.12 nm²/molecule) of monoglyceride.

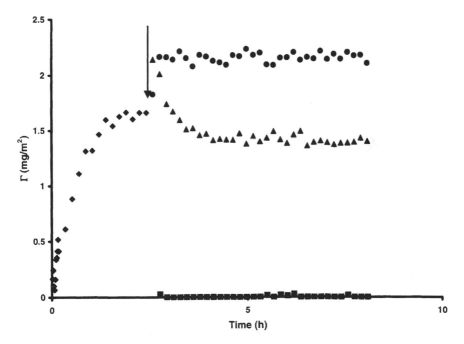

Figure 8 Time-dependent changes of Γ for 1 ppm [¹⁴C]β-lactoglobulin (c_b = 1 ppm) solution. At t = 150 min, (●) 24.5 μg (0.275 nm²/molecule), (▲) 71 μg (0.19 nm²/molecule), and (■) 113 μg (0.12 nm²/molecule) of 1-monopalmitoyl-rac glycerol (C16: 0) was spread at the surface.

IV. DISCUSSION

Spreading a lipid monolayer (lecithin or monglyceride) onto an adsorbing protein solution (BSA or β-lactoglobulin) resulted in an enhancement in the adsorption of the protein soon after the lipid surfactant was spread. This was evidenced by the increase in the initial rate of the protein adsorption. However, when the amount of lipid spread at the interface was increased, the steady-state surface concentration of the protein was observed to be lower than the one observed for protein when no surfactant was at the interface.

The initial enhancement in the protien adsorption, soon after spreading of the lipid, may be due to the formation of a complex between the hydrophobic patches of the protein and the hydrophobic chains of the lipid molecules [39,60]. Such a behavior can be explained in terms of the adsorption energy E_{ad} of a protein molecule [50], which can be written as

$$E_{ad} = E_{sp} + E_{el} + E_{hy}$$

where E_{sp} is the work that needs to be done by an adsorbing protein molecule to anchor itself at the interface and acts as the surface pressure energy barrier due to the steric interactions of the molecules already present at the interface. E_{el} is the electrostatic energy due to the formation of a electrical double layer subsequent to the adsorption of protein molecules at the interface. E_{hy} is the hydrophobic energy due to the exposure of the hydrophobic patches of the protein toward air [61,62].

E_{sp} and E_{el} are expected to be positive and will oppose the protein adsorption, whereas E_{hy} is negative and will promote adsorption. Protein may form a complex with lipid molecule if E_{hy} is greater than the sum $E_{sp} + E_{el}$. In the presence of a close-packed lipid monolayer at the interface, protein may be expelled from the interface because the contribution of E_{sp} is dominant.

Based on the average surface hydrophobicity and molecular dimensions of the protein molecules, the estimated value of E_{hy} is of the order of 60 kcal/mol for both BSA and β-lactoglobulin. For a loosely packed lecithin monolayer ($\pi \approx$ 6 mN/m), Γ was found to be 0.5 mg/m^2 and the calculated values of E_{sp} and E_{el} are of the order of 10 and 20 kcal/mol, respectively. Consequently, E_{hy} is higher than the sum $E_{sp} + E_{el}$ and protein adsorption is enhanced. On the other hand, if a close-packed monolayer of lecithin is spread ($\pi \approx 40$ mN/m), the estimated value of E_{sp} is of the order of 70 kcal/mol; E_{el} is negligible because the interface is not charged. Consequently, $E_{sp} + E_{el} > E_{hy}$ and BSA is expelled from the interface. For β-lactoglobulin, the estimated values of E_{el} were found to be negligible at any protein surface concentration (1.4 kcal/mol at the highest value of $\Gamma \approx 2.18$ mg/m^2). E_{sp} was estimated to be of the order of 105 and 275 kcal/mol when π was 18 and 47.5 mN/m, respectively (corresponding to a spread monolayer of monoglyceride of 0.27 and 0.12 nm^2/molecule, respectively). Consequently, E_{sp} is much higher than E_{hy} when a close-packed monolayer of monoglyceride is spread at the surface and, therefore, β-lactoglobulin molecules are expelled from the interface.

V. CONCLUSIONS

The π–A isotherms of spread monolayers of egg-yolk lecithin, 1-monopalmitoyl-rac glycerol, BSA, β-lactoglobulin, and mixtures of lecithin–BSA and monoglyceride–β-lactoglobulin were determined with a Langmuir minitrough. The isotherms of lecithin–BSA mixtures were closer to that of BSA alone for weight ratios of lecithin to BSA of 0.5–2.0. All mixed 1-monopalmitoyl-rac glycerol–β-lactoglobulin isotherms exhibited a phase transition between a liquid-expanded state and a liquid-condensed state, which indicated that the monoglyceride had a strong influence on the properties of the mixed monolayer during compression. In both systems studied, the isotherms of the mixed monolayers asymptotically approached that of the lipid monolayer, thus indicating that the protein was squeezed out from the monolayer at very small areas of the monolayer. Lipid-rich monolayers exhibited hysteresis and a smoother phase transition than protein-rich monolayers.

The dynamics of adsorption of [^{14}C] protein (BSA or β-lactoglobulin) for different spread amounts of lipid (lecithin or monoglyceride) and time delays after the initiation of protein adsorption was investigated through the measurements of surface pressure and protein surface concentration using a radiotracer technique. Spreading a lipid onto an adsorbing protein solution was found to enhance the rate of adsorption of the protein. Synergism was observed between BSA and lecithin for short times, leading to surface pressures higher than the sum of the surface pressures developed by lecithin and BSA individually. For both systems, spreading a low amount of lipid at the interface resulted in an enhancement of the protein adsorption for short times, leading to a protein surface concentration higher than the steady-state value in the absence of the lipid. Spreading higher amount of lipids decreased the amount of protein adsorbed, and spreading a close-packed lipid monolayer caused eventual desorption of protein from the interface, possibly because of the surface pressure and steric exclusion effects produced by the spread lipid monolayer.

REFERENCES

1. PJ Halling, CRC Crit Rev Food Sci Nutr 15:155, 1981.
2. E Dickinson, G Stainsby. Colloids in Food. London: Oxford Press, 1982.
3. DC Clark, AR Mackie, PJ Wilde, DR Wilson. Faraday Discuss 98:253, 1994.

4. DC Clark, In: AG Gaonkar, ed. Characterization of Food: Emerging Methods. Amsterdam: Elsevier, 1995, p. 23.
5. WE Ewers, KL Sutherland. Aust J Sci Res Ser A 5:697, 1952.
6. DJ McClements. Food Emulsions: Principles, Practice and Techniques. Boca Raton FL: CRC Press, 1998.
7. E Dickinson. An Introduction to Food Colloids. Oxford: Oxford University Press, 1992.
8. J Chen, E Dickinson, G Iveson. Food Struct 12:135, 1993.
9. DC Clark, AR Mackie, PJ Wilde, DR Wilson. In: Foods Proteins: Structure and Functionality. KD Schwenke, R Mothes, eds. New York: VCH, 1993, p. 263.
10. DC Clark, PJ Wilde, DR Wilson. Colloids Surfaces 59:209, 1991.
11. DC Clark, PJ Wilde, DR Wilson, R Wüstneck. Food Hydrocolloids 6:173, 1992.
12. M Coke, PJ Wilde, EJ Russel, DC Clark. J Colloid Interf Sci 138:489, 1990.
13. M Cornec, AR Mackie, PJ Wilde, DC Clark. Colloids Surfaces A: Physicochem Eng Aspects 114:237, 1996.
14. M Cornec, PJ Wilde, PA Gunning, AR Mackie, F Husband, ML Parker, DC Clark. J Food Sci 63:39, 1998.
15. JL Courthaudon, E Dickinson, WW Christie. J Agric Food Chem. 39:1365, 1991.
16. JL Courthaudon, E Dickinson, DG Dalgleish. J Colloid Interf Sci 145:390, 1991.
17. JL Courthaudon, E Dickinson, Y Matsumara, DC Clark. Colloids Surfaces 56:293, 1991.
18. JL Courthaudon, E Dickinson, Y Matsumara, A Williams. Food Struct 10:109, 1991.
19. JA De Feijter, J Benjamins, M Tamboer. Colloids Surfaces 27:243, 1987.
20. E Dickinson. Colloids Surfaces 42:191, 1989.
21. E Dickinson. In: M. El Nokaly, D Cornell, eds. Microemulsions and Emulsions in Foods. ACS Symposium Series 448. Washington, DC: American Chemical Society, 1991, p. 114.
22. E Dickinson, CM Woskett. In: RD Bee, P Richmond, J Mingins, eds. Food Colloids. Cambridge: Royal Society of Chemistry, 1989, p. 74.
23. E Dickinson, S Tanai. J Agric Food Chem 40:179, 1992.
24. E Dickinson, SE Rolfe, DG Dalgleish. Food Hydrocolloids 3:193, 1989.
25. DK Sarker, PJ Wilde, DC Clark. Colloids Surfaces B 3:349, 1995.
26. PJ Wilde, DC Clark. J Colloid Interf Sci 155:48, 1993.
27. M Bos, T Nylander, T Arnebrant, DC Clark. In: GL Hasenhuettl, W Hartel, eds. Food Emulsifiers and Their Applications. New York: Chapman & Hall, 1997, p. 95.
28. T Nylander, B Erickson. In: SE Friberg, K Larsson, eds. Food Emulsions. New York: Marcel Dekker, 1997, p. 189.
29. T Nylander. In: D Möbius, R Miller, eds. Protein at Liquid Interfaces. Amsterdam: Elsevier, 1998, p. 385.
30. DG Cornell, DL Patterson. J Agric Food Chem. 37:1455, 1989.
31. Z Kozorac, A Dhathathreyan, D Möbius. FEBS 229:372, 1988.
32. S Aynié, M Le Meste, B Colas, D Lorient. J Food Sci 57:883, 1992.
33. PJ Quinn, RMC Dawson. Biochem J 113:791, 1969.
34. PJ Quinn, RMC Dawson. Biochem J 115:65, 1969.
35. DG Cornell, DL Patterson, N Hoban. J Colloid Interf Sci 140:428, 1990.

36. E Brewkink, RA Demel, G De Korte-Kool, B De Kruijff. Biochemistry 31:1119, 1992.
37. R Grimard, P Tancrede, C Gicquaud. Biochim Biophys Res Commun 190:1017, 1993.
38. MA Bos, T Nylander. Langmuir 12:2791, 1996.
39. Y Du, J An, J Tang, L Jiang. Colloids Surfaces B 7:129, 1996.
40. DG Cornell, RJ Carroll. J Colloid Interf Sci 108:226, 1985.
41. DG Cornell. J Colloid Interf Sci 88:536, 1982.
42. F MacRitchie. Adv Protein Chem 32:283, 1978.
43. JA Ibdah, MC Phillips. Biochemistry 27:7155, 1988.
44. MR Rodriguez Nino, PJ Wilde, DC Clark, JM Rodriguez Patino. Langmuir 14:2160, 1998.
45. A Rahman, P Sherman. Colloid Polym Sci 260:1035, 1982.
46. MR Rodriguez Nino, JM Rodriguez Patino. J Am Oil Chem Soc 75:1233, 1998.
47. E Boyle, JB German. CRC Crit Rev Food Sci Nutr 36:785, 1996.
48. C Henry. Cereal Foods World 40:734, 1995.
49. JR Hunter, PK Kilpatrick, RG Carbonell. J Colloid Interf Sci 142:429, 1991.
50. D Cho, G Narsimhan, EI Franses. Langmuir 13:4710, 1997.
51. DE Graham, MC Phillips. J Colloid Interf Sci 70:403, 1979.
52. HJ Trurnit. J Colloid Sci 15:1, 1960.
53. D Cho, EI Franses, G Narsimhan. Colloids Surfaces A 117:45, 1996.
54. S Xu, S Damodaran. J Colloid Interf Sci 157:485, 1993.
55. F MacRitchie. J Colloid Interf Sci 79:461, 1981.
56. F MacRitchie, L Ter-Minassian-Saraga. Colloids Surfaces 10:53, 1984.
57. U Gehlert, G Weidemann, D Vollhardt. J Colloid Interf Sci. 174:392, 1995.
58. F Boury, I Ivanova, I Panaiotov, JE Proust. Langmuir 11:599, 1995.
59. F Boury, I Ivanova, I Panaiotov, JE Proust. Langmuir 11:2131, 1995.
60. MJ Castleden. J Food Sci 38:756, 1973.
61. D Cho, G Narsimhan, EI Franses. J Colloid Interf Sci 191:312, 1997.
62. G Narsimhan, F Uraizee. Biotechnol Prog 8:187, 1992.

18
Surfactant Adsorption Kinetics and Exchange of Matter for Surfactant Molecules with Changing Orientation Within the Adsorption Layer

Reinhard Miller
Max-Planck-Institut für Kolloid- und Grenzflächenforschung, Golm, Germany

Valentin B. Fainerman
Institute of Technical Ecology, Donetsk, Ukraine

Eugene V. Aksenenko
Institute of Colloid Chemistry and Chemistry of Water, Kiev, Ukraine

Alexander V. Makievski
Max-Planck-Institut für Kolloid- und Grenzflächenforschung, Golm, Germany, and Institute of Technical Ecology, Donetsk, Ukraine

Jürgen Krägel
Max-Planck-Institut für Kolloid- und Grenzflächenforschung, Golm, Germany

Libero Liggieri
Institute of Physical Chemistry of Materials, National Research Council, Genoa, Italy

Francesca Ravera
Institute of Physical Chemistry of Materials, National Research Council, Genoa, Italy

Rainer Wüstneck
Institute of Solid State Physics, University of Potsdam, Teltow, Germany

Giuseppe Loglio
Institute of Organic Chemistry, University of Florence, Florence, Italy

ABSTRACT

Recently developed thermodynamic adsorption isotherms are used to derive an improved diffusion controlled adsorption kinetics model for surfactants at liquid interfaces. The isotherms consider molecular reorientation processes within the adsorption layer, controlled by the actual surface pressure. It will be shown that the consideration of interfacial reorientation explains apparently too high diffusion coefficients. Moreover, a simple model for the matter exchange for sinusoidal surface area changes is presented which considers the molecular reorientations as the only relaxation process.

As shown previously, the adsorption kinetics of alkyl dimethyl phosphine oxides is influenced by interfacial reorientation. While the lower homologs follow the classical diffusion model, the higher homologs (C_{13}—C_{15}) yield diffusion coefficients several times larger than the physically expected values. Assuming two different adsorption states, again reasonable diffusion coefficients result.

New adsorption kinetics data for oxyethylated alcohols $C_{10}EO_8$ in terms of dynamic and equilibrium surface (interfacial) tensions at the water–air and water–hexane interfaces are analyzed with the reorientation model. The data agree well with the model derived for molecules, which can exist in two (or more) orientation states.

I. INTRODUCTION

The modeling of dynamic interfacial processes requires a knowledge of the equilibrium state of the adsorption layer. Although there are many adsorption isotherms available to describe the equilibrium behavior of adsorbed surfactant layers at liquid interfaces, in almost all cases the Langmuir isotherm is used as a standard. However, it had been shown in [1] that the isotherm had a significant effect on the quantitative course of the adsorption kinetics. For example, using the Langmuir isotherm, higher initial values for the rate of surface tension changes are obtained as compared to the Frumkin isotherm [2]. This makes it important to consider an exact isotherm for a surfactant under study in order to understand quantitatively its adsorption kinetics and exchange of matter.

The fundamental relationship needed for a quantitative description of adsorption/desorption processes is the well-known Ward and Tordai equation [3]. This nonlinear Volterra integral equation represents a general relationship between the dynamic adsorption $\Gamma(t)$ and the subsurface concentration $c(0, t)$ for freshly formed and nondeformed surfaces as

$$\Gamma = 2\sqrt{\frac{D}{\pi}}\left(c_0\sqrt{t} - \int_0^{\sqrt{t}} c(0, t - \tau)\, d\sqrt{\tau}\right) \tag{1}$$

where D is the diffusion coefficient, c_0 is the bulk concentration, t is the time, and τ is a dummy integration variable. From Eq. (1) dependencies, $\Gamma(t)$ can be obtained using respective isotherms as additional boundary condition for the diffusion-controlled adsorption model [4].

An adsorption rate slower than expected from a diffusion model was often ascribed to the existence of adsorption barriers [5–8]. The nature of a nondiffusional adsorption mechanism is not clear yet. One of the most frequently discussed reasons is the existence of energy barriers for adsorption or desorption. It is also possible that, at least for concentrated solutions, the nonequilibrium character of a surface layer can play a major role, as demonstrated elsewhere [9]. Recently, it has been shown impressively that aggregation processes set in at interfaces after a certain critical surface pressure is reached [10]. This process leads also to a deceleration of the adsorption rate.

For some surfactants, it was observed that their adsorption rate was higher than expected by a diffusion model. The reorientation of surfactant molecules within the adsorption layer can be one of the explanations for such a faster adsorption. In the theoretical models discussed in the literature so far, reorientation is described by a first-order reaction equation resulting in a variation of the molecular state at the surface [11–13]. Recent experimental and theoretical results provide a completely different understanding of reorientation effects on dynamic surface tension [14,15]. Apparently, the partial molar area of asymmetric surfactant molecules depends on their orientation at the surface so that a reorientation is possible [16]. This, however, is true also for surfactants without any oxyethylene groups which show a "super diffusional" kinetics as well [17,18]. Recently, the dynamic surface tension of aqueous solutions of alkyl dimethyl phosphine oxides (C_nDMPO) with different alkyl chain lengths was studied [19]. The lower homologs (C_8–C_{12}) adsorb according to the diffusion model based on the Langmuir isotherm. For higher homologs (C_{13}–C_{15}) however, the measured dynamic surface tensions are lower than the theoretically predicted ones and the application of the standard diffusion model would lead to diffusion coefficients that are much larger than the physically reasonable values. To understand these differences, again the theory of diffusion-controlled adsorption kinetics considering two states of C_nDMPO at the surface with different partial molar areas [20] is able to explain the results with typical diffusion coefficients.

The theory proposed in Ref. 19 is a quasiequilibrium model and assumes an equilibrium composition of the surface layer at all times. However, the reorientation of molecules within the surface layer may require some time, which can affect the adsorption rate [21].

The present overview will provide insight into the recently derived adsorption kinetics models, which consider molecular reorientation processes in the adsorption layer. Also, the transfer kinetics between the various states of adsorbed molecules within the monolayer is presented, which has nothing in common with a barrier-controlled adsorption mechanism. Dynamic surface tensions obtained for model surfactants are discussed in order to validate the theoretical models.

II. THEORETICAL MODELS OF ADSORPTION KINETICS AND EXCHANGE OF MATTER

The integrodifferential equation (1) proposed by Ward and Tordai [3] represents a general relationship between the dynamic adsorption Γ and the subsurface concentration $c(0, t)$ for freshly nondeformed surfaces. It is the starting point for many models of diffusion-controlled adsorption. According to the literature, these models are called diffusion controlled. If we assume an extra time step, for example, for the molecular reorientation in the adsorption layer, we deal with a mixed-kinetics model [4].

A. Orientation Changes in a Quasiequilibrium Adsorption Layer

The adsorption isotherm for surfactant molecules that can adsorb in two states (1 and 2) with different partial molar areas ω_1 and ω_2 ($\omega_1 > \omega_2$) can be expressed by [14]

$$bc(0, t) = \frac{\Gamma_1 \omega_\Sigma}{(1 - \Gamma \omega_\Sigma)^{\omega_1/\omega_\Sigma}} \tag{2}$$

where b is the adsorption equilibrium constant, $\Gamma = \Gamma_1 + \Gamma_2$ is the total adsorption, ω_Σ is the mean partial molar area,

$$\omega_\Sigma = \frac{\omega_2 + \omega_1 \beta \exp[\Pi(\omega_2 - \omega_1)/RT]}{1 + \beta \exp[\Pi(\omega_2 - \omega_1)/RT]} \tag{3}$$

where R is the gas constant and T is the absolute temperature. The parameter $\beta = (\omega_1/\omega_2)^\alpha$ is a constant which considers that the adsorption activity of surfactant molecules is higher in state 1 than in state 2 (α is a constant). An important relationship, which includes the experimentally available dynamic surface pressure $\Pi = \gamma_0 - \gamma$ (γ_0 is the surface tension of the solvent and γ the surface tension of the solution which is a function of time and surfactant concentration), is defined for this two-state model by

$$\Pi = -\frac{RT}{\omega_\Sigma} \ln (1 - \Gamma\omega_\Sigma) \qquad (4)$$

It was demonstrated in Ref. 19 that a finite-difference method was suitable for solving the respective set of equations numerically. The simultaneous solution of this set of equations is much more complicated than that using simpler isotherms such as Henry or Langmuir, as several iterations are required in each successive time step because Γ depends not only on $c(0, t)$ but also on ω_Σ and Π, which, in turn, are functions of Γ.

B. Orientational Changes as an Extra Kinetic Process

For a dynamic adsorption layer where adsorption from or desorption to the solution bulk take place continuously, one cannot generally assume an equilibrium composition at the interface. If the transition between the two different adsorption states 1 and 2 is not instantaneous, an additional rate constant has to be introduced. The transition between the two states can be described by the following kinetic scheme:

$$\Gamma_1 \underset{k_{21}}{\overset{k_{12}}{\rightleftarrows}} \Gamma_2 \qquad (5)$$

The rate constants k_{ij} characterize the transition from state i into state j. The transition process can be described by the first-order equation

$$-\left(\frac{d\Gamma_1}{dt}\right)_s = \left(\frac{d\Gamma_2}{dt}\right)_s = k_{12}\Gamma_1 - k_{21}\Gamma_2 \qquad (6)$$

The subscript s denotes that the transition between the two states takes place within the surface layer [21]. Variation in the molar area takes place both due to the diffusional flux of the surfactant molecules from the bulk to the surface layer, and the reorientation at the interface; that is,

$$\frac{d\Gamma_1}{dt} = \left(\frac{d\Gamma_1}{dt}\right)_b + \left(\frac{d\Gamma_1}{dt}\right)_s \qquad (7)$$

$$\frac{d\Gamma_2}{dt} = \left(\frac{d\Gamma_2}{dt}\right)_b + \left(\frac{d\Gamma_2}{dt}\right)_s \qquad (8)$$

The subscript b denotes the flux of each 'species' from or to the bulk phase. The total change in the surfactant amount at the interface is given by the sum of Eqs. (7) and (8):

$$\frac{d\Gamma}{dt} = \frac{d\Gamma_1}{dt} + \frac{d\Gamma_2}{dt} = \left(\frac{d\Gamma_1}{dt}\right)_b + \left(\frac{d\Gamma_2}{dt}\right)_b \tag{9}$$

where Eq. (6) was taken into account. If we assume that the fluxes between the solution bulk and the interface for the two states are equal to each other, that is,

$$\left(\frac{d\Gamma_1}{dt}\right)_b = \left(\frac{d\Gamma_2}{dt}\right)_b = \frac{1}{2}\frac{d\Gamma}{dt} \tag{10}$$

the equations describing the balance between states 1 and 2 read

$$\frac{d\Gamma_1}{dt} = \frac{1}{2}\frac{d\Gamma}{dt} - k\left[\Gamma_1 - \beta\Gamma_2 \exp\left(-\frac{\Pi}{RT}(\omega_1 - \omega_2)\right)\right] \tag{11}$$

$$\frac{d\Gamma_2}{dt} = \frac{1}{2}\frac{d\Gamma}{dt} + k\left[\Gamma_1 - \beta\Gamma_2 \exp\left(-\frac{\Pi}{RT}(\omega_1 - \omega_2)\right)\right] \tag{12}$$

where $k = k_{12}$.

C. Exchange of Matter at Sinusoidal Area Changes

The time dependence of the area A for sinusoidal surface deformation can be defined as

$$A(t) = A_0[1 + \delta \exp(i\overline{\omega}t)] \tag{13}$$

where δ is the relative amplitude of oscillations, $\overline{\omega}$ the angular frequency, and t is the time. Surface deformation gives rise to an additional transfer of surfactant between the bulk and the surface layer, which can be described by an ordinary molecular diffusion equation, as discussed in detail elsewhere [22,23]. The boundary condition at the interface ($z = 0$) for the diffusion problem should account for both the diffusion of surfactant toward the surface and the transitions between the states of adsorbed molecules. Equations (11) and (12) can be used to describe the mass balance between states 1 and 2. However, the change in area with time $A(t)$ has to be considered simultaneously and, thus, the equations have to be modified, respectively, as

$$\frac{d\Gamma_1}{dt} + \Theta\Gamma_1 = \frac{1}{2}\left(\frac{d\Gamma}{dt}\right)_b - k\left[\Gamma_1 - \beta\Gamma_2 \exp\left(-\frac{\Pi}{RT}(\omega_1 - \omega_2)\right)\right] \tag{14}$$

$$\frac{d\Gamma_2}{dt} + \Theta\Gamma_2 = \frac{1}{2}\left(\frac{d\Gamma}{dt}\right)_b + k\left[\Gamma_1 - \beta\Gamma_2 \exp\left(-\frac{\Pi}{RT}(\omega_1 - \omega_2)\right)\right] \tag{15}$$

The value of $\Theta = d \ln A / dt$ is determined from Eq. (13). For small oscillation amplitudes ($\delta < 0.1$),

$$\Theta = \frac{d}{dt}\left(\frac{\delta \sin \overline{\omega} t}{1 + \delta \sin \overline{\omega} t}\right) \cong \delta \, \overline{\omega} \cos \overline{\omega} t \tag{16}$$

To link the present model to experimentally available data, Eq. (4) can be used for any state of the adsorption layer. From the change in surface pressure Π or surface tension γ with time due to the sinusoidal area changes $A(t)$, an important characteristic of the surface layer (i.e., the viscoelasticity ε) can be obtained as

$$\varepsilon = \frac{d\gamma}{d \ln A} \tag{17}$$

The viscoelasticity is a complex parameter determined by the dilatational elasticity and viscosity [24,25]. The viscoelasticity modulus (or surface dilatational modulus) incorporates the real and imaginary constituents, elasticity and viscosity, respectively, as

$$|\varepsilon| = (\varepsilon_r^2 + \varepsilon_i^2)^{1/2} = \frac{\varepsilon_0}{(1 + 2\zeta + 2\xi^2)^{1/2}} \tag{18}$$

ε_0 is the Gibbs elasticity modulus, which can be determined from the equation-of-state of the adsorption layer via $\varepsilon_0 = d\gamma / d \ln \Gamma$. The value ζ characterizes the ratio of the so-called diffusion relaxation frequency $\overline{\omega}_0$ to the angular frequency of area oscillations $\overline{\omega}$ that is,

$$\zeta = \left(\frac{\overline{\omega}_0}{2\overline{\omega}}\right)^{1/2} \tag{19}$$

The diffusion relaxation frequency is determined by the diffusion coefficient, adsorption Γ, and bulk concentration c of the surfactant as

$$\overline{\omega}_0 = D\left(\frac{dc}{d\Gamma}\right)^2 \tag{20}$$

For high frequencies, the diffusional exchange between the solution bulk and the surface layer can be neglected as compared with the processes which take place in the surface layer. Assuming the approximate relation $dc/d\Gamma \approx c/\Gamma$ and noting that for usual surfactants $D = 4 \times 10^{-6}$ cm^2/s and $\Gamma = 4 \times 10^{-10}$ mol/cm^2, one obtains the condition of a so-called insoluble monolayer: $\overline{\omega} \geq 10^{-3}$ s^{-1} for $c = 10^{-6}$ mol/L and $\overline{\omega} \geq 10$ s^{-1} for $c = 10^{-4}$ mol/L. These frequencies are not extremely high and can be easily achieved with known experimental methods. On the other hand, the rate of exchange between the adsorption states 1 and 2 of a surfactant molecule can be quite high; Therefore, experiments at high frequencies of surface deformation would significantly contribute to the understanding of the phenomenon of interfacial reorientations.

Neglecting the diffusion fluxes from the solution bulk, Eqs. (14) and (15) reduce to [26]

$$\frac{d\Gamma_1}{dt} + \Theta\Gamma_1 = -k\left[\Gamma_1 - \beta\Gamma_2\exp\left(-\frac{\Pi}{RT}(\omega_1 - \omega_2)\right)\right] \tag{21}$$

$$\frac{d\Gamma_2}{dt} + \Theta\Gamma_2 = k\left[\Gamma_1 - \beta\Gamma_2\exp\left(-\frac{\Pi}{RT}(\omega_1 - \omega_2)\right)\right] \tag{22}$$

Model calculations have been performed recently which demonstrate that interfacial reorientations can significantly contribute to the overall relaxation process at sinusoidal interfacial expansions and compressions.

III. MODEL CALCULATIONS FOR THE REORIENTATION MODEL

To illustrate the effect of the difference in the molar areas of an adsorbing surfactant, the theoretical dependencies $\Pi(t)$ are shown in Figure 1. The surface pressure reaches a minimum at $\omega_1 = \omega_2$ (i.e., when there is no reorientation). With increasing ω_1 up to a value of $3\omega_2$, Π increases monotonously within the entire time range. Further increase to values of $6\omega_2$, however, results in an additional growth of Π only for short times, whereas in the region of intermediate and large Π, the surface pressure is lower than that obtained for $\omega_1 = 3\omega_2$. For $\omega_1 = 6\omega_2$, the dependence $\Pi(t)$ for large surface pressures is the same as that calculated for $\omega_1 = \omega_2$. This can be easily understood from Eq. (6). If the partial molar areas differ considerably from each other, then there will be almost no molecules adsorbed in state 1 even at small surface pressures.

Note that the calculated curves in Figure 1 are very similar to those calculated without the consideration of reorientation effects; however, diffusion coefficients there to five times larger than those assumed in Figure 1 have to be used. Thus, the pure diffusion-controlled adsorption kinetics model for surfactants with surface reorientation predicts a significant increase in the adsorption rate when the maximum partial molecular area is larger than two times the minimum value. This change in the adsorption rate is formally equivalent to an increase in the apparent diffusion coefficient by a factor of 4.

Some results of the calculated dynamic surface tensions for the reorientation model with an extra interfacial rate constant are presented in Figure 2. One can see that the two-state quasistationary model predicts faster kinetics of the

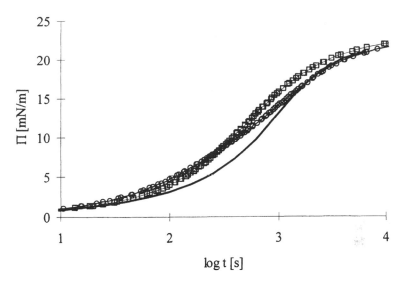

Figure 1 Dynamic surface pressure of an adsorption layer calculated from the diffusion model assuming two states of adsorbed surfactant molecules at the surface. For all curves, $\omega_2 = 2.5 \times 10^9$ cm^2/mol, $D = 5 \times 10^{-6}$ cm^2/s, and $c_0 = 5 \times 10^{-9}$ mol/cm^3, $\Pi(t^{\to\infty}) = 23.7$ mN/m; $\omega_1 = 2.5 \times 10^9$ cm^2/mol (solid line), 7.5×10^9 cm^2/mol ($\square$), and 1.5×10^{10} ($\bigcirc$) cm^2/mol.

surface tension decrease compared to the Langmuir model. However, for low-transition-rate constants ($k = 0.01$ s), the pressure increases less rapidly. The effect of slow-transition kinetics becomes more significant for large α values characteristic of, for example, oxyethylated surfactants.

As the rate constant of interfacial reorientation can be rather small, experiments on fast interfacial perturbations should be suitable for their determination, and exchange of matter functions at higher frequencies should be sensitive to this parameter. Thus, it is not unexpected that the kinetics of the transitions between the two states significantly affects the value of the elasticity modulus resulting from experiments based on sinusoidal area changes. The dependence of the elasticity modulus on $\lambda = k/\overline{\omega}$ is shown in Figure 3. In these calculations, the equilibrium surface pressure was assumed to be equal to 11 mN/m. One can see for large values of λ [i.e., for rapid transitions between the states and slow deformation rates (small frequencies $\overline{\omega}$)], the elasticity modulus attains a minimum value. The decrease of λ leads to an increase in viscoelasticity ε. For very small values ($\lambda < 0.01$), the viscoelasticity modulus reaches a value which corresponds to the Gibbs elasticity. Note that it was assumed in the calculations that prior to the imposition of the area oscillations, the composition of the surface

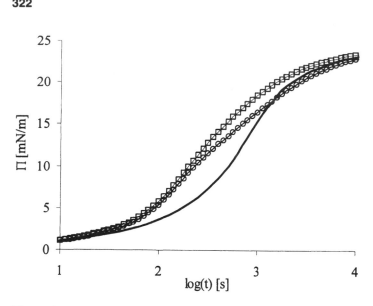

Figure 2 Dynamic surface pressure calculated from different models: reorientation quasiequilibrium model ($\square$); kinetic model ($\bigcirc$); Langmuir model (solid line), with $\omega_2 = 2.5 \times 10^9$ cm^2/mol, $\omega_1 = 7.5 \times 10^9$ cm^2/mol, $\alpha = 2$, $k = 0.01$ s, and $D = 5 \times 10^{-6}$ cm^2/s.

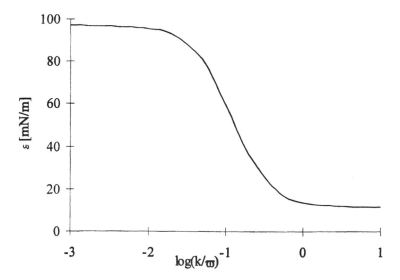

Figure 3 The dependence of the elasticity modulus ε on $k/\overline{\omega}$ at $\Pi = 11$ mN/m calculated according to the model given by Eqs. (21) and (22); the parameters used are $\omega_1 = 1 \times 10^{10}$ cm^2/mol, $\omega_2 = 4 \times 10^9$ cm^2/mol, and $\alpha = 4$.

layer was at equilibrium, corresponding to Eq. (4). The viscoelasticities obtained for the two-state model exceed significantly the values calculated from a Langmuir isotherm. This demonstrates impressively that the values of the viscoelasticity calculated in the framework of different models can be drastically different.

IV. COMPARISON WITH EXPERIMENTAL DATA AND DISCUSSION

In order to evaluate to capacity of the new adsorption kinetics models, experimental data are compared to the theoretical dependencies. It turns out that a good agreement with theoretical models can be obtained only when molecular processes within the adsorption layer are considered.

The members of the homologous series of alkyl dimethyl phosphine oxides C_nDMPO are excellent nonionic model surfactants and the results obtained at the water–air interface belong to a wide range of surface phenomena [19,27–29]. Although the equilibrium surface tension isotherm for C_8DMPO, C_{10}DMPO, and C_{12}DMPO can be described quite well by a Langmuir isotherm, the longer alkyl chains require the reorientation model for better agreement. For C_8DMPO, the molar areas for states 1 and 2 differ only slightly (i.e., $\omega_1/\omega_2 = 1.5$), whereas for C_{14}DMPO, we obtain $\omega_1/\omega_2 = 3$. According to the model calculations presented in Figure 1, one can expect a significant effect of the reorientation on $\gamma(t)$. This prediction is supported by the experimental data. Figure 4 shows the experimental and theoretical dependencies of $\gamma(t)$ for a C_{14}DMPO solution. The reorientation model describes the experimental data much better than the model neglecting the transition between the two orientation states. However, the shape of the experimental curve is somewhat different from that obtained from the reorientation model. The sharp decrease in the experimental γ values can possibly be explained by the nonequilibrium composition of the surface layer due to a finite transition rate of the C_{14}DMPO molecules between the two states. This point will have to be further studied by relaxation experiments, as discussed earlier. For the moment, no specific results exist.

Another widely used group of nonionic model surfactants is the n-decyl polyoxyethylene glycol ethers C_nEO$_m$. Due to their molecular structure, these surfactants are also candidates for reorientations at liquid interfaces [14,19,21,30–32].

The experimental dynamic surface tensions for C_{10}EO$_8$ at the water–air interface, measured in Ref. 2 using the pendent bubble method, are compared with the calculations for the two models in Figure 5. The adsorption characteristics used were those obtained from the surface tension isotherm [29]. The diffusion coefficient $D = 5 \times 10^{-6}$ cm^2/s was calculated from the equation proposed by Wilke and Chang [33]. The value of D required to match the experimental

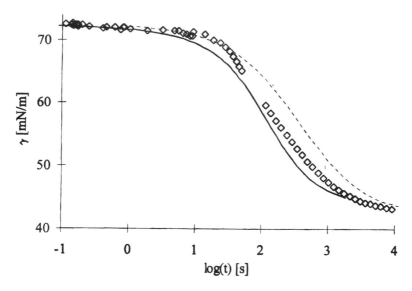

Figure 4   Dynamic surface tension γ of C_{14} DMPO solutions for $c_0 = \;\;.5 \times 10^{-9}$ mol/ cm³, experimental data ($\Diamond$) according to Ref. 19; model calculations: $\;$ reorientation effects according to a Langmuir isotherm (dotted line, $\omega_1 = \omega_2 = 2.2 \times 10^9$ cm²/mol) and considering reorientations (solid line, $\omega_1 = 6.6 \times 10^9$ cm²/mol, $\omega_2 = 2.2 \times 10^9$ cm²/ mol); $D = 4.1 \times 10^{-6}$ cm²/s.

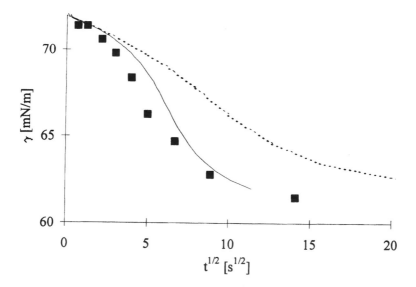

Figure 5   Dynamic surface tension $\gamma(t)$ for an aqueous solution of 6×10^{-9} mol/cm³ of $C_{10}EO_8$ at the air–water interface [2]; calculations with the Langmuir mc $\;$ dotted line) and the two-state model (solid line), $D = 5 \times 10^{-6}$ cm²/s.

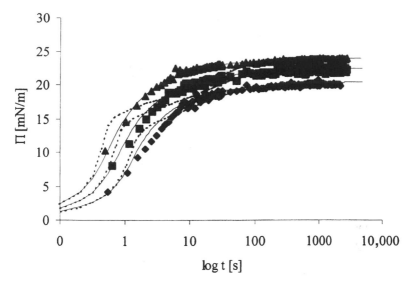

Figure 6 Dynamic surface pressure $\Pi(t)$ of aqueous solutions of $C_{10}E_5$ at three concentrations: 6×10^{-8} mol/cm^3 (▲), 8×10^{-8} mol/cm^3 (■), and 10^{-7} mol/cm^3 (♦); theoretical curves calculated with the two-state model without (solid line) and with a reorientation step with $k = 0.1$ s^{-1} (dashed line), $D = 4.3 \times 10^{-6}$ cm^2/s.

data would be 1.5×10^{-5} cm^2/s (i.e., three times higher than the expected value). At the same time, the two-state model agrees well with the experimental data using the typical diffusion coefficient of $D = 5 \times 10^{-6}$ cm^2/s. Thus, we conclude that the "superdiffusion" adsorption kinetics of $C_{10}EO_8$ at the water–air interface is not real but is caused by the self-regulation of surfactant molecular states within the surface layer.

The results for another member of the group of EO surfactants are shown in Figure 6. The mechanism assuming diffusion and reorientation with $k = 0.1$ s^{-1} is not adequate to describe the data. The best agreement with the experimental data is obtained when assuming a pure diffusion-controlled process and infinitely fast surface reorientation [34].

V. CONCLUSIONS

A consideration of reorientation explains the too large diffusion coefficients obtained when only the classical diffusion model is applied. Examples for which a considerable effect of the orientational changes on the adsorption kinetics has been analyzed are the longer-chain alkyl phosphine oxides and ionic and nonionic surfactants containing a certain number of EO groups. The kinetics of the reorientation process itself can have additional considerable effects and can be described by an extra rate constant. Adsorption kinetics data obtained as dynamic surface tensions do not allow the exact orders of magnitude of the rate constant of this molecular process to be deduced. A discussion of the conditions under which reorientations happen allows one to expect that for C_nEO_m with short alkyl groups (small n) and long EO parts (large m) such rate processes should become significant in dynamic surface tension data. For other surfactants capable of changing the orientation at an interface, relaxation experiments at higher frequencies should be more suited. For this purpose, studies with the oscillating bubble methods are underway.

ACKNOWLEDGMENTS

The work was financially supported by projects of the European Community (INCO ERB-IC15-CT96-0809, ESA Topical Team FAST), and the DFG (Mi418/9, Wu 187/6, and Wu 187/8).

REFERENCES

1. R. Miller, Colloid Polym Sci 259:375, 1981.
2. S-Y Lin, T-L Lu, W-B Hwang. Langmuir 10:3442, 1994.
3. AFH Ward, L Tordai. J Chem Phys 14:543, 1946.
4. SS Dukhin, G Kretzschmar, R Miller. In: D Möbius, R Miller, eds. Dynamics of Adsorption at Liquid Interfaces: Theory, Experiment, Application. Studies in Interface Science Vol. 1. Amsterdam: Elsevier, 1995.
5. JF Baret. J Chem Phys 65:895, 1968.
6. R Defay, I Prigogine, A Sanfeld. J Colloid Interf Sci 58:498, 1977.
7. G Bleys, P Joos. J Phys Chem 89:1027, 1985.
8. L Liggieri, F Ravera, A Passerone. Colloids Surfaces A 114:351, 1996.
9. VB Fainerman, SA Zholob, R Miller, P Joos. Colloids Surfaces A 143:243, 1998.
10. EV Aksenenko, VB Fainerman, R Miller. J Phys Chem 102:6025, 1998.
11. AG Bois, II Panaiotov, JF Baret. Chem Phys Lipids 34:265, 1984.
12. G Serrien, P Joos. J Colloid Interf Sci 139:149, 1990.
13. VB Fainerman. Colloids Surfaces 57:249, 1991.
14. VB Fainerman, R Miller, R Wüstneck, AV Makievski. J Phys Chem 100:3054, 1996.
15. VB Fainerman, R Miller, R Wüstneck. J Phys Chem 101:6479, 1997.
16. P Joos, G Serrien. J Colloid Interf Sci 145:291, 1991.
17. R Miller, K Lunkenheimer. Colloid Polym Sci 264:357, 1986.
18. J-P Fang, K-D Wantke, K Lunkenheimer. J Phys Chem 99:4632, 1995.

19. EV Aksenenko, AV Makievski, R Miller, VB Fainerman. Colloids Surfaces A 143: 311, 1998.
20. VB Fainerman, EH Lucassen-Reynders, R Miller. Colloids Surfaces A 143:141, 1998.
21. R Miller, EV Aksenenko, L Liggieri, F Ravera, M Ferrari, VB Fainerman. Langmuir 15:1328, 1999.
22. R Miller, G Kretzschmar. Adv Colloid Interf Sci 37:97, 1991.
23. R Miller, G Loglio, U Tesei, K-H Schano. Adv Colloid Interf Sci 37:73, 1991.
24. J Lucassen, M Van den Tempel. Chem Eng Sci 17:1283, 1972.
25. J Lucassen, M Van den Tempel. J Colloid Interf Sci 41:41, 1972.
26. R Miller, EV Aksenenko, VB Fainerman. J Phys Chem (in press).
27. K Lunkenheimer, K Haage, R Miller. Colloids Surfaces 22:215, 1987.
28. J-P Fang, K-D Wantke, K Lunkenheimer. J Phys Chem 99:4632, 1995.
29. AV. Makievski, DO Grigoriev. Colloids Surfaces A 143:233, 1998.
30. VB Fainerman, SA Zholob, R Miller. Langmuir 13:283, 1997.
31. VB Fainerman, R Miller. Langmuir 13:409, 1997.
32. HC Chang, CT Hsu, S-Y Lin. Langmuir 14:2476, 1998.
33. CR Wilke, P Chang. AIChE J 1:264, 1955.
34. L Liggieri, M Ferrari, A Massa, F Ravera. Colloids Surfaces A. 156:455, 1999.

19
Dynamic Surface Tension of Aqueous Dispersions of SP-C and DPPC and Their Monolayer Behavior

Sun Young Park
International Paper Corporate Research Center, Tuxedo, New York

John E. Baatz
Medical University of South Carolina, Charleston, South Carolina

Robert E. Hannemann and Elias I. Franses
Purdue University, West Lafayette, Indiana

ABSTRACT

The measurements of dynamic surface tensions at 23 or 37°C of aqueous saline dispersions of SP-C, a protein component of lung surfactant, indicate that SP-C can adsorb at significant rates, which depend on the dispersion preparation protocol and the surface tension-measuring method used. Surface tension equilibration is slower with the pendant drop method than with the bubble method. The SP-C has moderate surface activity by itself, but acts synergistically with DPPC [dipalmitoylphosphatidylcholine] to reduce significantly the minimum surface tension observed under either constant or pulsating area conditions. Moreover, the adsorption is faster with the mixture than with either component alone, even though each component was dispersed separately and no mixing was expected in the bulk aqueous phase. At a DPPC–SP-C weight ratio of about 43:1, comparable to that of a commercial lung surfactant, SP-C appears to enter the monolayer and remain in it after area compression to an extent sufficient to affect the dynamic surface tension. However, preliminary FTIR-ATR (Fourier-Transform Infrared Attenuated Total Reflection) results on Langmuir–Blodgett (LB) transferred films were evidently not sensitive enough to detect the presence of SP-C at the surface.

I. INTRODUCTION

The dynamic surface tension of aqueous lung surfactant is important for the interface of alveolar lining layers and for stabilizing the lungs for proper breathing [1,2]. The natural lung or pulmonary surfactant consists of various lipids and proteins. The major lipid component DPPC [dipalmitoylphosphatidylcholine] is essential for proper biophysical function, but in order to generate low dynamic surface tensions, it requires the synergistic presence of other lipids and two hydrophobic proteins, SP-B and SP-C [3–43]. Two commercial surfactants are currently available in the United States for exogenous replacement therapy of premature infants suffering from the Respiratory Distress Syndrome (RDS) [2,43]. The RDS is thought to be caused primarily by a lack of sufficient amounts of certain or most components of the lung surfactant mixture [1,2]. This insufficiency apparently produces higher surface tensions than biophysically required at the alveolar lining layer.

One effective commercial lung surfactant replacement formulation, Survanta (from Ross Laboratories), which is derived primarily from biological sources, contains (1) DPPC, (2) the hydrophobic proteins SP-B and SP-C, and (3) various other lipids, including cholesterol, which are thought to play supplementary but relatively minor roles in the low surface tension behavior [1–3]. A synthetic commercial surfactant mixture, Exosurf (from Glaxo–Wellcome) contains DPPC, hexadecanol, and tyloxapol [2,43]. Hexadecanol is thought to play synergistic and supporting roles together with DPPC, although recent results indicate that it is quite surface active by itself, producing dynamic surface tensions as low as 10 mN/m under pulsating area conditions [44–48].

This chapter focuses on the behavior of one major hydrophobic lung-surfactant-associated protein, SP-C, and its possible interactions with DPPC. This model mixture is considered to have several essential features of the behavior of natural or biologically derived lung surfactants, including Survanta. Although the behavior of SP-C in solvent-spread or aqueous-dispersion-spread monolayers has been studied with pressure–area isotherms and detailed surface characterization techniques [2–9,12,14–17,19–23,25–34,38–42], questions remain about its surface activity under either equilibrium or dynamic conditions. Because both SP-C and DPPC are largely insoluble in water, the question is how the method of dispersion preparation and the technique of surface tension measurement affect the observed surface tension. The following questions also pertain to the mechanism(s) of the interaction of SP-C with DPPC under laboratory conditions: (1) Is there a synergistic transport to the surface, via formation of a mixed particulate structure, or simply interactions in the mixed monolayer after competitive adsorption? (2) Does SP-C remain at the surface after the surface is compressed or is it "squeezed out" [40,41], or preferentially excluded, as it is observed with certain supporting lipids? (3) Does SP-C remain and reabsorb on the surface after repeated cycles of area compression and expansion?

In the following sections, we present certain monolayer isotherm results for SP-C and certain dynamic surface tensions of dispersions, as measured with various techniques. This behavior is compared to that of DPPC–SP-C mixtures mostly at molar or mass ratios similar to those in Survanta, and at physiological conditions of temperature and NaCl concentration. Finally, some preliminary results of Fourier-Transform Infrared Attenuated Total Reflection (FTIR-ATR) on Langmuir–Blodgett (LB) films are presented, showing little spectral evidence of the presence of SP-C in the monolayer, even though indirect surface tension evidence points to some presence of SP-C in the monolayer.

II. EXPERIMENTAL SECTION

A. Materials

Synthetic L-α-dipalmitoylphosphatidylcholine (DPPC), 99+% (from Sigma Chemical Co., St. Louis, MO) was used without purification. L-α-DPPC with perdeuterated hydrocarbon tail groups or d-DPPC (>99%), was obtained from Avanti Polar Lipids. Surfactant Protein-C (SP-C) [molecular weight (MW) = 4200] was isolated from bovine lung extract and was purified by isocratic elusion of the sample through Lipophilic Sephadex LH-60 gel filtration beads [16]. The elution-purified SP-C was dialyzed against 1:1 chloroform–methanol using dialysis tubes with a molecular weight cutoff of 3500. The amino acid sequence of bovine SP-C, which is highly hydrophobic, is as follows [49–51]: N-Leu-Ile-Pro-Cys-Cys-Pro-Val-Asn-Ile-Lys-Arg-Leu-Leu-Ile-Val-Val-Val-Val-Val-Val-Leu-Leu-Val-Val-Val-Ile-Val-Gly-Ala-Leu-Leu-Met-Gly -Leu-COOH. The first 11 residues have some hydrophilicity, and the remaining region is quite hydrophobic. The adjacent cystein residues (Gys) are palmitoylated [52]. The purity of the SP-C was ascertained with sodium dodecyl sulfate–polyacrylamide gel electrophoresis (SDS–PAGE) utilizing tricine buffers [15–17]. Sodium chloride, analytical reagent (AR®) grade, was purchased from Mallinckrodt, Inc. (Paris, KY). High-performance liquid chromatography (HPLC)-grade hexane from (Aldrich, Milwaukee, WI), and absolute alcohol (from Midwest Grain Products, Co., Pekin, IL) were solvents used for making spread monolayers. Methanol with 99.85% purity and chloroform with 99.9% purity were used for preparing solutions of SP-C and its mixtures with DPPC. All the experiments were done with pure Millipore water, which was prepared by passing distilled water through a Milli-Q 4 bowl system. The water had an initial resistivity of 18 M$\Omega \cdot$ cm. All dispersions were prepared in saline solution (0.9 wt% aqueous sodium chloride) to mimic physiological conditions. A saline solution was also used as the subphase in the Langmuir trough experiments.

B. Preparation Protocols for Dispersions

The dispersions were prepared with sonication, which was done at 55°C for 25 min in a sonicator bath (Branson ultrasonic cleaner heated tank, model B52H, Branson Cleaning Equipment Co., Shelton, CT). Because the SP-C protein is difficult to disperse in water when dry, some SP-C solution in an organic solvent (1/1 v/v methanol–chloroform) was partially evaporated until a very small amount of solvent remained. Then, saline solution was added and the sample was sonicated for about 20 min in a sonicator bath at ~50°C. The 50-ppm dispersion, first a bit turbid, turned clear with sonication, without any visible particles. DPPC and DPPC–SP-C dispersions were prepared with the Protocol 2 procedure [44,45], during which they were stirred and then sonicated after they were heated above 50°C [44,45].

C. Apparatus and Procedures

1.   KSV Langmuir Trough

A computer-controlled minitrough from KSV Instruments, Finland with a roughened platinum Wilhelmy plate was used for studying the spread monolayer behavior and for depositing Langmuir–Blodgett (LB) films. The trough was set up in a class 100 clean room. The precision of the surface area was ± 0.5%, and that of the surface pressure measurement was ± 0.1 mN/m, with reproducibility of ± 1 mN/m. The maximum trough surface area was 250 cm^2 and the volume of the subphase was ~220 cm^3. Experiments were done at room temperature (23 ± 1°C), or at 37°C, by flowing thermostatted water through the hollow body of the trough. The subphase was not stirred, unlike the procedure used by other authors [5].

2.   Spread Monolayer Experiments and Langmuir–Blodgett Depositions

For surface pressure (π)–surface area per molecule ($\overline{A}$) monolayer isotherm studies and for fabrication of LB monolayer films, solutions of DPPC or d-DPPC in 9/1 v/v hexane–ethanol (0.5 mg/mL) were used [46,48]. For SP-C and DPPC–SP-C systems, the components were dissolved in 1/1 v/v methanol–chloroform to yield solutions of ~ 1 mg/mL. After monolayers were spread on a clean saline

surface in the Langmuir trough, about 7 min were allowed for solvent evaporation. LB monolayers were made with controlled surface pressures ranging from
5 to over 60 mN/m. The surface area was compressed at 15 cm^2/min (about 2–
7 Å^2/molecule/min) until the deposition pressure was reached. Then, the monolayer was transferred onto a germanium (Ge) ATR crystal at a deposition rate
of 0.5 cm/min. During deposition, the set surface pressure was maintained to
within 1 mN/m.

3. Joyce–Loebl Langmuir Trough

A computer-controlled Joyce–Loebl trough (Gateshead, U.K.) of larger size was
used for some monolayer studies. The Wilhelmy plate was made of filter paper
strip (Whatman #1). The surface area was controlled from 1000 to 200 cm^2 with
a constant perimeter barrier made of Teflon tape. The precision of the surface
pressure measurement was ± 0.5 mN/m and that of the area was ± 5 cm^2. The
trough temperature was controlled with a water circulator.

4. Bubble Method

A commercial, thermostatted, computer-controlled, pulsating Bubble Surfacto-
meter (PBS), from Electronetics Co. (Amherst, NY), was used for measuring
dynamic surface tensions $\gamma(t)$. The instrument, which is based on Enhorning's
design [53], measures the pressure difference across a bubble surface with a pressure transducer. The Laplace–Young equation, $\Delta P(t) = 2\gamma(t)/R(t)$, where ΔP is
the pressure difference and R is the radius of the bubble, is used for calculating
the surface tension [53–56]. The sample volume is approximately 25 μL. Under
constant-area conditions, the radius of the bubble is fixed at $R = 0.40$ mm and
the surface tension is measured every 50 ms after a 1-s initial delay or "dead"
time. Under the pulsating area, the radius varies nearly sinusoidally from $R =
0.40$ to 0.55 mm (area ratio = 1.89) by using a liquid volume displacer. The
bubble volume and area can be pulsated at various pulsating frequencies ranging
from 1 to 100 cycles/min. The bubble is quite small and nearly spherical. The
bulk rheology and other dynamic effects are generally insignificant at the conditions used [55,56]. Surface rheology effects may be important, but usually not
dominant at the conditions used [55,56]. Most of the experiments were performed
within a day after the dispersions were prepared, to eliminating possible effects
of particle coagulation or DPPC hydrolysis [8,57].

5. Pendant–Drop Apparatus

A Ramé–Hart apparatus with a video camera coupled to a Q570 image analyzer
(Cambridge Instruments) provided an efficient and accurate method for determining the surface tension. There is an x-y-z stage, a light source, a syringe support,

and a mounting post for a camera. The video camera, with the microscope adapter attached to the mounting post, is set up to acquire live images at specific time intervals and then to process them into binary images. Each image is sliced into 40 horizontal sections, the radii of which are measured. Then, the drop-profile data of radii, $r(z)$, and the distances from the apex of the drop, z, are acquired. These data are used for determining the surface tension through the method of profile matching or axisymmetric drop-shape analysis [58,59]. This involves optimizing the fit of the profile to the Laplace–Young equation and calculating the surface tension. In the pendant-drop method, as the surfactant particles in a dispersion floated due to buoyancy forces, they moved away from the air–water interface, in contrast to the bubble method, in which buoyancy caused some particles to move toward the interface. These effects, the bubble geometry, and possible convection or stirring in the liquid phase can affect significantly the rate of mass transfer to the interface and the rate of surface tension decrease.

6. Fourier-Transform Infrared Spectroscopy

A Nicolet (Madison, WI) 800 FTIR Spectrometer with a liquid-nitrogen-cooled MCT (mercury–cadmium–telluride) detector was used. It was continuously purged with dry air from a Balston air purifier. The ATR accessory was custom-built by Connecticut Instrument Co. (Norwalk, CT). An unpolarized incident beam was used. Germanium ATR crystals ($1 \times 10 \times 50$ mm, 45° trapezoids) were purchased from Wilmad Glass, Co. (Buena, NJ). The resolution of the FTIR spectra was 2 cm^{-1}; 256 scans were taken for each spectrum.

III. RESULTS AND DISCUSSION: AQUEOUS SP-C SYSTEMS

A. Spread Monolayer Isotherms

For SP-C, the areas per molecule are based on a presumed molecular weight of 4200. The surface pressure–surface area per molecule ($\pi–\bar{A}$) isotherm at 23°C (measured with the KSV trough) for SP-C monolayers shows substantial compressibility of the monolayer, no plateaus, and some hysteresis (Fig. 1A). For an average compression rate of 16 Å^2/molecule/min, the surface pressure changed gradually, with no detectable changes in the isotherm slope. The surface pressure increased to only 40 mN/m at the minimum surface area used, for an area compression ratio of about 4 of the initial area over the final area.

The behavior at 37°C was quite different from that at 23°C. For the same area per molecule, the surface pressure was significantly higher (Fig. 1B). For the area compression ratio of 6, π values as high as 50 mN/m were seen. Around $\pi = 40$ mN/m, the isotherm slope increased more slowly, indicating some

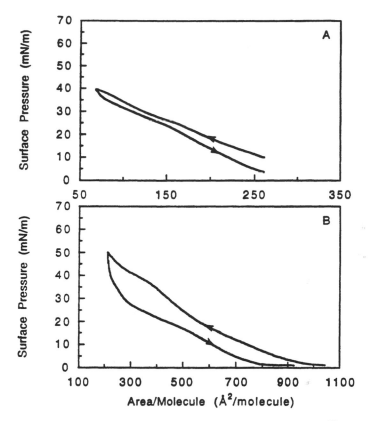

Figure 1   The surface pressure–surface area per molecule (π–$\overline{A}$) isotherm at 23°C (A) and at 37°C (B) for a pure SP-C monolayer on saline solution using 1/1 (v/v) methanol–chloroform as the organic solvent. The KSV trough was used for the measurements at 25°C (A) with a surface-area compression rate of 16 Å^2/molecule/min. The Joyce–Loebl trough was used for the measurements at 37°C (B) with a surface-area compression rate of 80 Å^2/molecule/min.

changes in the molecular packing or some transition in the monolayer structure. The maximum surface pressure observed was 50 mN/m at the minimum trough area. Hence, the SP-C has a significant capability to decrease the surface tension under surface compression.

B.   Dynamic Surface Tension of Aqueous Dispersions

When measured with the pendant-drop method at 23°C, the surface tension decreased quite slowly. For a 1-ppm dispersion, the surface tension changed little

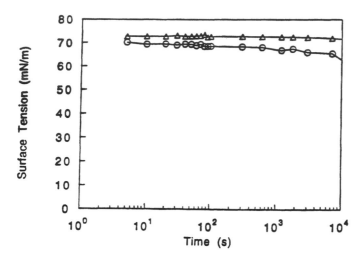

Figure 2 Dynamic surface tensions for SP-C dispersions in saline measured with the pendant-drop method at 23°C; concentrations of 50 ppm (O) and 1 ppm (△).

and remained around 72 mN/m, even after 3 h (Fig. 2). For the 50-ppm dispersion, the surface tension changed more substantially, but dropped to only ~62 mN/m after 3 h. The SP-C seems to have little surface activity by itself, in a time scale of hours, probably because of slow dissolution and apparently low equilibrium spreading pressure.

With the bubble method in saline and at 37°C (to mimic physiological conditions), the surface tensions decreased faster than at 23°C (Fig. 3). Another reason for the difference could be the different geometrical arrangements of the surface in the bubble and pendant-drop methods [44]. For the 50-ppm dispersion, the surface tension decreased to 49 mN/m after 40 min. For the 25-ppm dispersion, it decreased to 41 mM/m after 100 min. Hence, the SP-C is moderately surface active at 37°C and somewhat less active at 23°C. For the 1-ppm dispersion, the surface tension changed more slowly, as expected, and it reached 53 mN/m after 3 h. With the bubble method, the dynamic surface tension behavior of these dilute SP-C dispersions may be more susceptible to the presence of small amounts of impurities. Also, results with different measurement methods may differ if the presence of the dispersed particles and their transport toward or away from the surface are important [44–46]. Hence, our results could differ from those of literature sources, which use procedures of spreading a liquid dispersion on or under the surface under continuous stirring [5].

The dynamic surface tensions of the above three dispersions under pulsating area conditions (20 cycles/min) are higher than 38 mN/m (Fig. 4). Table 1

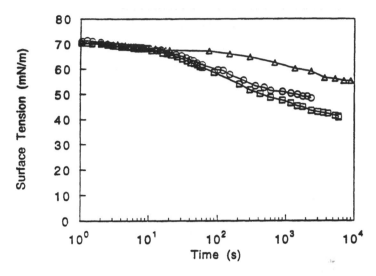

Figure 3 Dynamic surface tensions for SP-C dispersions in saline measured by the bubble method (PBS) at 37°C: (○) 50 ppm; (□) 25 ppm; (△) 1 ppm.

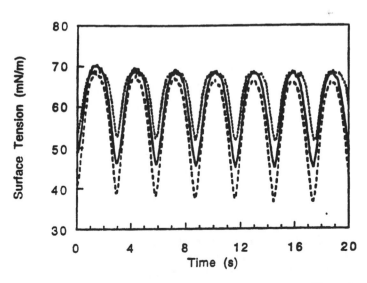

. igure 4 Dynamic surface tensions as the surface area oscillates sinusoidally at 37°C at 20 cycles/min for SP-C dispersions in saline: (—) 50 ppm; (---) 25 ppm, (...) 1 ppm.

Table 1 Surface Tensions of SP-C Aqueous Dispersions During Pulsation
Experiments

Concentration (ppm)	Frequency (cycles/min)	$\gamma_{initial}$ (mN/m)	γ_{max} (mN/m)	γ_{min} (mN/m)	$\Delta\gamma$ (mN/m)
50	10	47	63	41	22
	20	49	67	44	23
	30	46	69	42	25
	50	47	69	44	25
	80	46	67	43	24
1	1	54	66	52	14
	6	54	66	52	14
	10	54	68	51	17
	20	54	69	51	18
	40	55	69	52	17
	60	54	69	52	17
	80	55	68	53	15

lists the initial surface tension (possibly equilibrium) and the maximum and minimum surface tensions during pulsation. The surface tension amplitudes $\Delta\gamma \equiv \gamma_{max} - \gamma_{min}$ were from 22 to 25 mN/m at 50 ppm and from 14 to 18 mN/m at 1 ppm. For 20 cycles/min, the period of oscillation is much smaller than the adsorption time scale. Thus, the SP-C dispersions behave as having a nearly insoluble monolayer, or a slowly adsorbing, surfactant. Model calculations for soluble surfactants by Chang and Franses [54] show a maximum in the surface tension amplitude versus frequency ν when the dimensionless frequency is between the "soluble" limit or fast adsorption relative to oscillation rate and the "insoluble" limit. The SP-C dispersions fit this trend, indicating that their adsorption rate is significant. At 1 ppm, the maximum surface tension increases to the range 66–70 mN/m and the minimum surface tension is very close to the initial surface tension. Evidently, at the higher concentration or at slower rates ($\nu \leq 10$ cycles/min), there is time for some additional material to adsorb on the bubble surface, and the maximum surface tension is lower than that under a higher pulsating frequency. The surface tension amplitudes versus frequency for the 50- and 1-ppm dispersions show broad maxima at 30 and 20 cycles/min, respectively.

IV. RESULTS AND DISCUSSION: AQUEOUS DPPC–SP-C MIXTURES

The isotherm of a mixed-spread monolayer is quite different from that of either pure monolayer (Fig. 5). The weight ratio of 43:1 was chosen because it is about

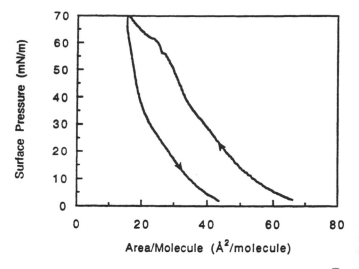

Figure 5   The surface pressure–average area per molecule (π–$\bar{A}$) at 37°C for DPPC–SP-C 43/1 by weight mixed monolayer on saline solution using 1/1 (v/v) methanol–chloroform solvent. The Joyce–Loebl trough was used. The surface area was compressed at 6 Å²/molecule/min.

the same as the ratio of DPPC to total protein (SP-C and SP-B) in Survanta. The isotherm shows no distinctive changes in the slope (no apparent transitions in the monolayer phases), except at a π value of about 55 mN/m, where the isotherm curve indicates some squeeze-out or other monolayer-collapse phenomena. The surface pressure increased gradually to 70 mN/m when the minimum trough area was reached. As the surface area was expanded, a sizable hysteresis in the surface pressure was observed, indicating some reorganization or rearrangement in the monolayers. The isotherm of a DPPC monolayer shows a high collapse pressure of about 70–72 mN/m (or a low surface tension $\gamma \cong 0$–2 mN/m) and a collapse area of about 40 Å²/molecule [2]. This is a typical area of a close-packed monolayer of a double-chain surfactant. The DPPC–SP-C mixed monolayer isotherm does reach $\pi \approx 70$ mN/m, but at a smaller average area per molecule. Because the SP-C alone cannot achieve very high surface pressures and because of the large weight ratio of DPPC to SP-C, we infer that the behavior of the mixed monolayer, especially at high π's, is dominated by DPPC. However, the SP-C interacts with DPPC sufficiently to alter the monolayer isotherm. Longo et al. [11] reported a collapse pressure of ~66 mN/m for a monolayer mixture of DPPC–PG–PA–SP-B1-25, with composition of 69/21/9/3 w/w. PA is palmitic acid, PG is phosphatidylglycerol, and SP-B1-25 is a positively charged amino-terminal peptide. These authors concluded that a specific interaction between the cationic peptide and the anionic lipid was responsible for the stabilization of the DPPC-rich monolayers. Other authors have confirmed the interaction of SP-C with DPPC in the monolayer with a variety of techniques [3,20,23,27,31,37,40].

The question now is whether a mixed monolayer forms by adsorption from aqueous dispersion. The dynamic surface tensions of aqueous DPPC–SP-C mixtures of weight ratio 43:1 (13,000 ppm DPPC and 300 ppm SP-C) were measured with the bubble method at 37°C. The weight ratio was the same as in the mixed monolayers of Figure 1. The dispersions were prepared with the Protocol-2-S procedure (detailed in Refs. 44 and 45), in which heating above 60°C with stirring is followed by a 1-h sonication in a sonicator bath and finally cooling to 37°C. Under constant area (Fig. 6), the surface tension decreased within 1 s from 72 mN/m (of pure saline) to ~50 mN/m. Then, it decreased more slowly to a value of 29 mN/m after about 1000 s. By contrast, the surface tension after 1 s for the pure 13,500-ppm DPPC dispersion with the same preparation protocol was ~65 mN/m, and it decreased to 30 mN/m after 10^4 s [45,46]. Even though the equilibrium surface tension of the DPPC–SP-C mixtures was similar to that of DPPC dispersions prepared by sonication, the surface tension of the former decreased much faster in the first few seconds than that of the latter.

With the bubble area pulsated at 20 cycles/min (Fig. 7), the minimum surface tension at the sixth cycle was 4 mN/m. At the 10th cycle (not shown), γ_{max} was 49 mN/m and the minimum surface tension was 1 mN/m. It was observed

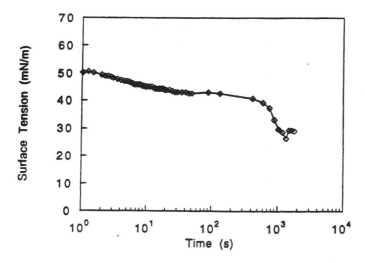

Figure 6   Dynamic surface tensions for 13,500/300 ppm/ppm DPPC–SP-C mixed dispersions in saline measured with the bubble method (PBS) at 37°C.

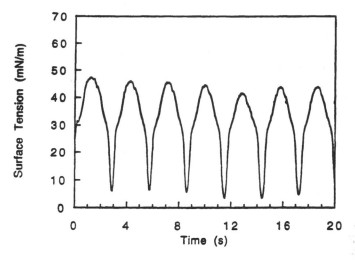

Figure 7 Dynamic surface tensions at 37°C as the surface area oscillates sinusoidally at 20 cycles/min for 13,500/300 ppm/ppm DPPC–SP-C dispersions (see Fig. 6).

that the bubble was deformed (flattened) at the time of the minimum surface tension, evidently due to the gravity force dominating the surface tension forces [47,55,56]. For such a shape of the bubble, the actual minimum surface tension may be even less than 1 mN/m at the minimum bubble area [55,56]. The results at other frequencies were not recorded because it was difficult to maintain the bubble size without deforming the bubble and the method was deemed unreliable. The bubble deformation indicated low surface tension, however. Smith et al. [12] reported a minimum surface tension of ~7 mN/m for 9/1 w/w DPPC–SP-B/SP-C mixture, with a concentration of 2.0 mg/mL. Although the method of sample preparation and surface tension measurement were different, their results indicate the feasibility of producing low minimum surface tensions with such a mixture.

The maximum surface tension for pure DPPC dispersions was ~61 mN/m and the minimum surface tension was <1 mN/m if the dispersion was prepared by a Protocol-2 followed by sonication [46–48]. The maximum surface tension for DPPC–SP-C mixtures was much lower than that of pure DPPC. At constant area, the surface tension decreased faster and the adsorption of molecules to the air–water interface was faster for the DPPC–SP-C mixtures than for pure DPPC. As the bubble expanded to the maximum area in 1.5 s (half of the pulsating period), the surface tension increased only to 49 mN/m. The maximum surface tension is close to the surface tension (50 mN/m) obtained after 1 s under the constant-area conditions. Similarly to hexadecanol in the DPPC–hexadecanol mixtures [44–48], the SP-C seems to play an important synergistic role in lowering the dynamic surface tension behavior of pure DPPC to ~50 mN/m quicker, within a few seconds after a new air–water interface is created.

The weight ratio of DPPC to SP-C at the surface could be quite different from the weight ratio of 43 : 1 in the bulk dispersion, because of selective competitive adsorption. Moreover, after surface compression, the spread mixed monolayer may show a "squeeze-out" effect in which the weight ratio may change by preferential collapse of some domains. For probing the dynamic surface compositions of DPPC–SP-C mixtures, some preliminary experiments were done on spread monolayers which were transferred to an ATR crystal plate and analyzed with FTIR-ATR spectroscopy. Perdeuterated DPPC (d-DPPC) was used for separating their infrared hydrocarbon bands from those of SP-C. To increase the sensitivity of the analysis for SP-C, the following mixed monolayer rich in SP-C was examined: 1 : 1 d-DPPC–SP-C by mol or 0.16 : 1 by weight. The π–$\overline{A}$ isotherm at 23°C for such a mixed monolayer (Fig. 8) is different from that of 99.6/0.4 mol/mol (43/1 by weight) mixtures (Fig. 5), showing some evidence of monolayer phase change at ~14 mN/m, from a liquid condensed to a more-ordered and less compressible monolayer state. The monolayer collapsed at ~64 mN/m and showed substantial hysteresis.

Figure 9 shows the spectra (in select regions) of pure SP-C and pure d-DPPC cast films obtained with the FTIR-ATR technique [46,60]. The spectrum of

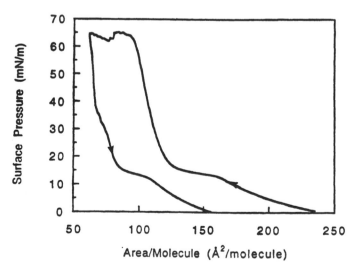

Figure 8   Surface pressure–average surface area per molecule (π–$\overline{A}$) isotherm at 23°C for a 1/1 mol/mol d-DPPC–SP-C mixed monolayer on saline solution using 1/1 (v/v) methanol–chloroform as the organic solvent. The KSV trough was used.

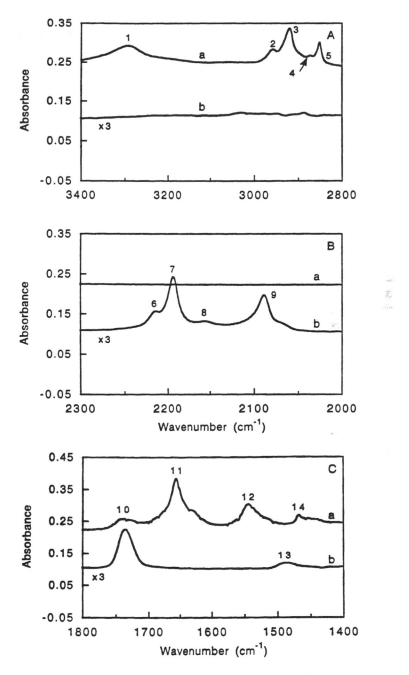

Figure 9 FTIR-ATR spectra of cast films of pure SP-C (a) and d-DPPC (b). Three different spectral regions, (A, B, and C) are plotted. Band assignments and peak positions are given in Table 2. The spectra are shifted vertically for clarity. The spectrum of d-DPPC is multiplied by a factor of 3 for clarity. The incident beam was unpolarized.

SP-C shows the hydrocarbon band region and also the secondary protein structure region [61]; see Table 2. The peak at 1656 cm^{-1} is assigned to the amide I peak, which indicates an α-helical secondary structure. It consists primarily of C=O vibration ($\sim$80%) and, in small part ($\sim$10% each), of CN vibrations and CH bending [40,57]. The peak at 1544 cm^{-1} is assigned to amide II vibrations. It consists of CN vibrations (40% of the intensity) and NH vibrations (60% of the intensity) [61]. For perdeuterated DPPC, the CD stretching vibration bands are in the region between 2300 and 2000 cm^{-1}. One would expect that the two components would be easily distinguished in the spectrum of a mixture.

Figure 10 shows the spectra of Langmuir–Blodgett films deposited at 23°C at surface pressures of 10 and 60 mN/m. Even though the mixture's initial mole ratio was 1:1, for either deposition pressure no SP-C was detected and only d-DPPC was detected. A larger signal was observed at 60 mN/m, as expected from the isotherm (Fig. 8). It remains unclear whether the lack of detection of SP-C is due to low experimental sensitivity or indicates the absence of SP-C from the deposited monolayer. The latter possibility seems unlikely, because indirect surface tension evidence indicates the presence of SP-C in the monolayer at the conditions used here. External reflection infrared data for different protocols of monolayer formation support the presence of SP-C at the surface [40]. Under in

Table 2 Assignments of Major Infrared Band Peaks for SP-C and d-DPPC (1/1 mol/mol) Cast Films

Peak[a]	Assignment	SP-C position (cm^{-1})	d-DPPC position (cm^{-1})
1	Bonded NH$_2$	3290	—
2	CH$_3$-v_a	2957	2950
3	CH$_2$-v_a	2919	—
4	CH$_3$-v_s	2871	2887
5	CH$_2$-v_s	2851	—
6	CD$_3$-v_a	—[b]	2214
7	CD$_2$-v_a	—	2193
8	CD$_3$-v_s	—	2156
9	CD$_2$-v_s	—	2088
10	C=O	1737	1736
11	Amide I	1656	—
12	Amide II	1544	—
13	CH$_3$-v_a, overtone	—	1485
14	CH$_2$-δ	1467	—

Note: v-stretch; δ-scissors; subscript a-asymmetric; subscript s-symmetric.
[a] Unpolarized incident light was used.
[b] Peaks are small or not observed.

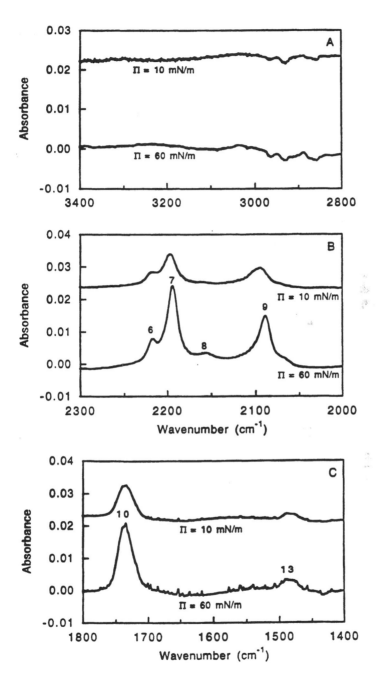

Figure 10 FTIR–ATR spectra of Langmuir–Blodgett monolayers of an initial bulk mole ratio of d-DPPC to SP-C of 1:1, deposited at the indicated surface pressures. In all spectral regions, the spectrum at 10 mN/m has been shifted vertically for clarity. Band assignments and peak positions are given in Table 2. The incident beam was unpolarized.

vivo conditions, the SP-C and other proteins may also play additional roles in forming a lipoprotein complex, which is more easily transportable to the surface. In those cases, there may be a coupled transport of lipids and proteins in addition to synergistic adsorption. Under the conditions used in this chapter, no coupled transport is expected.

V. CONCLUSIONS

The SP-C is surface active on its own, and its detailed dynamic surface tension behavior depends on the method used for measuring the surface tension and probably on its dispersion state. Because the adsorption of SP-C onto the air–water interface from the bulk dispersions is slow compared to the period of area pulsation, the SP-C surface layers behave as essentially insoluble monolayers during pulsation. The mixtures of DPPC–SP-C, however, show faster adsorption behavior, especially in the first few seconds after the air–water interface is created. Under pulsating area conditions (20 cycles/min), they show a smaller range of surface tensions, between 42 mN/m and 1 mN/m. The presence of SP-C in the dispersion leads to faster surface tension equilibration and lowers the maximum tension under pulsating area conditions, evidently without the need for a coupled transport, as with a lipid–protein complex, which may exist in vivo. Because both DPPC and SP-C are dispersed as solid particles, little interaction (or mutual dissolution) is expected in the bulk aqueous phase. Their synergistic behavior is inferred to be due to favorable interactions in the surface layer. The surface activity (minimum surface tension or maximum surface pressure under surface compression) of SP-C molecules depends on the temperature. The evidence of the presence of SP-C molecules in the monolayer at 23°C is inferred from the surface tension behavior but not confirmed from the LB deposition–FTIR analysis, perhaps for reasons of low experimental sensitivity.

ACKNOWLEDGMENTS

We are greateful to the National Science Foundation (grants CTS 93-04328 and 96-15649) for patial support of this work. We are also thankful to Dr. D. J. Ahn for useful discussion and advice on the FTIR method.

REFERENCES

1. JA Clements. Am Rev Respir Dis 115:67, 1977.
2. KMW Keough. In: B Robertson, LMG Van Golde, JJ Batenburg, eds. Pulmonary

Surfactant: From Molecular Biology to Clinical Practice. Amsterdam: Elsevier, 1992, p. 109.

3. J Johansson, T Curstedt. Eur J Biochem 244:675, 1997.
4. K Shiffer, S Hawgood, N Düzgünes, J Goerke. Biochemistry 27:2689, 1988.
5. SB Hall, AR Venkitaraman, JA Whitsett, BA Holm, RH Notter. Am Rev Respir Dis 145:24, 1992.
6. GA Simatos, KB Forward, MR Morrow, KMW Keough. Biochemistry 29:5807, 1990.
7. K Shiffer, S Hawgood, HP Haagsman, B Benson, JA Clements, J Goerke. Biochemistry 32:590, 1993.
8. J Johansson, T Curstedt, B Robertson, H Jörnvall. Biochemistry 27:3544, 1988.
9. W Seeger, A Günther, C Thede. Am J Physiol 261:L286, 1992.
10. T Curstedt, J Johansson, P Persson, A Eklund B Robertson, B Löwenadler, H Jörnvall. Proc Natl Acad Sci USA 87:2985, 1992.
11. ML Longo, AM Bisagno, JAN Zasadzinski, R Bruni, AJ Waring. Science 261:453, 1993.
12. GB Smith, HW Taeusch, DS Phelps, KMW Keough. Pediatric Res 23:484, 1988.
13. Y Tanaka, T Takei, T Aiba, K Masuda, A Kiuchi, T Fujiwara. J Lipid Res 27:475, 1986.
14. J Johansson, T Curstedt, B Robertson. Acta Pediatricia 85:642, 1996.
15. AD Horowitz, B Elledge, JA Whitsett, JE Baatz. Biochim Biophys Acta 1107:44, 1992.
16. JE Baatz, KL Smyth, JA Whitsett, C Baxter, DR Absolom. Chem Phys Lipids 63: 91, 1992.
17. AD Horowitz, JE Baatz, JA Whitsett. Biochemistry 32:9513, 1993.
18. DA Evans, RW Wilmott, JA Whitsett. Pediatric Pulmonol 21:328, 1996.
19. AJ Davis, AH Jobe, D Hafner, M Ikegami. Am J Respir. Crit Care Med 157:553, 1998.
20. Z Wang, O Gurel, RH Notter. J Lipid Res 37:1749, 1996.
21. Z Wang, O Gurel, JE Baatz, RH Notter. J Biol Chem 271:191, 1996.
22. Z Wang, SB Hall, RH Notter. J Lipid Res 37:790, 1996.
23. J Perez-Gil, K Nag, S Taneva, KMW Keough. Biophys J 63:197, 1992.
24. S Taneva, KMW Keough. Biophys J 66:1137, 1994.
25. S Taneva, KMW Keough. Biophys J 66:1149, 1994.
26. S Taneva, KMW Keough. Biophys J 66:1158, 1994.
27. SG Taneva, KMW Keough. Biochemistry 33:14,660, 1994.
28. R Qanbar, S Cheng, F Possmayer, S Schurch. Am J Physiol 271:L572, 1996.
29. K Nag, J Perez-Gil, A Cruz, NH Rich, KMW Keough. Biophys J 71:1356, 1996.
30. S Taneva, KMW Keough. Biochemistry 36:912, 1997.
31. K Nag, J Perez-Gil, A Cruz, KMW Keough. Biophys J 71:246, 1996.
32. K Nag, SG Taneva, J Perez-Gil, A Cruz, KMW Keough. Biophys J 72:2638, 1997.
33. AS Dico, S Taneva, MR Morrow, KMW Keough. Biophys J 73:2595, 1997.
34. A Post, A von Nahmen, M Schmitt, J Ruths, H Riegler, M Sieber, HJ Galla. Mol Membrane Biol 12:93, 1995.
35. A von Nahmen, M Schenk, M Sieber, M Amrein. Biophys J 72:463, 1997.
36. M Amrein, A von Nahmen, M Sieber. Eur Biophys J 26:349, 1997.

37. A von Nahmen, A Post, HJ Galla, M Sieber. Eur Biophys J 26:359, 1997.
38. B Pastrana, AJ Mautone, R Mendelsohn. Biochemistry 30:10,058, 1991.
39. CR Flach, JW Brauner, JW Taylor, RC Baldwin, R Mendelsohn. Biophys J 67:402, 1994.
40. B Pastrana-Rios, S Taneva, KMW Keough, AJ Mautone, R Mendelsohn. Biophys J 69:2531, 1995.
41. A Gericke, CR Flach, R Mendelson. Biophys J 73:492, 1997.
42. C Schroder, A Gunther, W Seeger, W Volter. Biomed Peptide Proteins Nucleic Acids 1:1994, 1998.
43. HD Modanlou, K Beharry, G Padilla, K Norris, S Safvatis, JV Aranda. J Perinatol 17:455, 1997.
44. SY Park, C-H Chang, DJ Ahn, EI Franses. Langmuir 9:3640, 1998.
45. SY Park, SC Peck, C-H Chang, EI Franses. In: DO Shah, ed. Dynamic Properties of Interfaces and Association Structures. Champaign, IL: American Oil Chemists Society Press, 1996, p. 1.
46. SY Park, Ph.D. thesis, Purdue University, West Lafayette, IN, 1995.
47. C-H Chang, KA Coltharp, SY Park, EI Franses. Colloids Surfaces A 114:185, 1996.
48. EI Franses, C-H Chang, JB Chung, KA Coltharp, SY Park, DJ Ahn. In: DO Shah, ed. Micelles, Microemulsions, and Monolayers: Science and Technology. New York: Marcel Dekker, 1998, p. 417.
49. J Johansson, H Jornvall, A Eklund, N Christensen, B Robertson, T Curstedt. FEBS Lett 232:61, 1988.
50. RW Olafson, U Rink, S Kielland, S-H Yu, J Chung, PGR Harding, F Possmayer. Biochim Biophys Res Commun 148:1406, 1987.
51. S-H Yu, W Chung, RW Olafson, F Possmayer. Biochim Biophys Acta 921:437, 1987.
52. JT Stults, PR Griffin, DD Lesikav, A Naidu, B Moffat, BJ Benson. Am J Physiol 261:L118, 1991.
53. GJ Enhorning. J Appl Physiol: Respir Environ Exercise Physiol 43:198, 1977.
54. C-H Chang, EI Franses. Chem Eng Sci 49:313, 1994.
55. C-H Chang, EI Franses. J Colloid Interf Sci 164:107, 1994.
56. SB Hall, MS Bermel, YT Ko, HJ Palmer, G Enhorning, RH Notter. J Appl Physiol. 75:468, 1993.
57. JB Chung, PC Shanks, RE Hannemann, EI Franses. Colloids Surfaces 43:223, 1990.
58. Y Rotenberg, L Boruvka, A Neumann. J Colloid Interf Sci 93:169, 1983.
59. AH Alexopoulos, EI Franses. Colloids Surfaces 43:263, 1990.
60. SY Park, EI Franses. Langmuir 11:2187, 1995.
61. H Susi. In: SN Timasheff, GD Fasman, eds. Structure and Stability of Biological Macromolecules. New York: Marcel Dekker, 1969.

20

Monolayer Penetration as a Means of Understanding Systems of Mixed Amphiphiles

Sekhar Sundaram
Air Products and Chemicals, Inc., Allentown, Pennsylvania

Kathleen J. Stebe
The Johns Hopkins University, Baltimore, Maryland

ABSTRACT

The fundamental understanding of adsorption of single surfactants or amphiphiles onto interfaces in terms of the equilibrium and dynamic transport properties is fairly well developed. However, mixed surfactant systems are less well understood. Here, a surfactant penetrating into a pre-existing insoluble monolayer is discussed as a means of studying mixed surfactant systems in a controlled manner.

We have developed models to study the equilibrium and dynamic behavior of the surface pressure. From the equilibrium data, interactions between the insoluble and soluble components can be quantified. The dynamic surface pressure evolution can be analyzed in terms of the surfactant diffusion and ad-desorption kinetics. The insoluble monolayer is predicted to reduce the diffusion timescales strongly as the surface coverage of the insoluble component is increased. Thus, at high surface coverage of insoluble component there may be a shift in mechanism controlling the mass transfer from diffu-

We would like to thank Darsh Wasan for his scientific insight and for his leadership in the Surface Science Community.

sion to kinetic control. This model is valid for any two amphiphiles, provided that one is insoluble, and the other is soluble.

Experimentally, the dynamics of monolayer penetration is studied using a penetration trough which allows the insoluble monolayer to be exposed to the soluble amphiphile with a well-defined initial condition. Experiments were performed on the system of the hard protein lysozyme penetrating into an insoluble monolayer of DPPC. The diffusion timescale is shown to vary as predicted. Furthermore, the surface phase change apparent in the lysozyme dynamic surface tension trace is disrupted by the insoluble component, also in agreement with theory.

Future directions of this research are briefly described. These include the demonstration of a shift in mechanism from diffusion to kinetic control with increasing surface coverage of an insoluble component and an investigation of ionic effects in monolayer penetration.

I. INTRODUCTION

Surfactants adsorb onto fluid interfaces, where they reduce the surface tension. They are used as additives to regulate the surface tensions between immiscible fluids in a wide variety of applications. The fundamental understanding of dynamic surface tension in single surfactant systems is well developed in terms of the surfactant mass transfer and the thermodynamic relationship between the surface concentration and the surface tension (as reviewed in Ref. 1). However, in most applications, surfactants are added as mixtures. These mixtures often have desirable properties caused by interactions among the components that are not achieved by a single surfactant. For example, the phenomenon of synergistic surface tension reduction is well documented; when two surfactants interact attractively, the net amount of surfactant adsorbed at the interface of the mixture is greater than the sum of the surface concentrations for the two components separately. As a result, the surface tension is strongly reduced (see, for example, Ref. 2).

Here, the system of a soluble surfactant penetrating into a preexisting insoluble monolayer is used to study surfactant interactions in a controlled manner. Monolayers are formed using a surfactant that is virtually insoluble in water. Typically, they are spread from a volatile solvent on the surface of water contained within a Langmuir trough. Teflon barriers that sweep along the interface are used to compress the monolayer to the desired area/molecule [3]. The surface pressure is commonly monitored by the Wilhelmy plate technique. To study interactions with a soluble amphiphile, the soluble molecule may simply be injected into the subphase. However, the disturbances caused by the resulting convection makes it difficult to study the transport dynamics. A less disruptive means of studying this system involves the use of a penetration trough [4] in which the monolayer is formed on a surfactant-free subphase, and transferred laterally to a surfactant-containing subphase. This can be achieved with little bulk convection, as evidenced by the surface pressure, which can be monitored throughout the transfer process. Thus, the evolution of the surface pressure as the soluble surfactant penetrates the monolayer can be studied with a well-defined initial condition. (See Figure 1 for a schematic of a penetration trough.)

Many previous monolayer penetration studies have focused on mixtures of soluble and sparingly soluble amphiphiles which occur in nature. For example, lung surfactant is comprised of lipids that form insoluble monolayers and soluble apoproteins; the search for a lung surfactant replacement has motivated lipid–protein penetration studies (see, for example, Ref. 5). The membranes of cells

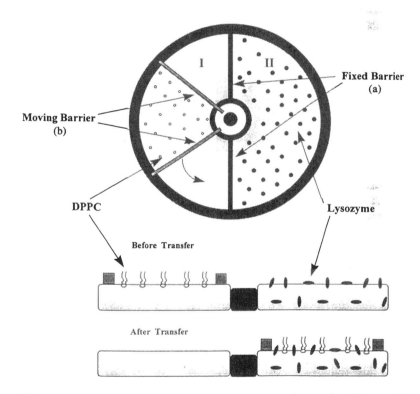

Figure 1 A penetration trough for monolayer penetration studies. The trough is divided into two regions by fixed barriers (a). There are also two mobile barriers (b) which compress the monolayer and transfer it from one region to the other. The insoluble monolayer is spread and compressed over the surfactant-free subphase in region I. It is transferred to the surfactant-containing subphase in region II.

are comprised of lipid bilayers with integral and peripheral proteins. Monolayer penetration studies provide a means for understanding how membrane fluidity (as inferred from the ability to form surface-condensed phases and from compressibilities calculated from the slopes of compression isotherms) is altered by a variety of drugs, proteins, and surfactants (see, for example, Ref. 6).

Optical techniques are also used for studying monolayer penetration. Recently, the surface concentration of a monolayer penetrated by a soluble amphiphile has been studied using surface-reflectometry techniques to infer the adsorbed mass [7]; Brewster angle and fluorescent microscopies have been used to directly visualize the domains which form in the evolving mixed monolayers [8].

We have developed a framework to analyze these systems that allows a quantitative analysis of the surface pressure in terms of the (fixed) surface concentration of the insoluble component and the instantaneous concentration of the soluble component [9–11]. Our treatment parallels the approach commonly adopted to study the dynamic surface tension of soluble surfactants [12–16]. Typically, these studies are structured as follows. The equilibrium surface tension is measured as a function of the bulk concentration. These data are analyzed using a surface equation-of-state based on the Gibbs adsorption equation and an assumed adsorption isotherm. They are used to find the best-fit isotherm parameters. The dynamic surface tension reduction is then measured. The surface tension is assumed to be in local equilibrium with the instantaneous surface concentration. The surface concentration evolution is modeled as being controlled by bulk diffusion, or the kinetics of adsorption–desorption, or both. The surface tension evolution is then regressed to find the best-fit diffusivities and adsorption–desorption kinetics constants. Thus, using the surface equation-of-state to relate the surface tension to the surface composition, dynamic surface tension studies are recast as mass-transfer problems.

In the monolayer penetration problem, the insoluble component (indicated by the subscript 1) is spread at the interface to a surface coverage x_1 on a subphase that is free of a soluble component (indicated by subscript 2). The initial surface pressure is given by $\pi_0 = \pi(x_1, x_2 = 0)$, where x_2 is the surface coverage of the soluble component. The monolayer is then transferred at a constant surface area onto a subphase which contains the soluble component at bulk concentration $C_{\infty 2}$. Once the monolayer is transferred onto the surfactant-containing subphase, the surface pressure evolves to a final value $\pi_f = \pi(x_1, C_{\infty 2})$. The surface pressure increase ($\Delta \pi$) caused by surfactant penetration of the monolayer is

$$\Delta \pi = \pi_f - \pi_0 = \pi(x_1, C_{\infty 2}) - \pi(x_1, x_2 = 0) \tag{1}$$

In this chapter, the derivation of the Gibbs adsorption equation which is properly constrained to account for the presence of the insoluble layer is reviewed. Assuming an adsorption isotherm $x_2(x_1, C_{\infty 2})$, an equation-of-state for $\Delta \pi$ $(x_1, C_{\infty 2})$ is found. The equilibrium change in surface pressure as a function of bulk concentration can then be regressed to find the best-fit isotherm parameters. Assuming a local state, $\Delta\pi(t)$ can be analyzed in terms of its instantaneous composition. In the following, the equilibrium and dynamic analyses are reviewed. Some supporting experimental data are presented, and future directions are discussed. Because our approach underscores the similarities in the dynamic adsorption of a single surfactant onto an initially surfactant-free interface and the dynamic adsorption (or penetration) of a soluble surfactant onto an initially surfactant populated interface, the terms *adsorption* and *penetration* are used interchangeably throughout the remainder of this text.

II. EQUILIBRIUM

A. The Monolayer Penetration Equation

An equation is determined governing the increase in surface pressure $\Delta\pi$. The Gibbs convention of a mathematical dividing surface located so that the surface excess of the solvent is zero is adopted. The Gibbs adsorption equation at constant temperature is

$$d\gamma = -\Gamma_1 \, d\mu_1 - \Gamma_2 \, d\mu_2 \tag{2}$$

where Γ_i is the surface excess and μ_i is the chemical potential of the ith component. This equation can be rearranged to show

$$d\Phi = d\left(\mu_1 + \frac{\gamma}{\Gamma_1}\right) = -\frac{\Gamma_2}{\Gamma_1}\left(\frac{\partial\mu_2}{\partial C_2}\right)_{T,P} dC_2 + \gamma \, d\left(\frac{1}{\Gamma_1}\right) \tag{3}$$

where Φ is defined by Eq. (3). The total differential of μ_2 at constant T and P is

$$d\mu_2 = \left(\frac{\partial\mu_2}{\partial C_2}\right)_{T,P} dC_2 \tag{4}$$

By equating mixed partial derivatives of Φ, Eq. (3) can be manipulated to show

$$\left(\frac{\partial\gamma}{\partial C_2}\right)_{T,P,1/\Gamma_1} = -\left(\frac{\partial\mu_2}{\partial C_2}\right)\left(\frac{\partial(\Gamma_2/\Gamma_1)}{\partial(1/\Gamma_1)}\right)_{T,P,C_2} \tag{5}$$

Integrating Eq. (5) over C_2 for an ideal bulk solution of uniform concentration $C_{\infty2}$ gives

$$\pi = \gamma(C_{\infty2})=0,\Gamma_1) - \gamma(C_{\infty2},\Gamma_1) = -\int_0^{C_{\infty2}} -\left(\frac{RT\,\partial\ln C_2}{\partial C_2}\right)\left(\frac{\partial(\Gamma_2/\Gamma_1)}{\partial(1/\Gamma_1)}\right)_{T,P,C_2} dC_2 \tag{6}$$

Given an adsorption isotherm [i.e., $\Gamma_2(C_{\infty2}, \Gamma_1)$], Eq. (6) can be integrated to find the corresponding equation for $\Delta\pi$. (This parallels the approach taken for soluble surfactants adsorbing onto a surfactant-free interface. The Gibbs adsorption equation is cast into an appropriate form for single-surfactant adsorption. An adsorption isotherm is assumed. The Gibbs adsorption equation is integrated to give the corresponding surface equation-of-state. For an example of such a manipulation, see the work of Frumkin [17].)

B. The Equation for $\Delta\pi$ for a Two-Component Frumkin Model

In adsorption-site models, the adsorption flux of the soluble component to the interface is assumed to be first order in the sublayer concentration C_{2s} and proportional to the space available for surfactant adsorption on the interface. The desorption flux is assumed to be first order in the surface concentration. The Frumkin model [17] is the simplest of such models that accounts for interactions among the adsorbing molecules.

The maximum packing of the soluble surfactant on the interface is given by $\Gamma_{\infty2}$; the equilibrium surface concentration is given by Γ_{eq2}. The maximum packing of the soluble surfactant $\Gamma_{\infty1}$ is given by the inverse minimum area/molecule $(1/A_{1min})$, where A_{1min} is the minimum area/molecule to which the insoluble component can be quasistatically compressed. Therefore, the fraction of the interface occupied by each component at equilibrium x_i is

$$x_i = \frac{\Gamma_{eqi}}{\Gamma_{\infty_i}} = \frac{A_{imin}}{A_{ieq}} \tag{7}$$

where A_{ieq} is the mean area per molecule to which the insoluble monolayer is compressed. The net flux to the interface is

$$\frac{\partial \Gamma_2/\Gamma_{\infty2}}{\partial t} = \beta C_{2s}\left(1 - \frac{\Gamma_2}{\Gamma_{\infty2}} - x_1\right) - \frac{\alpha\Gamma_2}{\Gamma_{\infty2}} \tag{8}$$

where β is the adsorption kinetic constant and α is the desorption kinetic constant.

The kinetic constants α and β can be expressed with an Arrhenius dependence:

$$\begin{aligned}\beta &= \beta_0 e^{E_a/RT}\\ \alpha &= \alpha_0 e^{E_d/RT}\end{aligned} \tag{9}$$

where E_a and E_d are the activation energies for adsorption and desorption, respectively. If the activation energies are independent of the surface concentration of either species (i.e., $E_a = E_a^\circ$ and $E_d = E_d^\circ$), the Langmuir adsorption isotherm for penetration into an insoluble monolayer results. If there are interactions between the surfactants, the activation energy for adsorption–desorption depends on surface concentration. As yet, the functional form of these activation energies in terms of the surface concentrations of each component has not been established. Commonly, these energies are expanded to retain the linear terms in a Taylor series in the surface concentrations. E_a and E_d can become

$$E_a = E_a^\circ + \sum_{i=1,2} v_{a_i} \Gamma_i$$

$$E_d = E_d^\circ + \sum_{i=1,2} v_{d_i} \Gamma_i \tag{10}$$

where v_{a_i} (v_{d_i}) is the partial derivative of the adsorption (desorption) energy with respect to the surface excess of component i. At equilibrium, the rate of change in Eq. (8) is zero, and the Frumkin adsorption isotherm is found:

$$\frac{x_2}{1 - x_1} = \frac{k}{k + e^{K_1 x_1 + K_2 x_2}} \tag{11}$$

where x_i is defined in Eq. (7). In Eq. (11), the adsorption number k is a weighted concentration,

$$k = \frac{C_{\infty 2}}{a} \tag{12}$$

and a is a measure of the surface activity of the molecule,

$$a = \frac{\alpha_0}{\beta_0} \exp\left(\frac{E_{a0} - E_{d0}}{RT}\right) \tag{13}$$

The greater the tendency of a surfactant or amphiphile to adsorb, the greater is $1/a$. The interaction parameters K_i are defined as

$$K_i = \frac{(v_{a_i} - v_{d_i}) \Gamma_{\infty 1}}{RT} \tag{14}$$

where $i = 1$ or 2 for 1–2 or 2–2 interactions, respectively. Attractive interactions cause the energy for desorption to increase faster than that for adsorption, resulting in $K_i < 0$. Conversely, repulsion yields $K_i > 0$. For $K_i = 0$, the Langmuir isotherm is recovered. Using Eq. (6), the equation for $\Delta\pi$ is given in closed form as

$$\Delta\pi = -RT\Gamma_{\infty 2}\left[\ln\left(1 - \frac{x_2}{1 - x_1}\right) - K_1 x_1 x_2 - \frac{K_2}{2} x_2^2\right] \tag{15}$$

Equation (15) is a two-component Frumkin model for the interface; it can also be derived from a regular solution theory for monolayers [18]. The interaction parameters K_i, the surface activity a and maximum packing of the soluble component $\Gamma_{\infty 2}$ are found by fitting Eq. (15) to equilibrium surface pressure data.

C. Equilibrium Model Results

The effects of surface saturation and intermolecular interactions on $\Delta\pi$ as a function of k are discussed in detail in Ref. 9. The trends can be understood by considering Eqs. (11) and (15). These equations show that in the absence of interactions, x_2 is reduced by x_1 by the factor $(1 - x_1)$, but $\Delta\pi$ is unaltered by x_1. If there are 1–2 attractive interactions, x_2 increases relative to the ideal case, yielding a higher $\Delta\pi$ as x_1 increases. The opposite trend is found when there are 2–2 attractive interactions, because x_1 reduces the amount of 2 that can adsorb and, therefore, the ability of 2 to attract itself. These trends are shown in Figures 2a–2c, where

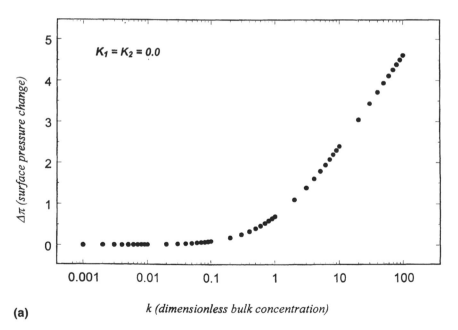

(a)

k (dimensionless bulk concentration)

Figure 2 A family of curves showing the dimensionless surface pressure increase $\Delta\pi$ (scaled by $RT\Gamma_{\infty 2}$) versus the dimensionless concentration k for $x_1 = 0.1, 0.5,$ and 0.9 (a) For no interactions (i.e., $K_1 = K_2 = 0$), the three curves superpose. (b) For 1–2 attractive interactions (shown here for $K_1 = -2.0$ and $K_2 = 0$), $\Delta\pi$ increases with x_1. (c) For 2–2 attractive interactions (shown for $K_2 = -2.0$ and $K_1 = 0$), $\Delta\pi$ decreases with x_1.

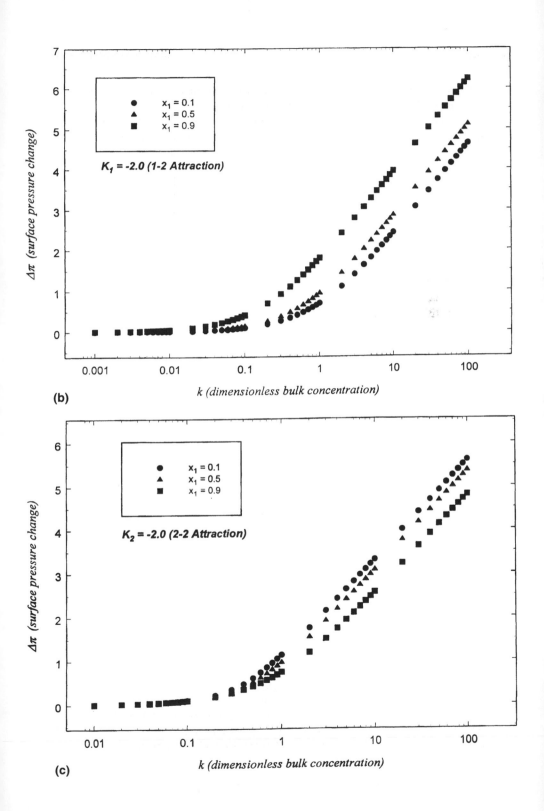

(b)

(c)

the dimensionless $\Delta\pi$ (scaled with $RT\Gamma_{\infty 2}$) versus the dimensionless concentration k is graphed as a function of interactions and x_1.

The results in Figure 2 show that intermolecular interactions have a pronounced effect. The trends in $\Delta\pi$ with interactions are summarized in Figure 3 for the Frumkin model. Consider a surfactant solution at a fixed concentration (i.e., k fixed). If that soluble surfactant were to adsorb onto a monolayer-free interface, it would decrease the surface tension by the amount π, where π is given by the single-component Frumkin equation:

$$\pi = -RT\Gamma_{\infty 2}\left(\ln(1 - x_2) - \frac{K_2}{2}x_2^2 \right) \tag{16}$$

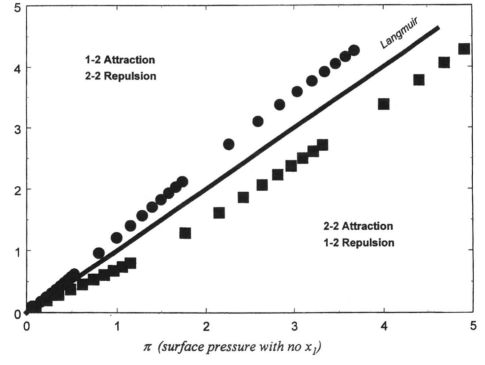

Figure 3 A comparison of the dimensionless surface pressure increase $\Delta\pi$ (scaled by $RT\Gamma_{\infty 2}$) determined by Eqs. (11) and (15) to the increase in surface pressure with no insoluble component π defined in Eq. (16). For no intermolecular interactions, $\Delta\pi = \pi$. For 1–2 attractive interactions (shown for $K_1 = -2.0$, $K_2 = 0$) or 2–2 repulsive interactions (shown for $K_1 = 0$ and $K_2 = 2.0$), $\Delta\pi > \pi$. For 1–2 repulsion ($K_1 = -2.0$ and $K_2 = 0$) and 2–2 attraction ($K_1 = 0$ and $K_2 = -2.0$), $\Delta\pi < \pi$. All curves here have $x_1 = 0.5$.

where x_2 is the surface coverage in Eq. (11) with $x_1 = 0$. If that same surfactant solution were placed in contact with an insoluble monolayer with fractional coverage of the insoluble component x_1, the surface pressure would increase at equilibrium to $\Delta\pi$ as given by Eq. (15), where x_2 is the surface coverage in Eq. (11) for the x_1 value that described the monolayer.

In the ideal case (i.e., no intermolecular interactions), π is equal to $\Delta\pi$ regardless of the value of x_1. This is represented by the 45° line in Figure 3. However, if there are 1–2 attractive interactions, $\Delta\pi$ is greater than π. The difference between these two quantities increases with x_1. The same trends are obtained if there are 2–2 repulsive interactions. Thus, in Figure 3, the region above the 45° line corresponds to either 1–2 attractive or 2–2 repulsive interactions. The opposite trends with x_1 are obtained for either 1–2 repulsive or 2–2- attractive interactions, so the region beneath the 45° line corresponds to either 1–2 attractive or 2–2 repulsive interactions. Differentiation between 1–2 and 2–2 interactions can be made by comparing the equation-of-state in Eq. (16) to equilibrium surface pressure data for component 2 separately, from which K_2 can be independently obtained.

III. MASS-TRANSFER MODEL FOR MONOLAYER PENETRATION

In this section, the diffusion-controlled penetration of a soluble surfactant into a preexisting insoluble monolayer is modeled. The analysis is a direct extension of the diffusion-controlled model of Ward and Tordai [19]. One significant difference between the single-surfactant and monolayer penetration cases is in the diffusion timescale that emerges from a dimensional analysis of the governing equations. In the following subsections, it is shown that because the insoluble amphiphile occupies space at the interface, it reduces the adsorption of the soluble component at equilibrium. Because less of the soluble component must diffuse to the interface to establish equilibrium, the timescale for equilibration is reduced. [In this discussion, intermolecular interactions are neglected. Therefore, Eqs. (11) and (15) reduce to a two-component Langmuir model for monolayer penetration. This model has been extended to include interactions [11]; the dynamic behavior of this system including strong 2–2 attractive interactions is compared to the experiment in Section IV.B.]

A. Diffusion-Controlled Adsorption

As shown schematically in Figure 4, a preexisting insoluble monolayer with coverage x_1 is exposed at time $t = 0$ to a solution of soluble surfactant of initially

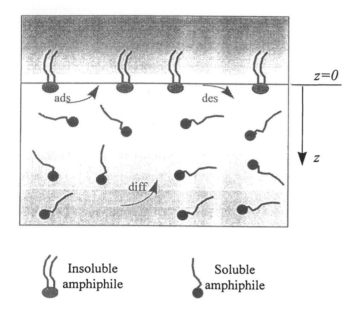

Figure 4 A schematic of the mass-transfer process. An insoluble monolayer is exposed at t = 0 to a solution of soluble surfactants. The coordinate z indicates the depth beneath the interface.

uniform concentration $C_{2\infty}$. The monolayer is initially free of component 2. Bulk diffusion establishes the sublayer concentration $C_{2s}(t)$:

$$C_{2s}(t) = C_2(z = 0, t) \tag{17}$$

where $z = 0$ locates the interface, z pointing into the solution. Local equilibrium between $C_{2s}(t)$ and $\Gamma_2(t)$ is maintained according to the adsorption isotherm, which, in the absence of interactions, is simply

$$\frac{\Gamma_2(t)/\Gamma_{\infty 2}}{1 - x_1} = \frac{(\beta/\alpha)C_{2s}}{(\beta/\alpha)C_{2s} + 1} \tag{18}$$

The bulk concentration $C_2(z, t)$ is determined by Fickian diffusion:

$$D\frac{\partial^2 C_2}{\partial z^2} = \frac{\partial C_2}{\partial t} \tag{19}$$

where D is the bulk diffusion coefficient of component 2. The initial conditions are

$$\Gamma_2(t=0) = 0$$
$$C_2(z, t=0) = C_{2\infty} \qquad (20)$$
$$C_{2s}(t=0) = 0$$

The boundary conditions for a diffusion-controlled flux to the interface are

$$\lim_{z\to\infty} C_2 = C_{2\infty}$$
$$\frac{d\Gamma_2}{dt} = D\frac{\partial C_2}{\partial_z}\bigg|_{z=0} \qquad (21)$$

The solution of Eqs. (17) and (19)–(21) is the Ward and Tordai solution:

$$\Gamma_2(t) = 2\sqrt{\frac{D}{\pi}}\left(\sqrt{t} - \int_0^{\sqrt{t}} C_{2s}(t - \tau)d\sqrt{\tau}\right) \qquad (22)$$

which is solved numerically simultaneously with the instantaneous adsorption isotherm Eq. (18) to give $\Gamma_2(t)$. Given $\Gamma_2(t)$, $\Delta\pi(t)$ is given by

$$\Delta\pi(t) = -RT\Gamma_{2\infty}\ln\left(1 - \frac{\Gamma_2(t)x_2}{\Gamma_{2\,eq}(1 - x_1)}\right) \qquad (23)$$

These equations can be recast in dimensionless form using the following definitions:

$$C_{2s}^* = \frac{C_{2s}}{C_{2\infty}}; \qquad \Gamma_2^* = \frac{\Gamma_2}{\Gamma_{2\,eq}}; \qquad t^* = \frac{t}{\tau_D}; \qquad \Delta\pi^* = \frac{\Delta\pi}{RT\Gamma_{2\infty}} \qquad (24)$$

where the asterisk superscript denotes a dimensionless variable. The characteristic diffusion timescale τ_D in Eq. (24) is

$$\tau_D = \frac{h^2}{D} \qquad (25)$$

where h is defined as

$$h = \frac{\Gamma_{2\,eq}}{C_{2\infty}} = \frac{x_2\Gamma_{\infty 2}}{C_{2\infty}} = \frac{(1 - x_1)\Gamma_{\infty 2}}{C_{2\infty}}\left(\frac{k}{k + 1}\right) \qquad (26)$$

This is the adsorption depth, which is the characteristic distance below the interface depleted by surfactant adsorption.

The governing equations in dimensionless form become

$$\Gamma_2^*(t^*) = \frac{2}{\sqrt{\pi}}\left(\sqrt{t^*} - \int_0^{\sqrt{t^*}} C_{2s}^*(t^* - \tau^*)\,d\sqrt{t^*}\right) \qquad (27)$$

$$\Gamma_2^*(t^*) = (1 + k)\left(\frac{C_{2s}^*(t^*)}{1 + C_{2s}^*(t^*)k}\right) \tag{28}$$

$$\Delta\pi^*(t^*) = -\ln\left(1 - \frac{\Gamma_2^*(t^*)x_2}{1 - x_1}\right) = -\ln\left(1 - \frac{\Gamma_2^*(t^*)k}{1 + k}\right) \tag{29}$$

Note that at equilibrium, both Γ_2^* and C_{2s}^* are unity.

By choosing τ_D to scale time, Eq. (27) is free of any physicochemical parameters. Because $\Gamma_2^*(t^*)$ results from the simultaneous solution of Eqs. (27) and
(28), $\Gamma_2^*(t^*)$ depends only on the dimensionless bulk concentration k. Although
x_1 appears explicitly in Eq. (29), recall that in the absence of interactions, x_2 $(1 -
x_1) = k/(1 + k)$. Therefore, the surface pressure change $\Delta\pi$ is also independent
of x_1, since interactions among amphiphiles has been neglected.

B. Diffusion-Controlled Adsorption: Numerical Results

Figures 5a and 5b are dimensionless plots of $\Gamma_2^*(t^*)$ and $\Delta\pi^*(t^*)$, respectively.
As predicted by Eqs. (27)–(29), the results for fixed k and differing x_1 superpose
in dimensionless form. Thus, the role of the insoluble component in the surface
pressure evolution for diffusion-controlled adsorption of a noninteracting surfactant is captured by scaling time with τ_D. This timescale accounts for the fact that
the insoluble component occupies space at the interface, thereby reducing the
amount of soluble surfactant that can adsorb onto the interface. Because less
soluble surfactant can adsorb into the monolayer, the time required to deliver the
surfactant by diffusion is reduced.

Because τ_D goes as $(1 - x_1)^2$, the dimensional dynamic quantities evolve
to equilibrium more rapidly as x_1 increases. Thus, in the absence of interactions,
the dynamic behavior of the interface is altered by x_1, even though the equilibrium
behavior is not.

The definition of τ_D in Eq. (25) is general. However, the result that τ_D
reduces strictly proportionally to $(1 - x_1)^2$ is valid only for the no-interactions
case, because x_2 does not vary simply as $1 - x_1$ when interactions are finite.
However, regardless of the magnitude of the interactions, τ_D diminishes with x_1.
These results are discussed in greater detail in Ref. 10.

IV. EXPERIMENTS

The experiments were performed in a penetration trough (used by S.S. at the
Max Planck Institute for Colloid and Surface Science, Berlin), which is described

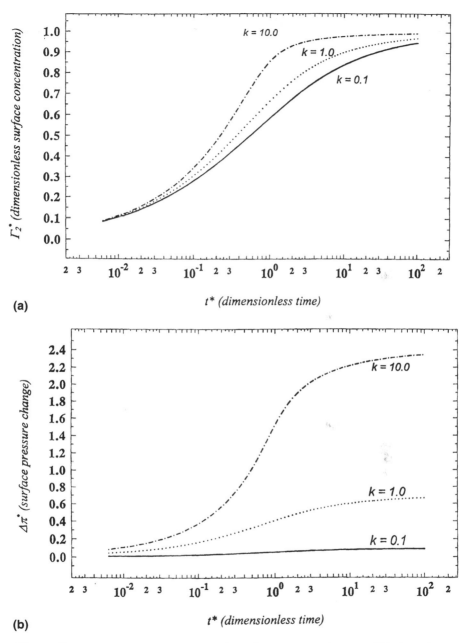

(a)

(b)

Figure 5 (a) The dimensionless dynamic surface concentration trace $\Gamma_2^*(t^*)$ and (b) the dimensionless evolution of the surface pressure change $\Delta\pi^*(t^*)$. Results for fixed k and differing x_1 superpose in dimensionless form. This demonstrates that in the absence of interactions, the role of the insoluble component in altering the surface pressure evolution has been isolated in the timescale τ_D used to define the dimensionless time t^*.

in detail elsewhere [10,11]. The adsorption of the soluble protein lysozyme (a hard protein) into an insoluble monolayer of the lipid dipalmitoylphosphatidyl-choline (DPPC) was studied. The details can be found in Refs. 10 and 11. Here, some of the main results are reviewed.

The penetration trough is depicted in Figure 1. A circular trough was partitioned by fixed barriers (A) into two semicircular regions (labeled I and II in the figure). Region I was filled with pure buffer solution; region II was filled with buffered lysozyme solution. The interfaces in both regions were aspirated before depositing DPPC on region I. In addition to the fixed divid-ing barriers, there are two more barriers which sweep the interface (B). The DPPC monolayer is deposited and compressed by the two sweeping barriers in region I. The monolayer is then moved to region II, where it is exposed to the surface-active lysozyme. The transfer takes roughly 20 s. During the transfer process, the monolayer is kept at constant area. The surface pressure is moni-tored by a Wilhelmy plate, which also moves from one region to the other with the barriers. Surface pressure data are recorded continuously for up to 1600 min.

A. Experimental Results and the Diffusion Timescale

In these experiments, the surface pressure evolution timescale reduced with x_1, in rough agreement with the factor $(1 - x_1)^2$ suggested by the diffusion timescale argument.

In Figure 6, the dynamic $\Delta\pi$ results for lysozyme at a fixed bulk concentra-tion of 8.3×10^{-5} wt% are presented as a function of x_1. In Figure 6a, $x_1 = 0$ (i.e., there was no DPPC at the interface); in Figures 6b and 6c, $x_1 = 0.45$ and $x_1 = 0.55$, respectively. The x_1 values are calculated by taking the ratio of the minimum area per molecule for DPPC (or the collapse area, assumed to be 0.4nm^2/molecule) to the mean area per DPPC molecule in the experiment. (Simi-lar behavior was found for bulk lysozyme concentrations of 3×10^{-5} wt% and 1.35×10^{-4} wt% [11]). The time at which the surface pressure has attained 90% of its final value is indicated as a bold dot in the figures. For each experiment, this is reported as $t_{90\%}$ in Table 1. In all cases, the time required to attain equilib-rium surface pressure is reduced with increasing x_1. In the third column of Table 1, the time required for the lipid-free case to attain 90% of its final value is indicated by $t_{90\%}|_{x_1=0}$. The reduction in equilibration time predicted by the Lang-muir argument for τ_D is the product $t_{90\%}|_{x_1=0}(1 - x_1)^2$. Comparing this calculated value to the time attained in experiment, the τ_D argument is in fair agreement with the data, although the time required in experiment is always longer than the predicted time.

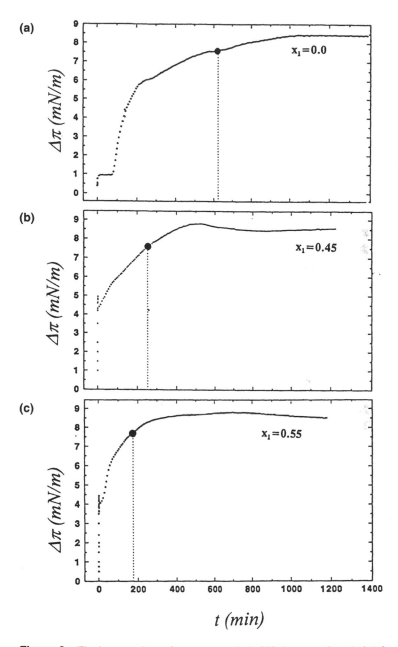

Figure 6   The increase in surface pressure $\Delta\pi$(mN/m) versus time (min) for a lysozyme at a concentration of 8.3×10^{-5} wt% adsorbing onto a clean air–water interface from buffer solution at pH 7.4 and ionic strength 0.15. The time at which the surface pressure has attained 90% of its final value, $t_{90\%}$, is indicated as a bold dot. The graphs correspond to (a) $x_1 = 0.0$ (no DPPC), (b) $x_1 = 0.45$, and (c) $x_1 = 0.55$.

Table 1 Experimental Data

x_1	$t_{90\%}$ (min)	$\dfrac{t_{90\%}\vert_{x_1=0}}{(1-x_1)^2}$ (min)
	$C_{2\infty} = 3 \times 10^{-5}$ wt%	
0.0	1075	
0.44	630	337
	$C_{2\infty} = 8.3 \times 10^{-5}$ wt%	
0.0	630	—
0.45	250	191
0.55	170	128
	$C_{2\infty} = 1.35 \times 10^{-4}$ wt%	
0.0	670	
0.44	360	270

Note: $t_{90\%}$ is the time required to achieve 90% of the equilibrium surface pressure. The first column indicates the fractional coverage of the insoluble component. The second column indicates the time required to attain 90% of the final surface pressure in experiment. The third column represents the time as predicted by the diffusion timescale for diffusion-controlled adsorption.

B. Interactions and Surface Phase Behavior: Numerical and Experimental Results

If attractions among the soluble molecules are sufficiently pronounced, a soluble amphiphile can undergo a surface phase change from a surface gaseous to a surface liquid-expanded state when it adsorbs at a clean interface. The critical points for the single-surfactant system can be determined by equating to zero the first and second derivatives with respect to x of a surface equation-of-state. Performing this calculation for Eq. [16], $K_2 = -4$ and $x_2 = \frac{1}{2}$ are the critical interaction parameter and the critical coverage for $x_1 = 0$, respectively (see Ref. 17). For interactions more pronounced than these (i.e., $K_2 < -4$), the interface phase separates into surface gaseous and surface liquid-expanded states. By performing a Maxwell construction on the interface, binodal concentrations can be found, locating the endpoints to the coexistence region [17]. The signature of a surface phase change in a dynamic surface tension evolution is a prolonged induction period before the surface pressure rises [20]. This induction period is predicted for a single surfactant undergoing a surface phase transition. The results are shown as the solid line in Figure 7 using Eq. (16) and the Maxwell construc-

tion [16]. Such an induction period is shown in the surface tension trace in Figure
6a for a lysozyme at the buffer–air interface.

The insoluble layer has a pronounced effect on the phase behavior of the
adsorbing surfactant. For example, for $K_2 = -6$, x_1, dilutes the interface, thereby
disrupting the self-interactions in the soluble compound, and can prevent the induction period. In Figure 7, the dashed curve is the result for $x_1 = 0.4$. Comparing the
dashed and solid curves, the induction period that is evident in the dynamic surface
pressure trace for $x_1 = 0$ is eliminated for finite x_1. This compares favorably to the
data for lysozyme penetrating into DPPC in Figure 6b and 6c [11].

V. FOCUS OF CURRENT RESEARCH

This monolayer penetration theory has only been compared to the lysozyme–
DPPC data thus far. However, the framework is general and should allow the

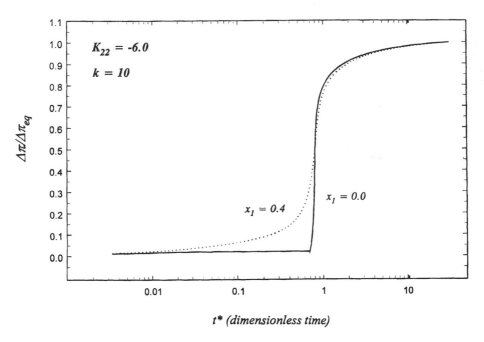

Figure 7 The predicted $\Delta\pi$ (scaled with its equilibrium value) versus dimensionless
time t^* for a diffusion-controlled surfactant which undergoes a gaseous to liquid-expanded
surface phase change when adsorbing on a monolayer-free interface with $K_2 = -6$. For
$x_1 = 0$ (the solid curve), a prolonged induction period occurs before the surface pressure
rises. As x_1 increases (the dashed curve, shown for $x_1 = 0.4$), the induction period is
eliminated. These results compare favorably to the data for a lysozyme in Figure 6.

quantitative study of any two-component system containing both an insoluble and a soluble amphiphile. We are extending the model to include adsorption–desorption barriers and electrostatic effects. Our current projects concerning monolayer penetration are described in the following subsections.

A.   Shift of Controlling Mechanism with x_1

When an interface is exposed to a surfactant solution, the surfactant adsorbs, depleting the sublayer concentration and causing the surfactant to diffuse to the interface. The supply to the interface by adsorption and diffusion continues until the interface and the solution attain equilibrium.

For surfactants adsorbing to initially clean (surfactant-free) interfaces, it has been established that the kinetics of this process is diffusion controlled at dilute bulk concentrations, shifting to mixed kinetic–diffusion-controlled adsorption at higher bulk concentrations [12–14]. This shift in mechanism occurs because the interface cannot deplete the bulk solution significantly at elevated concentration. The characteristic adsorption timescale is

$$\tau_{ads} = \frac{1}{\beta C_\infty} \tag{30}$$

The ratio of the characteristic diffusion timescale to the adsorption timescale, Λ, determines whether diffusion or adsorption is the rate-limiting process:

$$\Lambda = \frac{\tau_D}{\tau_{ads}} = \frac{\beta \Gamma_{eq}^2}{C_\infty D} \tag{31}$$

For $\Lambda \gg 1$, diffusion controls the mass transfer; for $\Lambda \sim 1$, the system has mixed control; for $\Lambda \ll 1$, adsorption controls the system. For a monolayer-forming surfactant, the surface concentration approaches the maximum packing as the concentration increases. Therefore, Λ diminishes with bulk concentration, and the system shifts from diffusion-controlled to kinetic-controlled mass transfer.

We are extending the mass-transfer model for adsorption into a preexisting monolayer to include mixed kinetic–diffusion control [21]. Using this model, conditions in which the mass flux is governed by diffusion and/or kinetics can be predicted. These predictions will be used to guide experiments to measure the kinetics for adsorption–desorption.

Recall that in the presence of an insoluble monolayer in the absence of intermolecular interactions,

$$\tau_D = \tau_D|_{x_1=0} (1 - x_1)^2 \tag{32}$$

and the adsorption timescale remains unchanged. The ratio of timescales, Λ, becomes

$$\Lambda = \Lambda|_{x_1=0}(1 - x_1)^2 \tag{33}$$

This suggests that monolayer penetration is diffusion controlled for both dilute bulk concentrations and small x_1, whereas for either elevated bulk concentrations or for x_1 approaching unity, the system is kinetically controlled.

In Table 1, the equilibration timescales for the lysozyme–DPPC experiments were compared to the shift expected for pure diffusion control. The time required to equilibrate in the experiment was always longer than the diffusion-controlled prediction, suggesting that the kinetic barriers to adsorption–desorption may play a role. A more quantitative analysis of the protein system was not performed because of the extremely long timescales for these systems and the lack of a true equilibrium state for proteins at interfaces. It is our aim to probe these issues for a simple surfactant penetrating an insoluble monolayer, for which a thorough quantitative analysis can be performed.

B. Ionic Effects in Adsorption

The adsorption of an ionic surfactant establishes a surface charge density, causing an electrical potential gradient to develop which retards the surfactant flux toward the surface. In order for additional surfactant to adsorb, it must overcome this potential gradient to adsorb onto the interface. The instantaneous potential during the adsorption process is described by the nonlinear Poisson–Boltzmann equations for the instantaneous surface charge density. The potential creates a Boltzmann distribution that reduces C_{2s} and alters the equilibrium and dynamics of the system. For surfactants with valences z_1 and z_2, respectively, the surface charge density is given by $z_1 F \Gamma_1 + z_2 F \Gamma_2(t)$, where F is Faraday's constant. By studying the adsorption of ionic surfactants into neutral ($z_1 = 0$) and charged (z_1 finite) monolayers with fixed hydrophobic moieties, the role of surface charge in altering surfactant adsorption can be studied. The quasiequilibrium analysis described by MacLeod and Radke [22] has recently been extended to include adsorption–desorption barriers and initial surface charge densities, allowing the role of ionic interactions in the surface pressure evolution to be elucidated [23].

VI. CONCLUSION

We have outlined a quantitative approach to the interpretation of both dynamic and equilibrium monolayer penetration studies. Our approach parallels that for dynamic surface tension studies for single-surfactant systems. The Gibbs adsorption equation is constrained to account for the presence of the insoluble mono-layer. A surface equation-of-state for the surface pressure increase caused by monolayer penetration is derived for an assumed isotherm. By regressing equilibrium data, the isotherm parameters can be obtained. Assuming a local state, the surface equation-of-state can be used to understand the surface pressure evolution in terms of the soluble-surfactant mass-transfer kinetics. Results obtained on the protein lysozyme penetrating an insoluble monolayer of DPPC were reviewed.

The aim of this work is to improve our understanding of the manner in which interactions among amphiphiles alter the thermodynamic and dynamic behavior of amphiphile mixtures. The research is designed to extend recent fundamental studies in single surfactant systems to two components. Our current work focuses on understanding the mass-transfer kinetics, the surface phase behavior, and ionic effects in the monolayer penetration problem.

REFERENCES

1. CH Chang, EI Franses. Colloids Surfaces 100:1, 1995.
2. MJ Rosen, F Zhao. J Colloid Interf Sci 95:443, 1983.
3. GL Gaines. *Insoluble Monolayers at Liquid–Gas Interface*. New York: Interscience Publishers, 1966.
4. D Vollhardt. Mater Sci Forum. 25:541, 1988.
5. BA Holm, G Enhorning, RH Notter. Chem Phys Lipids 49:49, 1988.
6. G Colacicco. Lipids 5:636, 1968; M Karel. J Food Sci 38:756, 1973; KS Birdi J Colloid Interf Sci 57:228, 1976; SS Chen, HL Rosano. J Colloid Interf Sci 61:207, 1977.
7. JR Charron, RD Tilton. J Phys Chem 100:3179, 1996.
8. TK Vanderlick, H Mohwald. J Phys Chem 94:886, 1990; MA Mayer, TK Vanderlick. J Chem Phys. 100:8399, 1994.
9. S Sundaram, KJ Stebe. Langmuir 12:2028, 1996.
10. S Sundaram, KJ Stebe. Langmuir 13:1729, 1997.
11. S Sundaram, JK Ferri, D Vollhardt, KJ Stebe. Langmuir 14:1208, 1998.
12. R Pan, J Green, CJ Maldarelli. J Colloid Interf Sci 205:213, 1998.
13. SY Lin, HC Chang, EM Chen. J Chem Eng Japan 29:634, 1996.
14. DO Johnson, KJ Stebe. J Colloid Interf Sci 182:526, 1996.
15. SY Lin, RY Tsay, LW Lin, SI Chen. Langmuir 12:6530, 1996.
16. JK Ferri, KJ Stebe. J Colloid Interf Sci 209:1, 1999.
17. A Frumkin. Z Phys Chem 116:466, 1925.
18. R Defay, I Prigogine. Surface Tension and Adsorption. New York: John Wiley & Sons, 1966.
19. AFH Ward, L Tordai. J Chem Phys 14:453, 1946.
20. SY Lin, K McKeigue, CJ Maldarelli. Langmuir 7:1055, 1991.
21. S Datwani, S Sundaram, KJ Stebe. Unpublished data.
22. C MacLeod, CJ Radke. Langmuir 10:3555, 1994.
23. S Datwani, KJ Stebe. J Colloid Interf Sci 219:282, 1999.

21

Protein–Silicone Interactions at Liquid–Liquid Interfaces

Vasiliki Bartzoka, Michael A. Brook, and Mark R. McDermott
McMaster University, Hamilton, Ontario, Canada

ABSTRACT

The enzyme alkaline phosphatase and the globular protein human serum albumin (HSA) were emulsified with low molecular weight silicone oil and a silicone/polyether surfactant to give stable emulsions. The biological activity of the proteins in the emulsions was assayed both in vitro (enzyme kinetics of alkaline phosphatase) and in vivo (immune response to oral administration of the HSA emulsion). Depending on the assay, either a small decrease or large increase in biological activity was noted. The nature of the silicone–protein interactions is discussed.

I. INTRODUCTION

Silicone emulsions have found wide application in the medical and pharmaceutical fields due to their desirable physical and chemical properties. For instance, silicone oil emulsions have been used for the preparation of elastomers used in controlled drug-delivery systems [1]. Simple aqueous [2] and nonaqueous silicone oil emulsions (e.g., propylene glycol-in-silicone oil) and multiple emulsions [3] have become increasingly popular in the cosmetics industry mainly due to the desirable physical properties imparted from the silicone oils. These properties include a smooth, nongreasy feel and a wide range of available viscosities, which eliminates the need to use waxes in formulations. However, despite the extensive industrial use of silicone oil emulsions, the systematic study of their interfacial properties is only in its infancy.

One of the current challenges in drug delivery is the formulation of systems for the administration of proteins and proteinaceous materials [4]. We have an interest in the use of silicones as vehicles for the oral delivery of vaccines. It is clearly important for such an application to understand the effects, deleterious or otherwise, of the silicone on the immunogenic protein. Most of the knowledge of these interactions comes from studies of protein solutions at solid silicone elastomer surfaces [5,6]. However, such studies on solid–liquid interfaces may not apply to liquid–liquid interfaces. Although fundamentally the same, these interfaces differ in interfacial properties: The former offer specific, fixed sites for protein adsorption, whereas the latter are homogeneous and flexible. It is accepted in general that proteins at fluid interfaces are more mobile and thus free to penetrate the nonaqueous phase. Furthermore, it is believed that proteins at fluid interfaces exist in a state between the native and the denatured state, termed the "molten globule" state [7]. The only study of silicone oil–protein interactions at liquid–liquid interfaces of which we are aware suggests that proteins undergo (detrimental) conformational changes upon long-term mixing with silicone oil [8].

Stable silicone emulsions could serve as protein carriers for drug-delivery systems. In the present study, we are concerned with the degree of retention of the biological activity of proteins in silicone emulsions. We have, therefore, evaluated the biological activity, in vitro and in vivo, of two different proteins in silicone emulsions.

II. MATERIALS AND METHODS

A. Materials

Alkaline phosphatase type VII-S from bovine intestinal mucosa (EC 3.1.3.1), human serum albumin (HSA, Fraction V, 96–99% albumin), and p-nitrophenyl phosphate tablets 104–105 were obtained from Sigma. Phosphatase-conjugated goat anti-mouse IgG or IgA was purchased from Southern Biotechnology Associates. Octamethylcyclotetrasiloxane (D_4; viscosity 2.3 cS) was obtained from Gelest. The 3225C formulation aid (a silicone–polyether copolymer) was provided by Dow Corning. Trishydroxymethylaminomethane (Tris) and HCl were obtained from BDH. All aqueous solutions were prepared with Milli-Q-treated distilled water. Female BALB/c mice, age 6–8 weeks were purchased from Charles River Laboratories Inc., Montreal.

B. Instrumentation

A Hewlett-Packard 8452A UV-VIS Spectrophotometer with constant temperature circulating bath was used for monitoring the enzymatic reaction. A Caframo mixer was used for the emulsion formulation and a Leitz Wetzlar microscope was used for droplet-size determination.

C. Methods

1. Emulsion Formulation

(a). Emulsion Composition. Different emulsion compositions were used for the different studies. For in vitro studies, the emulsion comprised a 50 wt% aqueous phase (50/50 composition). The protein concentration in the aqueous phase ranged from 0.08 to 0.008 mg/mL. The surfactant formulation aid was 2.5 wt% in the emulsion, which provides 0.25 wt% of active surfactant (the silicone–polyether surfactant is supplied as a 10 wt% active solution in cyclic silicones).

For in vivo studies, the emulsion formulated for the immunological studies contained a 25 wt% aqueous phase and 3.75 wt% of surfactant solution (25/75 composition). The protein concentration in the aqueous phase was 0.5 mg/mL.

(b). Emulsification Conditions. Emulsification was achieved under turbulent mixing by slow admixing of the aqueous phase to the silicone oil phase containing the macrosurfactant. The mixing required to produce stable emulsions was obtained in a vessel equipped with two blades: a straight (90°) and a pitched (45°) one. The pitched blade was placed between one-half to two-thirds of the liquid level height; the straight blade was placed at a level equal to one-sixth to one-third the diameter of the vessel (both measured from the vessel bottom). The shaft was slightly tilted to minimize vortex formation. This system provides a mixing tip velocity (speed at the edge of the blade most remote from the mixing shaft) of 275 m/min at an agitator speed of 3440 rpm. After completion of the addition of the aqueous phase, the emulsion was left stirring for at least as long as the time of addition (20 min).

2. Demulsification

Emulsion inversion was induced by increasing the volume fraction of the dispersed phase by simple addition of buffer solution. Prior to inversion, the original emulsion was concentrated to a gel-like consistency by centrifugation at 2000 rpm for 1 h. The supernatant was a clear octamethylcyclotetrasiloxane solution. Emulsion inversion and phase separation was achieved by slowly adding Tris–HCl buffer solution into the gel-like protein-in-silicone-oil emulsion with dual-blade turbulent mixing at 3440 rpm.

3. Enzymatic Activity

(a). In Situ. The protein-in-silicone emulsions were mixed either with an aqueous phase containing the substrate or another silicone oil emulsion containing an aqueous substrate solution. The enzymatic activity was qualitatively

observed by the development of a yellow color, characteristic of the product formation (p-nitrophenoxide from p-nitrophenyl phosphate) in emulsions containing alkaline phosphate at concentrations as low as 0.008 mg/mL.

(b). In Solution (Recovered Aqueous Phase). The hydrolytic activity of alkaline phosphatase was quantified by following spectrophotometrically the formation of p-nitrophenoxide at 404 nm. The molar absorption coefficient of p-nitrophenoxide at 404 nm in Tris–HCl buffer at pH = 8.0 is 18,000 M^{-1}/cm.

The kinetic experiments were performed in disposable plastic 4-mL (1 $cm^2 \times 5$ cm) cuvettes, containing a magnetic stirrer. The cuvettes were filled with the substrate and were equilibrated at 25°C before the enzyme addition. The enzymatic reaction was started by the rapid addition of 10 μL of alkaline phosphatase to 2.99 mL of a range of p-nitrophenyl phosphate concentrations in Tris–HCl buffer solution at pH = 8.0. The concentration range of p-nitrophenyl phosphate was 1–6 μM. The absorption signal was recorded every 2 s for 30 s. Initial velocities were derived from the velocity values that corresponded to the linear part of the Michaelis–Menten equation by extrapolation to t_0.

4.   Immunological Studies

A group of five mice was immunized orally with HSA-containing silicone oil emulsions (25/75 wt% composition). The animals received 50 μg of protein per dose, three times in 7-day intervals. The collection of individual blood samples was performed as described elsewhere [9] at days 7, 21, 28, 35, and 48. An enzyme-linked immunosorbent assay (ELISA) was used to detect and quantify the HSA-specific antibody responses in the individual serum samples. The ELISA assays were performed as described elsewhere [9].

III.   RESULTS

A.   Emulsion Characterization

1.   Droplet-Size Measurements

Microscopic viewing of one part of the emulsions, diluted in three parts of silicone oil, showed that the average droplet size was 5 μm and 2 μm for the 50/50 and 25/75 emulsion compositions, respectively.

2.   Stability Observations

In 50/50 and 25/75 compositions of buffer solutions in D_4, sedimentation was observed due to viscosity differences. Emulsions made with other more viscous internal phases (such as glycerol–buffer) at 50/50 ratios were gel-like and no creaming was observed after several months.

B. Enzymatic Activity

Alkaline phosphatase retains at least some of its enzymatic activity while being entrapped in silicone oil over several days. This was clear from the development of the yellow color characteristic of p-nitrophenoxide upon addition of p-nitrophenyl phosphate-containing solutions or emulsions to alkaline phosphatase-containing emulsions. However, the quantification of the degree of the retention of the enzymatic activity was not possible *in situ*. Therefore, the aqueous phase was recovered from the emulsions and an enzymatic assay was run in solution. The alkaline phosphatase solution that had been exposed to silicone in the emulsion was compared to an aqueous solution that had not been emulsified. The Lineweaver and Burk equation [10] was used to determine the kinetic parameters, K_M and V_{max}, of the enzymatic reactions

$$\frac{1}{U} = \frac{K_M}{V_{max}[S]} + \frac{1}{V_{max}}$$

where U is the initial velocity, K_M is the Michaelis–Menten constant, V_{max} is the maximum velocity, and [S] is the concentration of the substrate p-nitrophenyl phosphate (PNP).

A graph of $1/U$ against $1/[S]$ gave a straight line of slope K_M/V_{max} and ordinal intercept $1/V_{max}$. The K_M values were 1.34×10^{-5} M and 1.99×10^{-5} M, and the V_{max} values were 60 and 74 μM/min for the emulsified and nonemulsified alkaline phosphatase solutions, respectively (Fig. 1).

The turnover number, which is a measure of the enzymatic efficiency, can be derived from the V_{max} values. The turnover number is defined as the ratio of V_{max} to the total enzyme concentration. Whereas the enzyme concentration is less for the solution recovered from the emulsion due to the buffer added during demulsification, the turnover numbers for the emulsified and nonemulsified enzyme solutions should be similar within experimental error if enzyme degradation does not take place, as was observed (60 and 74, respectively). The determination of enzyme concentrations is not trivial at such low concentrations and work is ongoing to ensure that the results presented here are accurate and reproducible.

C. Immune Response

Human serum albumin was emulsified in silicone oil using the same conditions as for alkaline phosphatase. The emulsion was orally administered to mice as previously described for HSA-containing starch microspheres [9]. The immunogenicity of the protein contained in the emulsion was reflected in the blood antibody titer assayed using ELISA. The results are shown in Fig. 2 along with the microparticle assays for the purposes of comparison. In both cases, the protein dose to the animal was identical: 50 µg.

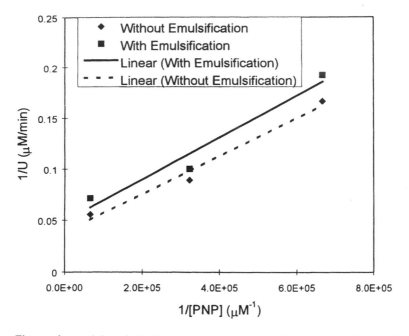

Figure 1   Activity of alkaline phosphatase without (◆) and with (■) emulsification in silicone oil.

It can be seen that the immunogenic activity of the HSA administered in a silicone emulsion is higher than the same amount of protein entrapped in a microparticle. We note that these results are preliminary and have not been repeated with the necessary controls in place.

IV. DISCUSSION

A. Emulsion Stability

Silicone emulsions formulated with the silicone–polyether emulsifiers used in the present study have been found to be stable [11]. This fact opens the door to a wide range of applications that exploit the desirable silicone oil properties. To date, protein (enzyme) entrapment in such emulsions has only been proposed [3]. In the current study, proteins were entrapped in silicone oil emulsions (Fig. 3) without any observable detrimental changes in the emulsion properties. We are, as yet, unable to comment on the ability of the protein to enhance emulsion stability.

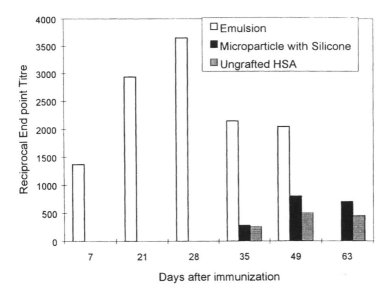

Figure 2 Immunogenic activity of HSA-in-silicone oil emulsions and in starch micro-particles coated with functionalized silicone.

Materials:

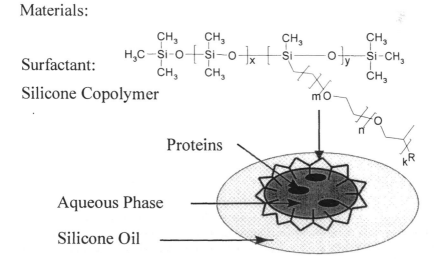

Figure 3 Structure of protein-in-silicone-oil emulsions.

B. Protein Stability

The focus of the present study was the determination of the degree of retention of biological activity of two different proteins, alkaline phosphatase (molecular weight ≈ 140 kDa) and HSA (molecular weight ≈ 70 kDa), after emulsification in silicone oil.

In vitro, alkaline phosphatase emulsified in silicone oil was found to be enzymatically active. Furthermore, the enzymatic activity was retained, even after breaking the emulsion, at levels similar to the enzyme solution that was not exposed to silicones.

In vivo, human serum albumin emulsified in silicone oil was found to lead to an immune response following oral administration to mice. This indicates that at least some of the protein was effectively protected through passage via the gastrointestinal tract of the animals (the oral administration of protein in water or in combination with unfunctionalized polydimethylsilicone oils and starch (ungrafted HSA), led to very minimal or no immune responses [9]); that is, the protein in the silicone oil emulsions was effectively protected from enzymatic degradation or decomposition due to the extreme pH conditions in the stomach. However, at least some of the protein was released from these emulsions (possibly in the intestine), in a form recognizable to the immune system, since HSA-specific antibodies were raised. A comparison of the magnitude of immune response to HSA entrapped in microparticles with the current HSA–silicone oil emulsion carrier system suggests the emulsion system to be promising direction for oral drug-delivery.

It should be noted that silicones can behave as adjuvants; that is, the immune response of a given protein can be enhanced when the protein is emulsified in silicones (or other hydrophobic materials) [12]. This effect is known for emulsions administered by injection. However, such an effect could also arise for the emulsions considered here; that is, the amount of protein required for the observed immune response may be less than normal as a result of the presence of the silicone emulsion. The question of silicone adjuvancy in these orally administered emulsions needs to be clarified.

These preliminary results demonstrate that the biological activity of proteins may be retained in silicone oil droplets even under the relatively harsh emulsification conditions used. This cannot be attributed to the conformational stability of the proteins studied, as the HSA is known to be a "soft" globular protein that can easily denature [13]. These findings are somewhat in contrast with previous work at silicone oil–protein solution interfaces [8], where protein denaturation was ascribed to its interaction with the silicone oil and with other work that attributed protein denaturation to emulsification processes [14]. There are many possible explanations for the contradiction between these studies and the current results.

The retention of enzymatic activity in the presence of silicone oil may be explained in four different ways. First, a significant amount of the protein present in the emulsion may remain in the water phase and not be affected by contact with silicone oil or the surfactant. Any degradation of protein at the interface would not affect the protein in the water bulk phase, from which the biological activity results.

Second, the proteins we have examined may not be affected by exposure to silicone oil, or conformational changes of the enzyme may indeed occur but not to the extent that the active site is completely and irreversibly altered. To test this possibility, silicone oil was mixed with protein in buffer solution in the absence of surfactant: Some enzymatic activity was observed but at a much lower level.

A third explanation is that the surfactant present in these emulsions plays an important role in the maintenance of biological activity. The surfactant we used, Dow Corning 3225C, is a rake copolymer consisting of a siloxane backbone and block polyethylene oxide–polypropylene oxide chains. It is possible that the protein never has the opportunity to interact directly with the silicone oil or with air that could also lead to conformational changes, due to steric protection at the interface provided by the polyethylene oxide–polypropylene oxide chains: It is well known that proteins are very compatible with such polyethers [15].

Finally, the protein may find its way to the interface where it acts as a cosurfactant. With favorable interactions with the surfactant, which restrict the flexibility/mobility of the protein at the interface, it is possible that denaturation is minimized. In this case, the orientation of the protein between the two phases may be very important. To investigate these possibilities, we are examining the biological activity of a series of different proteins over a range of concentrations.

V. SUMMARY

Preliminary in vitro and in vivo studies on protein–silicone interactions at fluid interfaces demonstrate that proteins entrapped in silicone oil emulsions retain their biological activity despite relatively harsh emulsification and demulsification conditions. Although still too early to reach any concrete conclusions, it can be said, based on these studies, that silicone emulsions could be developed into protein delivery systems.

ACKNOWLEDGMENTS

We gratefully acknowledge the financial support of the Natural Sciences and Engineering Research Council of Canada. We also wish to thank Dow Corning

Corporation for providing DC 3225C and Ken Kasprzak (Dow Corning) for helpful discussions.

REFERENCES

1. R Sutinen, V Laasanen, A Paronen, A Urtti, J Controlled Release 33:163, 1995.
2. RP Gee. U.S. Patent 5,300,286 (1994) (to Dow Corning).
3. G Dahms. Parfuem Kosmet 77:102, 1996.
4. SP Schwendeman, M Cardamone, A Klibanov, R Langer. In: S Cohen, H Bernstein, eds. Microparticulate Systems for the Delivery of Proteins and Vaccines Drugs and the Pharmaceutical Sciences (Vol. 77). New York: Marcel Dekker, 1996, p. 1.
5. BR Young, WG Pitt, SL Cooper. J Colloid Interf Sci 124:28, 1988.
6. V Bartzoka, MA Brook, MR McDermott. Langmuir 14:1887, 1998.
7. E Dickinson, Y Matsumura. Colloids Surfaces B: Biointerfaces 3:1, 1994.
8. L Sun, H Alexander, N Lattarulo, NC Blumenthal, JL Ricci, G Chen. Biomaterials 18:1593, 1997.
9. PL Heritage, ML Loomes, J Jianxiong, MA Brook, BJ Underdown, MR McDermott. Immunology 88:162, 1996.
10. RF Boyer, Modern Experimental Biochemistry. Menlo Park, CA: Benjamin/Cummings, 1986, p. 315.
11. GH Dahms, A Zombeck. Cosmet Toilet 109(11):91, 1995.
12. PC Klykken, KL White, Jr. In: M Potter, NR Rose, eds. Immunology of Silicones. (Current Topics in Microbiology and Immunology Vol. 210). Berlin: Springer-Verlag, 1996, p. 113.
13. E Blomberg, PM Claesson. In: TA Horbett, JL Brash, eds. Proteins at Interfaces II. Fundamentals and Applications. ACS Symposium Series, No. 602. Washington, DC: American Chemical Society, 1995, p. 296.
14. AL de Roos, P Walstra. Colloids Surfaces B: Biointerfaces 6:201, 1996.
15. P Alexandrididis, TA Hatton. Colloids Surfaces A: Physicochem Eng Aspects 96: 1, 1995.

22
Effect of a Nonadsorbing Polyelectrolyte on Colloidal Stability: An Overview

John Y. Walz
Yale University, New Haven, Connecticut

ABSTRACT

The effect of nonadsorbing polyelectrolytes (macromolecules) on the interaction between two charged particles was studied. First, a force-balance model was developed in which the particles and macromolecules were simulated as hard charged spheres. The model showed that the presence of charge can greatly increase both the magnitude and range of the depletion interaction between two particles. In addition, a significant, longer-range repulsion can arise due to an ordering of the macromolecules in the gap region.

Two types of experiments were then conducted. In the first, the interaction energy between a single colloidal particle and a flat plate in aqueous solutions of nonadsorbing macromolecules was measured using the optical technique of total internal reflection microscopy (TIRM). Second, the stability of a dispersion of charged particles in an aqueous solution of the macromolecules was studied using optical turbidity. Three different species were used to represent the non-adsorbing macromolecular material—hard charged silica spheres, cationic micelles (CTAB), and the anionic polyelectrolyte sodium polystyrene sulfonate (SPSS). With the silica sphere, reasonable agreement between the predicted and measured profiles was observed. In the stability experiments, both critical flocculation and higher critical restabilization concentrations were found. By comparison, the micelle and SPSS interaction energy profiles showed poor agreement with the model. In addition, both materials were able to induce flocculation of the particles, however no subsequent restabilization occurred. These experiments demonstrate the need for additional modeling work in these relatively complicated polymer-colloid systems.

I. INTRODUCTION

The depletion interaction arises whenever a dispersion of colloidal particles exists in equilibrium with a nonadsorbing macromolecular species. These macromolecules can be polymer molecules, micelles or other aggregate structures, or even other small particles. As two colloidal particles approach each other, such as through Brownian motion, the concentration of macromolecules in the gap region is altered relative to the bulk. At small separations, the macromolecules are excluded from the gap region, producing a depletion layer. This reduced concentration results in a lower osmotic pressure relative to the bulk, and the resulting attraction is termed the depletion force.

Although first observed experimentally in 1925, a satisfactory explanation of this phenomenon was not published until 1954 [1]. Asakura and Oosawa [1] showed that a volume exclusion mechanism could lead to a net attraction between two hard parallel plates immersed in a solution of rigid spherical macromolecules. Since then, numerous theoretical and experimental studies of the depletion interaction have been performed. Extensive reviews of prior work are given by Seebergh and Berg [2], Napper [3], and Walz and Sharma [4]. As pointed out by Walz and Sharma, much of the prior work in the field has focused on the depletion interaction produced by neutral macromolecules. Because the interaction in such systems is relatively small and below the sensitivity level of the common force-measurement techniques, most of the experimental work has consisted of studies of colloidal stability or phase transitions.

Our goal over the past several years has been to understand the depletion interaction produced by charged macromolecules on the interaction between two charged colloidal particles. Although, clearly, more complex than the uncharged case, systems of this type are commonly found in aqueous solutions. Our work to date has consisted of the development of a force balance model for predicting the depletion interaction in charged systems, direct experimental measurement of the interaction using the optical technique of total internal reflection microscopy, and studies of dispersion stability. This chapter will present an overview of each of these three components.

II. THEORY

Here, we provide a brief overview of our depletion model, which is described in detail elsewhere [4]. The model system consists of two hard, charged spheres

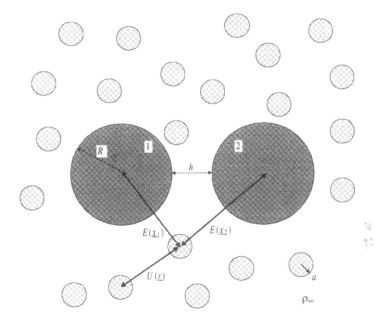

Figure 1   Schematic defining the variables used in developing the depletion force equation [Eq. (1)]. Two spherical particles of radius R are interacting across a gap distance of h in a solution containing a solute of spherical macromolecules of radius a and bulk number density ρ_∞.

of radius R in an aqueous dispersion of hard charged spherical macromolecules of radius a (see Fig. 1). For purposes of clarity, the term ''particle'' will be used when referring to the larger species and the smaller nonadsorbing material will be referred to as macromolecules. The bulk number density (number/volume) of the macromolecules is ρ_∞. The total force exerted *on* particle 1 by the surrounding macromolecules, $\mathbf{F}_1$, can be calculated as

$$\mathbf{F}_1 = \int\int_V\int \rho(\mathbf{x}_1)\nabla_1 E(\mathbf{x}_1)\, d\mathbf{x}_1 \tag{1}$$

where $\rho(\mathbf{x}_1)$ is the number density of macromolecules at position $\mathbf{x}_1$ and $\nabla_1 E(\mathbf{x}_1)$ is the spatial gradient of the *total* interaction energy of a macromolecule at $\mathbf{x}_1$ with particle 1. The distribution of macromolecules around the two particles is calculated using a virial expansion of the single-particle function correct to second order in bulk macromolecule density [5,6]. Thus,

$$\rho(\mathbf{x}_1) = \rho_\infty \exp\left(-\frac{E(\mathbf{x})}{kT}\right)[1 + b_1(\mathbf{x})\rho_\infty] \qquad (2)$$

where $E(\mathbf{x})$ is the *total* interaction energy of a single macromolecule located at position $\mathbf{x}$ and b_1 is the second virial coefficient, which is a function of the particle–macromolecule interaction, $E(\mathbf{x})$, and the macromolecule–macromolecule interaction, $U(\mathbf{r})$. The particles and macromolecules are modeled as hard-charged spheres, and the long-range particle–macromolecule and macromolecule–macromolecule interaction energies are calculated using the electrostatic expression given by Bell et al. [7].

An illustration of the effect of the long-range electrostatic repulsion can be seen in Figure 2. [The depletion energy is obtained by integrating the force predicted by Eq. (1) from infinite separation to any separation h, where h is the surface-to-surface distance between the two particles.] Here, the depletion energy

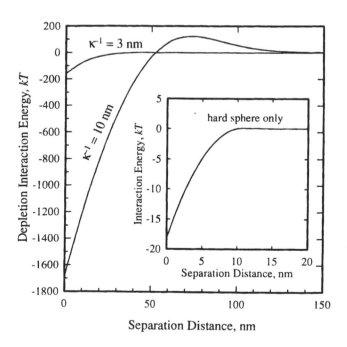

Figure 2   The depletion energy between two 10-μm-diameter spherical particles at different Debye lengths when long-range electrostatic interactions and second-order concentration effects are included. The particles are in a solution of 10-nm-diameter macromolecules at 1 vol% concentration and all surface potentials are assumed to be -50 mV.

between two 5-μm-radius particles in a 1 vol% solution of 5-nm macromolecules is given at different Debye lengths. The particles and macromolecules are assumed to possess a constant surface potential of -50 mV. Also shown is the interaction energy when the particles and macromolecules are modeled as simple hard spheres (no long-range interaction). As seen, the electrostatic repulsion not only increases the magnitude of the depletion attraction but also creates a significant repulsion at larger separations. As discussed by Walz and Sharma [4], this repulsion arises from layering of the macromolecules in the gap region.

In order to compare the depletion energies predicted with this force balance model with the experimental measurements of total internal reflection microscopy (TIRM), it was necessary to modify the model to account for the sphere–plate experimental geometry (as opposed to two spheres). This revision only required a simple modification to the volume integration in Eq. (1).

To calculate the total interaction energy between the particle and plate, we assume that the total interaction energy can be written as the sum of the electrostatic, van der Waals, and depletion energies. Thus,

$$E_{\text{tot}}(h) = E_{\text{el}}(h) + E_{\text{vdw}}(h) + E_{\text{dep}}(h) \tag{3}$$

III. EXPERIMENT

A. Overview of Experiments Conducted

Experiments were performed using three different types of nonadsorbing material—hard charged spheres, ionic micelles, and anionic polymers. In all cases, the particles were model polystyrene latex spheres that were negatively charged at the solution conditions. Two types of experiments were performed in each system. In the first experiment, direct measurement of the depletion energy between a single particle and a flat glass plate was obtained with the optical technique of TIRM. The measured profile was then compared to that predicted using the force balance model. In the second experiment, varying concentrations of macromolecules were added to stable dispersions of polystyrene particles. The stability of the dispersion was then monitored using optical turbidity. The sizes of the particles and macromolecules used in each of these experiments is listed in Table 1.

B. Direct Measurement of Depletion Interaction Energy

The depletion interaction energies between a polystyrene sphere and a glass plate were measured using the optical technique of TIRM, the details of which are given elsewhere [9,10]. A schematic of the setup is shown in Figure 3. In this technique, a laser beam is reflected from a solid–fluid interface at a high enough

Table 1   Sizes of the Various Materials Used in the TIRM and Dispersion Stability
Experiments

Macromolecule	Experiment	Polystyrene particle diameter	Macromolecule size (diameter or molecular weight)
Hard charged sphere (colloidal silica)	TIRM[a]	15 μm	12 nm
	Stability	0.41, 0.96 μm	5, 7 nm
Cationic micelle[b] (CTAB)	TIRM[a]	15 μm	6 nm[c]
	Stability	0.96 μm	6 nm[c]
Anionic polymer (137 kDa sodium polystyrene sulfonate)	TIRM[a]	15 μm	137 kDa
	Stability	0.47, 0.96 μm	137 kDa

[a] In the TIRM experiments, the particle was interacting with a polished BK-7 glass slide.
[b] In the CTAB experiments, the CTA$^+$ ion adsorbed onto the surface of both the particle and plate, giving all surfaces a net positive charge.
[c] The diameter of the CTAB micelle was taken from the literature [8].

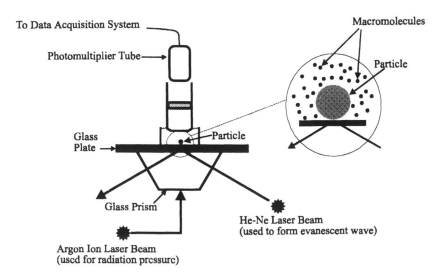

Figure 3   Schematic of the TIRM/radiation pressure experimental setup used to measure the depletion interaction energy between a single colloidal particle and a flat plate.

incident angle that an evanescent wave is formed in the fluid. A particle located sufficiently close to the interface will scatter the evanescent wave with a scattering intensity that varies exponentially with the particle–plate separation distance [11]. Measuring the scattering intensity provides a sensitive and instantaneous measure of the separation distance.

Interaction energy profiles are determined with TIRM by measuring the separation distances sampled by a Brownian colloidal particle for a statistically long period of time (the sphere–plate separation distance fluctuates around an equilibrium position due to Brownian motion). The probability of finding the particle at any separation distance, $p(h)$, is related to its total potential energy of interaction, $E_{tot}(h)$, at that separation through the Boltzmann equation

$$p(h) = A \exp\left(-\frac{E_{tot}(h)}{kT}\right) \tag{4}$$

where A is a normalization constant to ensure that all probabilities sum to 1. Given a sufficient number of independent separation distance measurements, the probability density profile, $p(h)$, can be determined, which can then be converted into an interaction energy profile using Eq. (4). Typically, 50,000 measurements of the scattered intensity were taken at a frequency of 100 measurements per second.

During the time required for data collection, the particle will also tend to migrate laterally (horizontal to the plate). Because there are no energy wells to hold the particle laterally, the particle can migrate to different parts of the surface or even totally out of the sampling area during the period of study. Obtaining accurate results then becomes difficult. To overcome this problem, an optical radiation pressure technique, similar in design to that used by Prieve and Walz [10], was employed. The principle of this technique lies in the fact that a laser beam made incident directly on the particle will exert a force on it that prevents its lateral movement. To ensure that the radiation pressure was not affecting the concentration of the macromolecules, we collected some data both with and without the radiation pressure beam. No discernible difference in the interaction energy profiles was found.

C. Dispersion Stability Measurements

The optical turbidity (absorbance) of dispersions was measured using a Shimadzu UV-VIS Recording Spectrophotometer UV-160 (Fig. 4). The spectrum was scanned in the wavelength range of 500–700 nm. Before scanning, each sample cuvette was inverted so that the amount of sample scanned was representative of the state of aggregation of the whole solution. This was done gently so as to prevent the flocs from breaking. The ultraviolet (UV) spectrum for each sample was taken at short intervals of time for the first 15–30 min to obtain the initial

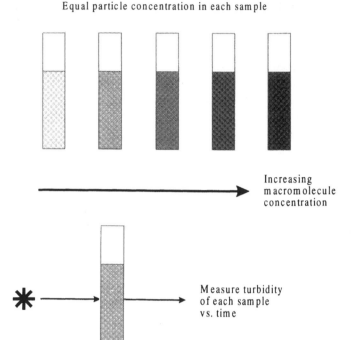

Equal particle concentration in each sample

Increasing
macromolecule
concentration

Measure turbidity
of each sample
vs. time

Figure 4 Illustration of the method by which the stability experiments were performed. Samples containing equal concentrations of particles, but varying concentrations of macro-molecules were prepared. The turbidity of each sample was then measured as a function of time.

rate of change of turbidity. Thereafter, the absorption was monitored periodically for the next 2–3 days, until there were no additional changes in floc size. The floc size of each sample was measured at the end of the experiment using the Coulter Sub-micron Particle Analyzer. Additionally, a set of the samples were placed in capped test tubes and let stand for visual analysis of flocculation.

The state of flocculation was determined using the procedure used by Snowden et al. [12]. Specifically, Heller et al. [13] give the following equation for the dependence of specific turbidity of nonabsorbing particles on the wavelength,

$$\left(\frac{\tau}{c}\right)_{c \to 0} = K\lambda^n \tag{5}$$

where τ is the turbidity at concentration c, λ is the wavelength in the medium, K is a constant which depends on the particle size and the relative refractive indices of the particle and the medium, and n is the exponent relating the turbidity of a particle dispersion to the wavelength. For Rayleigh scatterers (diameter $<$ $\lambda/10$), n has a constant value of -4. For our systems, the plots of log(turbidity) versus log(wavelength) for each sample were found to be almost linear in the range of wavelengths used, and the slope of the line, n, was obtained by linear regression. As the size of the floc increases, the value of n becomes less negative [12,14]; hence, plots of n versus silica concentration at various times provide a measure of the change in floc size in each sample.

IV. RESULTS

A. Hard Charged Sphere System

The first set of experiments were performed using the colloidal silica particles as the nonadsorbing macromolecules. A measured interaction energy profile is shown in Figure 5, where the silica concentration was 0.05 vol%. The $6kT$ second-

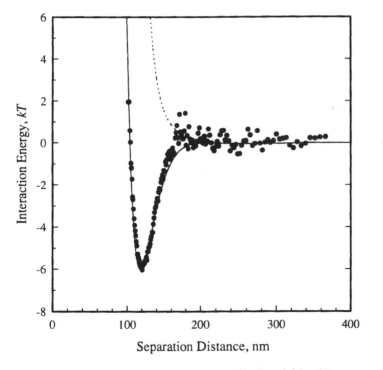

Figure 5   Measured interaction energy profile in a 0.05 vol% macromolecule (silica) solution. The dashed and solid lines give the model prediction before and after adjusting for screening due to the macromolecules, respectively. $\kappa^{-1} = 15.5$ nm; $\kappa^{-1}_{adj} = 13.4$ nm.

ary well is produced by the attractive depletion interaction coupled with the electrostatic repulsion between the particle and plate. The dashed line is the model prediction using no adjustable parameters. As seen, the model is not capturing the secondary well, which was observed consistently in numerous experiments. One possible explanation for this is that the macromolecules are screening the electrostatic repulsion between the particle and plate. To test this theory, the Debye screening length for the particle–plate interaction only was lowered very slightly (from 15.5 to 13.4 nm in this case). The resulting interaction is given by the solid curve, which can be seen to match the measured profile very closely.

A interaction energy profile measured at 0.21 vol% silica is shown in Figure 6. Now, a significant repulsion is observed, in addition to a long-range oscillatory profile. Again, these oscillations are thought to be due to an ordering of the silica spheres in the gap region. The dotted line shows the model prediction with no adjustable parameters, whereas for the solid line, the screening length for the particle–plate interaction was lowered from 11.6 to 10.9 nm. Again, very good agreement was achieved using this one adjustable parameter. It should be noted that

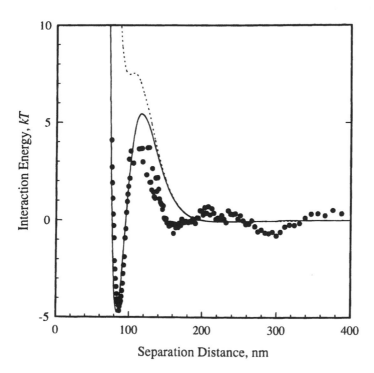

Figure 6 Measured interaction energy profile in a 0.21 vol% macromolecule (silica) solution. The dashed and solid lines give the model prediction before and after adjusting for screening due to the macromolecules, respectively. $\kappa^{-1} = 11.6$ nm; $\kappa^{-1}_{adj} = 10.9$ nm.

the long-range oscillations are not predicted by the model because it accounts for concentration effects only to second order in bulk concentration (including additional terms in the virial expansion would presumably produce these oscillations).

Some results from the flocculation experiments are shown in Figure 7. Shown is the turbidity parameter, n, which is an indication of floc size, versus silica concentration at various times. (After approximately 24 h, no additional changes in floc size were observed, indicating equilibrium had been reached.) As seen, no changes in floc size were observed up to a concentration of 2 vol%. At that point, a rapid increase was seen, indicating a critical flocculation concentration (CFC). At 5–6 vol%, the floc size begins to decrease again, indicating that the system is restabilizing.

This flocculation–restabilization behavior has been observed in other experiments involving nonadsorbing polymers [15,16] and is consistent with both the predicted and measured interaction energy profiles. At low macromolecule concentrations, the secondary depletion well develops and flocculation occurs

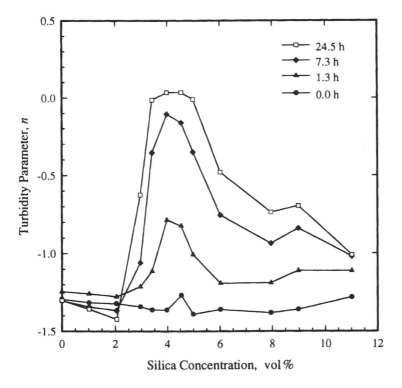

Figure 7   Variation in the turbidity parameter, n, with silica concentration at various times. The particles were 413-nm polystyrene latex spheres at 0.01 vol% and the macromolecules were 5-nm-diameter silica spheres.

once the well depth has reached an order of $1kT$. As the macromolecule concentration increases, the longer-range repulsive barrier begins to develop and the system restabilizes once this barrier becomes sufficient to prevent closer approach of the particles (of the order of $20kT$).

In summary, although a slight adjustment for screening by the nonadsorbing silica particles was required, both the measured interaction energy profiles and the stability behavior show reasonably good agreement with the model predictions.

B. Cationic Micelle System

The second system studied was that in which cationic CTAB (cetyltrimethylammonium bromide) micelles were used as the nonadsorbing species. In the CTAB system, the CTA^+ cations adsorb onto both the polystyrene particles and glass slide, giving all surfaces a net positive charge. A measured interaction energy profile from this system is presented in Figure 8. Shown are two profiles: one

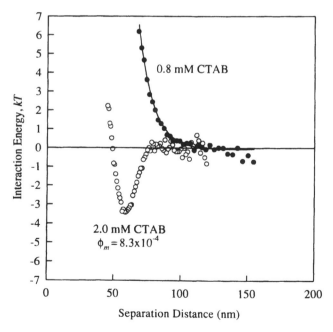

Figure 8   Measured interaction energy profile between a 15-μm polystyrene latex sphere and a flat glass plate both below and above the cmc for CTAB (cmc = 0.9 mM). The solid line is an exponential fit to the data at 0.8 mM, and the good agreement shows that the interaction can be well represented by the DLVO theory below the cmc. Above the cmc, however, a significant depletion attraction is evident. (ϕ_m is the bulk volume fraction of micelles.)

measured slightly below the cmc (cmc = 0.9 mM) and one measured above the cmc. Below the cmc, only an electrostatic repulsion between the particle and plate is observed, which can be described accurately with an exponential decay, as expected from the DLVO (Derjaguin–Landau–Verwey–Overbeek) theory. Above the cmc, however, a secondary depletion well is observed. Although not shown here, the depth of this well was found to increase with increasing CTAB concentration. A structural profile was also observed at micelle volume concentration above approximately 0.1 vol%.

A comparison of the measured profile to that predicted with the force balance model showed extremely poor agreement. Essentially, the model predicts a much larger structural contribution to the profile (larger repulsion). One possible explanation for this disagreement could be the effect of the strong electric fields near the charged wall and particle on the structure of the micelle (e.g., decreased degree of dissociation).

The stability behavior of the micelle system is shown in Figure 9 (only the

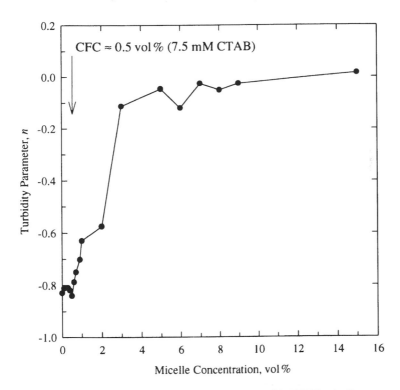

Figure 9 Variation in the turbidity parameter, n, with CTAB micelle concentration once equilibrium had been established in the system (long-time behavior). The particles here are 0.96-μm polystyrene latex spheres. Unlike the silica system, no restabilization is observed.

long-time equilibrium result is shown). Although a clear CFC was seen, no evidence of restabilization was observed, even out to concentrations as large at 15 vol% micelles. This behavior is fundamentally different from that seen with the silica particles and is currently not understood. One possibility is the much weaker than predicted structural contribution to the interparticle interaction observed in the TIRM experiments.

C. Anionic Polyelectrolyte System

The final system studied is that using a true polyelectrolyte, in this case 137 kDA sodium polystyrene sulfonate (SPSS), as the nonadsorbing macromolecule. The interaction energy profile measured at 0.001 wt% SPSS is shown in Figure 10. As seen, even at this low polymer concentration, a significant depletion well is observed. When the polymer concentration is increased by a factor of 10, the profile shown in Figure 11 is observed. Now, a substantial amount of structure

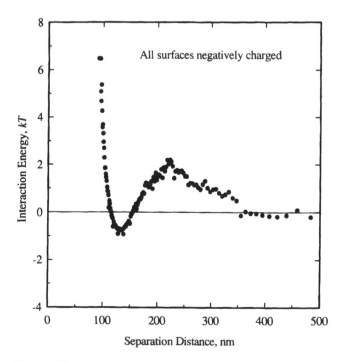

Figure 10 Measured interaction energy profile between a 15-μm polystyrene sphere and a glass slide in 0.001 wt% SPSS with no added electrolyte.

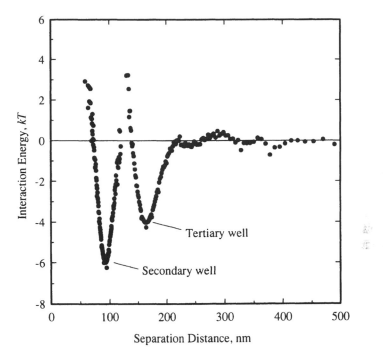

Figure 11   Measured interaction energy profile between a 15-μm polystyrene sphere and a glass slide in 0.01 wt% SPSS with no added electrolyte (10 × greater concentration than Figure 10).

is observed in the interaction energy profile, with both a secondary and a longer-range tertiary well present.

Several attempts were made to compare these measured profiles to model predictions. Poor agreement was found using the force balance model presented above, which is not unexpected because the model assumes a hard, charged sphere configuration. Because various researchers have suggested that SPSS assumes a rigid-rod conformation at low concentrations and low ionic strengths [17], comparisons were also made with the hard rod model of Mao et al. [18]. To account for the charged surfaces, the dimensions of the rod were increased by one Debye length in each direction. Once again, however, the agreement was poor. Essentially, the model was unable to capture the correct distance dependence of the interaction.

Finally, the results of the stability measurements are shown in Figure 12. As with the micelle system, although a clear CFC was found, no subsequent restabilization at higher polymer concentrations was observed.

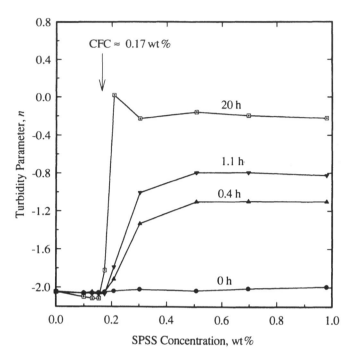

Figure 12   Variation in the turbidity parameter, n, with SPSS concentration at various times. The particles were 470-nm polystyrene latex spheres at 0.01 vol%. As with the micelles, although a clear CFC is seen, no restabilization at higher polymer concentrations is found.

V. CONCLUSIONS

The results of the experiments presented here clearly demonstrate the additional modeling work that is needed in these complex polyelectrolyte systems. On the one hand, for the simple hard charged sphere systems, the force balance model is capable of predicting the system behavior fairly accurately, especially at low macromolecule concentrations. In the micelle and polyelectrolyte systems, however, the model is clearly limited.

One very interesting yet puzzling phenomenon in this area is that of the restabilization at higher concentrations. Based on the substantial amount of structure observed in the interaction energy profile measured in the SPSS system, one would expect restabilization to occur—yet none was observed. Perhaps restabilization is highly dependent on the exact shape of the interaction energy profile in a manner that is currently not well understood. There is clearly a substantial amount of interesting research yet to be performed.

ACKNOWLEDGMENTS

The bulk of this research was conducted by Amber Sharma as part of his Ph.D. thesis at Tulane University. I would also like to personally acknowledge Professor Darsh Wasan who provided much assistance in understanding our observations, both through his manuscripts as well as several personal conversations with him.

REFERENCES

1. S Asakura, F Oosawa. J Chem Phys 22:1255, 1954.
2. JE Seebergh, JC Berg. Langmuir 10:454, 1994.
3. DH Napper. Polymeric Stabilization of Colloidal Dispersions. London: Academic Press, 1983, p. 13.
4. JY Walz, A Sharma. J Colloid Interf Sci 168:485, 1994.
5. ED Glandt. J Colloid Interf Sci 77:512, 1980.
6. JL Anderson, JH Brannon. J Polym Sci Polym Phys Ed 19:405, 1981.
7. GM Bell, S Levine, LN McCartney. J Colloid Interf Sci 33:335, 1970.
8. T Imae, R Kamiya, S Ikeda. J Colloid Interf Sci 108:215, 1985.
9. SG Bike, DC Prieve. Int J Multiphase Flow 16(4):727, 1990.
10. JY Walz, DC Prieve. Langmuir 8:3073, 1992.
11. DC Prieve, JY Walz. Appl Optics 32:1629, 1993.
12. MJ Snowden, SM Clegg, PA Williams, ID Robb. J Chem Soc Faraday Trans 87: 2201, 1991.
13. W Heller, H Bhatnagar, M Nakagaki. J Chem Phys 36:1163, 1962.
14. J Long, W Osmond, B Vincent. J Colloid Interf Sci 42:545, 1973.
15. S Rawson, K Ryan, B Vincent. Colloids Surfaces 34:89, 1988.
16. M Yasrebi, WY Shih, IA Aksay. J Colloid Interf Sci 142:357, 1991.
17. J Marra, ML Hair. J Colloid Interf Sci 128:511, 1989.
18. Y Mao, ME Cates, HNW Lekkerkerker. Physica A (Amsterdam) 222(1–4):10, 1995.

23
Viscosity of Colloidal Suspensions

Abbas A. Zaman and B. M. Moudgil
Engineering Research Center for Particle Science and Technology, University of Florida, Gainesville, Florida

ABSTRACT

The effect of different variables on the colloidal properties of concentrated aqueous dispersions has been studied. The electrostatic, steric, and hydrodynamic interactions are of significant importance in term of flow properties of the suspension and the role of interparticle interactions increases rapidly as the size of the particles is decreased. The role of polydispersity on suspension rheology has been studied by considering well-defined bimodal systems. The flow behavior of bidisperse aqueous silica suspensions was investigated at different volume fractions of the particles as a function of shear rate, and volume ratio of small to large particles. It is found that concentrated suspensions of particulates can show shear thickening behavior and the onset of shear thickening can be delayed by the addition of polymers to dispersion and by introducing polydispersity to the system.

I. OVERVIEW AND BACKGROUND

Control of the rheological behavior and stability of concentrated dispersions that are widely used as coatings, clays and ceramics, electronic pastes, paints, pharmaceuticals, sludges, and effluents is critical for the successful manufacturing of high-quality products of particulate suspensions. As industry moves toward higher solids loading to improve the material properties, control of the basic rheological behavior and physical stability of suspensions has become increasingly important. To achieve optimal process conditions, a basic understanding of the role of different variables that govern the suspension properties, such as particle size, particle size distribution, particle shape, volume fraction of the particles, temperature, shear rate, aging, use of polymers, and colloidal forces on flow properties of suspensions is required [1,2]. The effect of interparticle forces on suspension properties becomes increasingly important as the particle size is decreased. The range of interparticle interactions is of significant importance in terms of stability, flow properties, and packing of submicrometer-sized particles. In order to improve the material properties to minimize the inhomogeneities in the packing of particles and the distribution of the pores, interparticle forces in the suspension need to be manipulated during processing and forming [1,3,4]. The effect of nonhydrodynamic interactions between suspended particles on the behavior of the dispersion becomes more important at a higher volume fraction of the particles as the average interparticle spacing decreases [5]. Due to an increase in surface–surface separation at low volume fractions, the particle–particle interactions become less pronounced and the rheological properties of dilute suspensions are mainly affected by hydrodynamic forces and Brownian motion of the particles [6].

II. FLOW BEHAVIOR OF COLLOIDAL SUSPENSIONS

Concentrated suspensions of particles behave as non-Newtonian fluids and can exhibit different rheological behaviors such as shear thinning, shear thickening, thixotropy, rheopexy, and viscoelasticity depending on the general characteristics of the system [7,8]. Figure 1 is a plot of viscosity as a function of shear rate for suspensions of 0.25-μm silica particles at different solids loading (in an aqueous solution of 0.01 M NaNO$_3$ and suspension pH = 9.5). The viscosity shows changes with both shear rate and volume fraction of the particles. It can be observed that although the system shows Newtonian behavior for solids loading of up to 30 vol%, this behavior changes to non-Newtonian at a higher volume fraction of the particles at which the viscosity falls monotonically with increasing shear rate and approaches a final limiting value at high shear rates. There is at least four orders of magnitude increase in viscosity as the solids loading is increased from 10 to 53 vol%. It appears that the differences in viscosity at various volume fractions are more significant at low shear rates. The transition from a shear-rate-dependent region to a shear-rate-independent region occurs at higher shear rates as the solids loading is increased.

Suspensions of particles in shear flows experience different kinds of force such as hydrodynamic forces (including the viscous drag force and particle–particle interaction through a flow field induced by neighboring particles), colloid

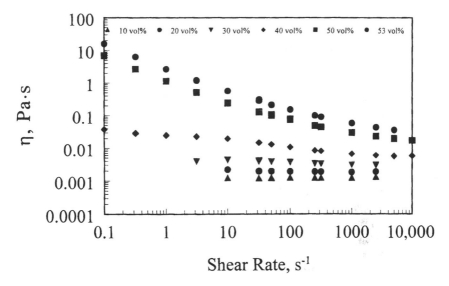

Figure 1 Variation of viscosity with shear rate and volume fraction for 0.25-μm silica particles dispersed in a solution of 0.01 M NaNO$_3$ at 25°C (suspension pH = 9.5).

chemical forces (including electrostatic, steric and London–van der Waals attractive forces), and forces due to gravitational, inertial, electroviscous, and thermal or molecular collisional effects. Thermal and electroviscous forces decrease as the size of the particles is increased and they become negligible for particles of larger size [9]. The effect of particle size on the higher limiting (1000 s^{-1}) relative viscosity of aqueous silica suspensions (1.5 μm and 0.25 μm (diameter) silica particles in a solution of 0.01 M NaNO$_3$ and suspension pH = 9.5) is shown in Figure 2, which is a plot of relative viscosity (with respect to the viscosity of the suspending fluid, μ) as a function of the core volume fraction, φ, of the particles. A decrease in particle size led to an increase in the apparent viscosity of suspension. The adsorbed layer of the fluid on the particle surface and double layers around the particles will increase the effective volume fraction of the particles. The relative increase in effective volume fraction is larger for smaller particles and, as a result, higher viscosities are observed for suspensions of smaller particles. Boersma et al. [10] have qualitatively shown that at a constant surface potential, the relative importance of hydrodynamic to electrostatic forces is decreased as the diameter of the particles is decreased. A reduction in viscosity with increasing particle size indicates that electroviscous effects are more significant for smaller particles, and colloid chemical forces become more dominant

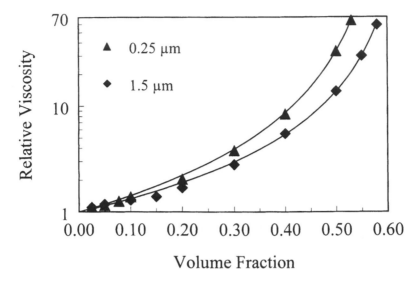

Figure 2 Higher limiting relative (1000 s^{-1}) viscosities for silica suspensions of two different sizes as a function of volume fraction of the particles at 25°C (dispersed in a solution of 0.01 M NaNO$_3$ and suspension pH = 9.5).

with decreasing particle size. This will affect the balance between the hydrodynamic and nonhydrodynamic forces acting on the particles and, therefore, the rheological properties will vary as the particle size is changed.

A significant scientific and technological issue in recent years has been improving the rheological properties of suspensions and increasing the maximum solids loading through broadening the size distribution of the particles [7,11–17]. The maximum solids loading can be increased significantly above the value of close packing of monosized spheres (0.64 in the low-shear-rate region and 0.74 in the high-shear-rate region) for bidisperse and polydisperse suspensions. In these systems, with the right size ratio of the particles, the smaller particles can be accommodated in the spaces between the larger particles, resulting in a more efficient packing. Although this concept has been employed by various industries to achieve lower viscosities at a higher volume fraction of the particles [16,18], systematic studies on rheological behavior of well-characterized polydisperse suspensions are not extensive and, to date, there is not a unique empirical or theoretical model to account for the effect of different variables on flow properties of polydisperse suspensions.

Figure 3 is a comparison between the higher limiting relative viscosities (1000 s^{-1}) of bimodal suspensions of 0.25- and 0.6-µm-diameter (at a volume

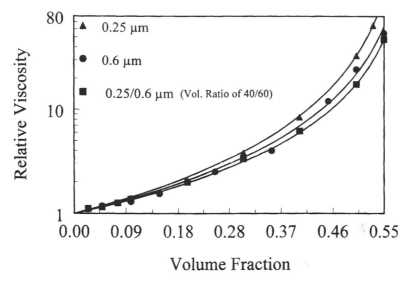

Figure 3 Comparison between the higher limiting (1000 s^{-1}) relative viscosities of bimodal silica suspensions with the relative viscosity of the related monosized systems as a function of volume fraction of the particles. ($25°C$, $0.01 \text{ } M \text{ NaNO}_3$, and suspension pH = 9.5).

ratio of 40/60) silica particles and the higher limiting relative viscosity of the related monosized systems (suspended in a solution of $0.01 \text{ } M \text{ NaNO}_3$ and suspension pH = 9.5) as a function of the core volume fraction of the particles. It can be observed that the bimodal suspension shows lower viscosity than the related monosized systems, indicating that the resistance to flow can be decreased significantly by using bimodal particle size distribution. This is the result of more efficient packing of polydisperse spheres, as indicated by earlier investigators [15,17]. In these systems, small particles can fit into the spaces between the larger particles; if they are small enough, they act along with the suspending fluid like a larger sea for the large particles [17]. The level of viscosity reduction in bimodal suspensions varies with particle size and particle size distribution, volume ratio of the two sizes, and total volume fraction of the particles [19,20].

III. EFFECT OF NONHYDRODYNAMIC INTERACTIONS

The rheological properties of suspensions can also be controlled by manipulating surface charge, ionic strength, and steric and electrosteric repulsive forces

[1,19,21,22]. The surface properties of the particles and the ionic strength of the supernatant solution strongly affect the colloid chemical interactions. Determination of the significance of the effect of the ionic strength (electrolyte concentration and type of the electrolyte) on the rheological properties of concentrated suspensions is necessary to establish criteria for controlling their rheological behavior so as to obtain highly loaded mixtures of desired fluidity while maintaining sufficient stability against sedimentation of particles.

The effect of pH (surface charge) on shear viscosity of silica suspensions prepared at a high volume fraction is shown in Figure 4, which is a plot of viscosity as a function of pH at different shear rates [21] for suspensions of 1-μm silica particles at 58 vol% (dispersed in a solution of 0.03 M NaNO$_3$). The pH values were adjusted by adding HNO$_3$ and NaOH (in one direction only) to the suspension. It can be observed that the viscosity of the suspension significantly decreases with increasing pH, reaches a minimum, and then increases with further increase in pH. The effect of the pH on the apparent viscosity of the suspension can be described in terms of the changes in the electrical charge of the particles. An increase in the viscosity at pH values above 10 is most probably related to the dissolution of silica at high pH values and formation of very large polymeric species at high silica concentrations. Partial dissolution may result in a hairy

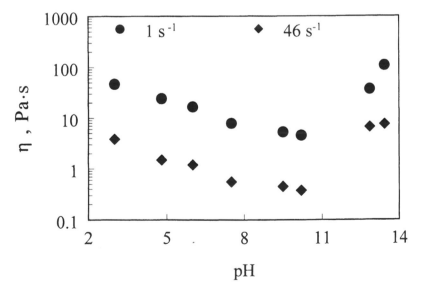

Figure 4 Variation of viscosity with pH at two different shear rates for 1-μm silica particles dispersed in a solution of 0.03 M NaNO$_3$ at 58 vol% (25°C). (From Ref. 21.)

surface, inducing a higher effective volume fraction and, therefore, an increase in the viscosity of the suspension.

Another method that is widely used to improve the rheological properties and stability of colloidal suspensions is the addition of polymers to the system. When water-soluble polymers are added to the suspension, the polymer may be adsorbed on the particle surface. The distribution of the polymer molecules in the solution is affected by particle–particle interaction, producing a force whose sign and magnitude depends on the nature of the particle–polymer interaction [1]. The adsorbed polymer can either stabilize or destabilize the suspension depending on the magnitude of the repulsive forces between the adsorbed polymer molecules. Steric stabilization occurs when the repulsive forces overcome the attractive van der Waals forces acting between the particle surfaces.

The effect of polyethylene oxide (PEO 6000) on the viscosity of silica suspensions is presented in Figure 5, which is a plot of viscosity as a function of PEO dosage for aqueous silica slurries at 58 vol% (dispersed in a solution of 0.03 M NaNO$_3$ and suspension pH = 9.5) at different shear rates [21]. It can be observed that the viscosity of the suspension initially decreases to a minimum with increasing the polymer dosage and then starts to increase with further addi-

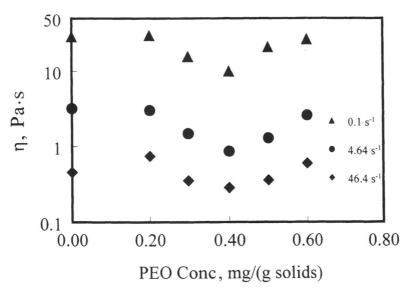

Figure 5 Variation of viscosity with shear rate for 1-μm silica particles dispersed in a solution of 0.03 M NaNO$_3$ and PEO (molecular weight = 6000) at 58 vol% (25°C and pH = 9.5). (From Ref. 21.)

tion of polymer to the suspension. There is a critical amount of polymer that must be added to the system to achieve minimal viscosities.

IV. SHEAR THICKENING PHENOMENON IN CONCENTRATED SUSPENSIONS

Particulate suspensions can show shear thickening behavior (increase in viscosity with shear rate) if subjected to the right conditions (combination of interparticle forces, particle size, and hydrodynamic interactions). In most of the industrial processes, shear thickening is a severe problem leading to difficulties in processing of the material and is to be avoided, if possible. It assists the development of excessive shear stress that causes fracture. Failure of the mixer motors and mixer blades due to development of high shear stress, roller bowing, problems in the atomization of these materials for spray applications [23], and rupture of shear-thickening coatings under the coating blade [24] are typical problems in the processing of shear-thickening materials that have been reported in the literature.

Apart from the above-mentioned problems, a few advantages for the shear-thickening behavior have been reported in a review paper by Barnes [25]. A certain amount of dilatancy may be helpful in maintaining small pore size and preventing bubble coalescing in the case of foamed poly(vinyl chloride) [26]. In the case of high-speed disk dispersers or roller mills, operating at the boarder line of dilatancy will provide the maximum effectiveness due to breakdown of aggregates [27]. Another direct application of the shear thickening is in the case of sport shoes cushioning, liquid couplings, shock-absorber fillings, and rotary-speed limiters [25].

Different explanations have been given for the phenomenon of shear thickening in monodisperse suspensions that are reviewed in details by Barnes [25]. At high volume fractions, the particles are very close together and the average separation of the particles is very small. Therefore, the particle–particle interaction forces are very large and will strongly affect the relative position of the particles. When shear flow is imposed on stabilized suspensions, repulsive interparticle forces tend to keep the particles in a layered-structure [10] form. However, when the magnitude of the hydrodynamic forces is larger than the interparticle forces, the electrical double layers overlap and particles will move from their equilibrium position and lead to a transition from a layered to a disordered structure, causing shear-thickening behavior.

The shear-thickening phenomenon has been described in terms of the formation of hollow enclosures due to the sudden onset of turbulent flow between the particles [28]. It is explained in terms of momentum interaction between lines of aligned particles at high shear rates by virtue of Brownian motion [29]. Light-scattering experiments performed by Hoffman [30] indicate that well-stabilized suspensions in shear flow show a transition from a two-dimensional ordered state to a disordered state at shear rates corresponding to a sudden jump in viscosity. The computer simulations performed on concentrated colloidal dispersions by Durlofsky et al. [31], considering hydrodynamic interactions and interparticle forces, indicate an increase in viscosity at high shear rates. The simulation results indicated a change from order to clusters of particles that were formed because of shear forces and were responsible for the increase in viscosity [10].

An example of the shear-thickening behavior in concentrated suspensions is given in Figure 6, which is a plot of viscosity as a function of shear rate for suspensions of electrostatically (●) and sterically stabilized suspensions (▲) of 1.5-μm-diameter silica particles (dispersed in a solution of 0.01 M NaNO3 and suspension pH = 9.5) at 58 vol%. Also, in the same figure, the viscosity of electrostatically stabilized bimodal suspensions of 0.6-μm and 1.5-μm (at a volume ratio of 30/70) silica particles at 58 vol% is presented. It can be observed from this figure that the onset of shear-thickening behavior occurs at higher shear rates for sterically stabilized suspension and the bimodal system does not show shear-thickening behavior over the shear rate range studied. In general, the onset of shear-thickening behavior is a function of particle size, volume fraction of the particles, and particles size distribution. The work of Boersma et al. [10] on

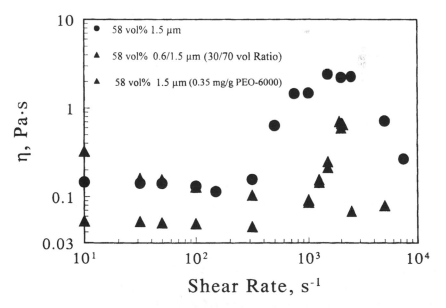

Figure 6 Effect of bimodality and PEO on the onset of shear thickening behavior in suspensions of 1.5-μm silica particles at 25°C (0.01 M NaNO$_3$ and suspension pH = 9.5).

rheological properties of dispersions of SiO_2 particles in glycerol–water (86.1/ 13.9 w/w) at 20°C indicate that the samples showed shear-thickening behavior at shear rates lower than 1 s^{-1} (at different ϕ values of 0.524, 0.562, and 0.598). Our results on the shear viscosity of aqueous silica suspensions (0.01 M NaNO$_3$ and suspension pH = 9.5) indicate that the samples do not show shear-thickening behavior at low shear rates. Therefore, interparticle forces play an important role in the onset of shear-thickening behavior in concentrated suspensions.

V. SUMMARY

The role of hydrodynamic, electrostatic, and steric forces, particle size, and poly-dispersity on flow properties of concentrated aqueous dispersions has been investigated. It is shown that the flow properties of suspensions of particulates can be controlled by manipulating hydrodynamic, electrostatic, and steric forces, particle size, and polydispersity. The role of polydispersity in suspension rheology was studied using well-defined bimodal suspensions of silica particles. The flow behavior of the system was investigated as a function of shear rate, total volume fraction of the particles, and volume ratio of small to large particles. The viscosity levels were lower in bimodal suspensions as compared with the viscosity of the related monosized systems at the same volume fraction of the particles. It is shown that highly concentrated suspensions can show shear-thickening behavior depending on the general condition of the system. The onset of shear thickening can be delayed through the addition of polymers to the suspension and also by introducing polydispersity in the system.

ACKNOWLEDGMENTS

The authors are grateful for the financial support provided by the University of Florida Engineering Research Center for Particle Science and Technology (NSF Grant No. EEC-94-02989) and the industrial partners of ERC. We like to express our appreciation to Ms. G. B. Basim and Mr. J. Adler for their assistance in performing some of the experiments.

REFERENCES

1. WB Russel, DA Saville, WR Schowalter. Colloidal Dispersions. New York: Cambridge University Press, 1991.
2. J Persello, A Magnin, J Chang, JM Piau, BJ Cabane. J Rheol 38:1845–1870, 1994.

3. IA Aksay. In: JA Mangels, GL Messing, eds. Advances in Ceramics, Vol. 9, Forming of Ceramics. Westerville, OH: American Ceramic Society, 1984, pp. 94–104.
4. FF Lange. J Am Ceram Soc 72:3–15, 1989.
5. WB Russel. J Fluid Mech 85:209–232, 1978.
6. IM Krieger. Trans Soc Rheol 7:101–109, 1963.
7. DJ Jeffrey, A Acrivos. AIChE J 22:417–432, 1976.
8. VV Jinescu. Int Chem Eng 14:397–420, 1974.
9. WB Russel. J Rheol 24:287–317, 1980.
10. WH Boersma, J Laven, HN Stein. AIChE J 36:321–332, 1990.
11. JS Chong, ED Christiansen, AD Baer. J Appl Polym Sci 15:2007–2021, 1971.
12. NJ Wagner, ATJM Woutersen. J Fluid Mech 278:267–287, 1994.
13. ATJM Woutersen, CG de Kruif. J Rheol 37:681–693, 1993.
14. BE Rodriguez, EW Kaler, MS Wolfe. Langmuir 8:2382, 1992.
15. RJ Farris. Trans Soc Rheol 12:281–301, 1968.
16. SC Goto, H Kuno. J Rheol 28:197–205, 1984.
17. RL Hoffman. J Rheol 36:947–965, 1992.
18. AP Shapiro, RF Probstein. Phys Rev Lett 68:1422–1425, 1992.
19. AA Zaman, BM Moudgil. J Rheol 42:21–39, 1998.
20. KM Wilson, AA Zaman, C-W Park. Proceedings of the 24th NOBCChE Conference, 1997, p. 34.
21. AA Zaman, BM Moudgil, AL Fricke, H El-Shall. J Rheol 40:1191, 1996.
22. AA Zaman, BM Moudgil. J Colloid Interf Sci 212:167–175, 1999.
23. JT Lazor. Mod Plast 42:149, 1965.
24. WD Todd. Fed Paint Varnish Prod Clubs 24:98, 1952.
25. HA Barnes. J Rheol 33:329, 1989.
26. DS Newton, JA Cronin. Br Plast 437 (October 1958).
27. TC Patton, TC Paint Flow and Pigment Dispersion. 2nd ed., New York: John Wiley & Sons, 1979.
28. EN Andrarde. Viscosity and Plasticity. Cambridge: Oil and Colour Chem Association, 1947.
29. AB Metzner, M Whitlock. Trans Soc Rheol 2:239, 1958.
30. RL Hoffman. Trans Soc Rheol 16:155, 1972.
31. L Durlofsky, JF Brady, G Bossis. J Fluid Mech 180:21, 1987.

24

Attachment of Gas Bubbles to Solid Surfaces from an Impinging Jet

Tadeusz Dabros, Hassan A. Hamza, and Q. Dai
Natural Resources Canada, Devon, Alberta, Canada

Jan Czarnecki
Syncrude Canada Ltd., Edmonton, Alberta, Canada

ABSTRACT

The attachment of small gas bubbles to solid surfaces was studied using a modified impinging jet technique. The average diameter of the bubbles generated by polarizing a platinum wire was around 40 μm. The solid surfaces studied were untreated glass, methylated glass, and glass with a thin layer of bitumen on it. A multistage experimental procedure was developed to provide information about the instantaneous bubble concentration in the impinging jet, bubble size distribution, and time dependence of the coating density, from which the experimental flux was estimated. As far as the theory is concerned, the characteristic feature of the bubble attachment is that Peclet and gravity numbers are much greater than one. For these conditions, it is demonstrated that the flux can be calculated using either the continuity equation or a limiting trajectory approach. A comparison of the experimental flux with the calculated value, which is based on an assumed model for the bubble-collector interaction, provides information about these interactions in dynamic conditions.

I. INTRODUCTION

The attachment of air bubbles to hydrophobic solid particles or liquid droplets is a critical step in flotation. Despite many years of commercial application of the process, some fundamental questions regarding the mechanism of bubble attachment remain unanswered. Factors such as solution chemistry, bubble size, particle or droplet size, mechanical energy, and flow conditions are considered important to bubble–particle attachment, but are very difficult to quantify.

The deposition of colloidal particles onto various solid surfaces has been investigated using the impinging jet technique developed by one of the authors [1]. In this chapter, we describe an application of the modified impinging jet cell, which allows one to study the process of gas-bubble attachment to solid surfaces. The advantage of this technique is that the dynamic process of bubble attachment

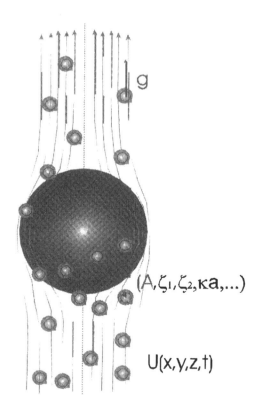

Figure 1 Factors determining the mass-transfer conditions for the attachment of gas bubbles to a solid particle.

can be studied experimentally in well-defined flow conditions that are similar to those experienced by particles and bubbles during flotation. As illustrated in Figure 1, the attachment of a gas bubble to a solid surface is a mass-transfer process controlled by (1) the flow field in the vicinity of the collector, (2) external forces acting on the bubble,and (3) colloidal (specific) interaction forces between the bubble and collector. The flow and mass-transfer conditions at the front of a

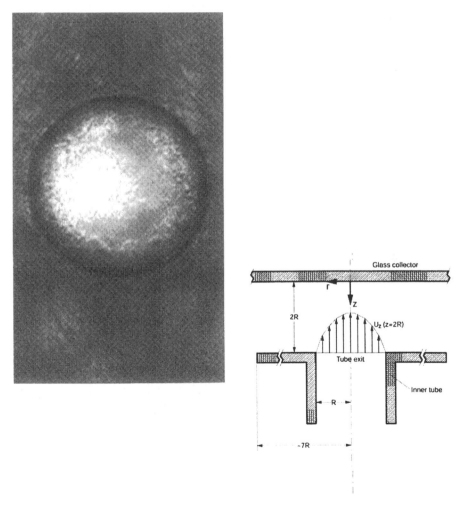

Figure 2 Spherical collector in a uniform flow (left) and an impinging cell setup (right) for creating flow conditions similar to those that exist near the stagnation point.

spherical collector in uniform flow are similar to those in the vicinity of the stagnation point of the impinging jet cell. In the impinging jet cell, the flowing fluid, which carries a dilute suspension of bubbles, is directed toward the transparent collector plate (see Fig. 2). The microscope allows one to observe the behavior of the bubbles brought by flow and gravity into the vicinity of the collector and also to count the number of (immobilized) bubbles attached to the collector surface.

The experimental results can be analyzed by solving the mass-transfer equations with colloidal, external, and hydrodynamic forces included in the analysis. Because the Peclet number is large for bubbles that are 10 μm in diameter or larger, the rate of deposition (flux) of gas bubbles onto a collector surface can be calculated by (1) solving the continuity equation or (2) using the limiting trajectory analysis. In any case, external forces such as gravity and specific colloidal interaction forces have to be specified. In this chapter, in order to provide some reference-calculated flux, we assume weak, dispersive-type attractive forces between bubbles and the collector. This simplified approach avoids the controversial, and not completely resolved, problem of hydrophobic interactions and repulsive–dispersive forces that can be expected for gas bubbles in water interacting with solid surfaces. This issue is discussed in detail elsewhere [2].

In this chapter, we first discuss the flow field and mass-transfer conditions in an impinging jet cell, with special attention being given to the characteristic for the gas-bubbles range of parameters, such as Peclet and gravity numbers. The wide range of these parameters and possible interaction is discussed in Ref. 3. After describing the experimental procedure, which also involves determination of the instantaneous concentration of gas bubbles, results are presented and discussed.

II. THEORY

A. Fluid Flow Field and Mass Transfer

The flow field in the gap between the collector and inner tube was found by solving the steady-state Navier–Stokes equation assuming constant viscosity and constant fluid density. This assumption is justified only for a sufficiently low bubble concentration. Nonslip boundary conditions were assumed at the solid walls of the tube, collector plate, and lower plate (see Fig. 2). For sufficiently large distances between the exit of the tube and the collector plate and for a laminar flow regime, the velocity profile at the exit can be assumed to be parabolic. The Navier–Stokes equation, with appropriate boundary conditions, was solved using the overrelaxation finite-difference method described generally in Ref. 1 and in more detail in Ref. 4.

The flow field close to the stagnation point can be expressed as

$$v_r = \alpha r z$$
$$v_z = -\alpha z^2 \tag{1}$$

where the flow intensity parameter α, which is a function of the average fluid velocity in the tube, fluid viscosity, and the distance between the tube exit and the plate, is calculated by solving the Navier–Stokes equation, as described in Ref. 1. Radial coordinates, z and r, have origin in the stagnation point and are directed toward the flow. As the distance from the stagnation point increases, Eq. (1) no longer describes the actual flow field. Analysis of numerical results shows that α does not change by more than 5–10% at distances from the symmetry axis less than 20% of the inner tube radius. Within this region, the flux of bubbles to the collector is independent of radial distance from the symmetry axis.

The flux of small spherical bubbles in a dilute suspension under steady-state conditions can be found by solving the continuity equation

$$\frac{1}{r}\frac{\partial(rj_r)}{\partial r} + \frac{\partial j_z}{\partial z} = 0 \tag{2}$$

where j_r and j_z are the radial and normal components of the flux, respectively. Because of the plate's horizontal orientation, the radial component of the flux is due to fluid convection, with corrections for the hydrodynamic wall–particle interaction [1]. The normal component, j_z, has additional terms due to gravity forces and specific colloidal forces and also includes a diffusion term, which cannot be neglected as was possible in the case of radial flux [5]. All normal flux terms have to be corrected for the hydrodynamic interactions.

The relative importance of convective and diffusive contributions to the bubble flux is measured by the Peclet number $\text{Pe} = 2\alpha a^3/D_0$, where a and D_0 are the bubble radius and diffusion coefficient, respectively. In this representation of the Peclet number, as it becomes much larger than 1, the thickness of the diffusion layer becomes much smaller than the radius of the bubbles and, at the same time, the bubble trajectory is essentially deterministic.

In the absence of an energy maximum, two other dimensionless numbers are important in defining the mass-transfer regime. One is the gravity number, $\text{Gr} = 2\Delta\rho g a^3/(9\mu D_0)$, where $\Delta\rho$ is the density difference between water and gas, g is the gravitational acceleration, and μ is the kinematic viscosity of the fluid. The second parameter describes the interaction between bubbles and collector at small separations, when colloidal interaction, such as electrical double layer, van der Waals, and other specific forces, become important. As the distance between the particle and collector becomes much smaller than its radius, particle mobility decreases to zero linearly with the gap width. Therefore, at a sufficiently small separation, the existence of an attractive force is necessary to ensure attachment. Usually, dispersion forces are believed to be responsible for overcoming decreasing mobility and immobilizing the particles. The detailed nature of interactions at very small separations, particularly in the case of heterogeneous system such as a solid collector and a gas bubble, is not known. Dispersion, electrical double layer, and other forces that are related to the structure of the medium in the vicinity of the interface, including so-called hydrophobic interactions, play a role in defining the total force between the collector and bubble. These issues are discussed in some detail elsewhere [2].

In this chapter, in order to obtain a reference value of the flux that would exist if there were an attractive force at small bubble–wall separations, a dispersion-type attractive interaction is assumed to exist between the bubble and the collector. This interaction is characterized by two parameters: (1) Hamaker constant A_{123} for the interaction between particle one and collector three suspended in medium two and (2) the London wavelength λ, which accounts for retardation of the dispersive interactions. It has been demonstrated [1] that in the absence of the energy barrier, when the ionic strength of water solution is not extremely low, the flux of particles to the collector is not very sensitive to the actual value of the Hamaker constants. In the calculations presented below, it was assumed that $\lambda = 0.1$ μm and $A_{123} = 4.2 \times 10^{-22}$ J.

B. Boundary Conditions

The most widely used boundary condition at the collector surface is derived from the so-called perfect sink model [6], which requires that the bubble concentration vanishes at small distances δ from the collector ($c = 0$ at $z = \delta$). The physical meaning of this boundary condition is that particles or bubbles are irreversibly captured at the collector surface when $z < \delta$ and, therefore, they are no longer part of the solution. In order to treat the particle or bubble as being captured, its tangential movement has to vanish. In the case of colloidal particles, the problem of boundary conditions at the collector surface has been discussed in Refs. 6 and 7. The particles are immobilized at the solid surface as a result of tangential forces, which are usually attributed to physical or chemical heterogeneity of the surface. Note that even small surface roughness (as compared to colloidal particle dimensions) can produce sufficient static friction to prevent particle movement by hydrodynamic forces. It is usually assumed that these forces start to operate when the particle reaches the primary energy minimum. The existence of strong attractive forces at sufficiently small particle–collector separations is essential in order to immobilize the particles. In the case of gas-bubble attachment to a solid surface, two complications arise. First, the bubble can be deformed by the flow, particularly near the contact region, leading to uncertainty where the boundary conditions should be applied. Second, the bubble is smooth down to the molecular level and permanent immobilization can be difficult to achieve, particularly when attractive forces are weak.

Regarding the first point, both elastic and nonelastic deformations in the contact region are anticipated even for a solid particle due to dispersive forces operating in the contact area after the particle is captured. As a result, the force of adhesion can increase substantially. For bubbles, this effect should play a minor role. More important could be the deformation due to hydrodynamic forces exerted on the bubble by the fluid flow. The ratio of surface tension forces to the hydrodynamic forces is of the order of $\sigma/\mu\alpha a^2$, where σ is the surface tension. Using this formula, it can be shown that even for 100-μm bubbles, the interfacial forces predominate. Also, for similar reasons, it can be argued that small bubbles behave like solid particles. This relates to the ongoing discussion of whether the Rybczynski–Hadamard formula is confirmed by experiment [5,8]. According to Leal [8], in purified water, gas bubbles with diameters larger than 1 mm obey the Rybczynski–Hadamard formula, indicating that tangential stresses at the bubble surface vanish. However, for smaller bubbles, the terminal velocity follows the Stokes formula, which requires that nonslip boundary conditions be satisfied. The most likely explanation of this phenomenon was given by Levich, who associated it with the surface tension stresses caused by the surface tension gradients arising from an uneven distribution of surface-active components at the bubble surface in dynamic conditions created by flow [5]. The order of magnitude of the surface tension stress, τ_0 is $\Delta\sigma/a$, where $\Delta\sigma$ is the variation in surface tension over the bubble surface, which, in turn, can be associated with the variations in the surface concentration of the active components. The hydrodynamic stresses, τ_h, in the system being considered are of the order of $\mu\alpha a$. The ratio $\tau_0/\tau_h = \Delta\sigma/\mu\alpha a^2$ is indeed much higher than unity for 1-mm bubbles even for minute changes in the surface tension. Bubbles 100 μm in diameter can certainly be treated as objects with nonslip boundary conditions.

The second point refers to the mechanism of immobilization of bubbles at the solid surface. Probably the most effective mechanism of immobilization would be through formation of the three-phase contact line (Fig. 3). Contact-angle hysteresis due to the surface roughness or chemical inhomogeneity could assure permanent immobilization of the bubble even if it is subjected to significant hydrodynamic drag forces. However, in order to create the three-phase contact line, the liquid film between the bubble and the collector surface has to break. The problem of the stability of thin liquid films on solid substrates was studied by Ruckenstein and Jain [9]. Their approach is based on the hydrodynamic stability analysis of a thin layer of liquid, the surface of which has a given interfacial tension and is subject to a potential energy field having contributions from London–van der Waals and electrical double-layer forces. According to this analysis, in order to destabilize the films, the liquid film energy has to be a nonincreasing

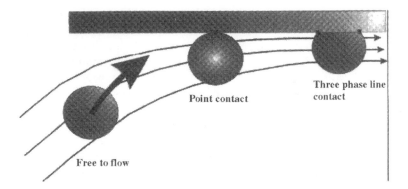

Figure 3 Different modes of attachment of a bubble to a solid collector surface. The behavior of the bubble in contact with the collector determines the boundary conditions for the mass-transfer problem.

function of the film thickness; therefore, the Hamaker constant has to be positive. In the case of water, $A_{ww} = 3 \times 10^{-20}$ J. The generally accepted values of Hamaker constants for almost all substances, including plastics and paraffins, are higher than for water. Therefore, according to the theory presented, the films should not rupture. However, it is well known from experimental observations that liquid films at nonwetting solid surfaces break spontaneously after reaching a thickness of the order of 10^2 μm. This phenomenon cannot be explained by the introduction of nonlinear terms to the flow equations or by stretching the theory of interactions to its limit. Recently, Sharma and Ruckenstein [10] developed so-called energetic criteria for the breakup of liquid films on nonwetting solid surfaces. According to this theory, the film breaks when the free energy of the system, consisting of the film with a hole in it and the exposed solid surface, reaches the energy of the undisturbed film. Prior to the formation of the hole, the energy necessary to develop a sufficiently large hole has to be supplied from external sources such as vibrations. Analysis of the accessible data on equilibrium contact angles and surface tensions led Sharma and Ruckenstein to conclude that the critical film thickness was 510–787 μm for water on Teflon, 310 μm for water on paraffin, and 420 μm for water on polyethylene. These values were confirmed by experiments. Therefore, the liquid film can break on a nonwetting solid surface after reaching a thickness of hundreds of micrometers, provided that an external (not very well specified) force starts to create the hole. If the film breaks, a three-phase contact line is created. Contact-angle hysteresis can immobilize the bubble if the hydrodynamic forces are smaller than some critical values. Therefore, the perfect sink boundary condition can be applied at the critical film thickness. Another mode of attachment, without film breaking, is also illustrated in Figure 3. Both modes, with the formation of a three-phase line and a "point" attachment, have been observed experimentally in this work.

The dimensionless flux of bubbles can be expressed in terms of the Sherwood number $Sh = ja/D_0c_0$, where $j = |j_n|$ and c_0 is the bulk concentration of the bubbles. Solving the continuity equation for specific values of the gravity number and the adhesion number, $Ad = A_{123}/kT$, with appropriate boundary conditions provides a general relationship between the Sherwood and Peclet numbers. Figure 4 shows the Sherwood number as a function of Peclet number for different values of the gravity number. For Peclet numbers smaller than the gravity number, the flux to the collector is mainly due to the buoyancy. It is interesting

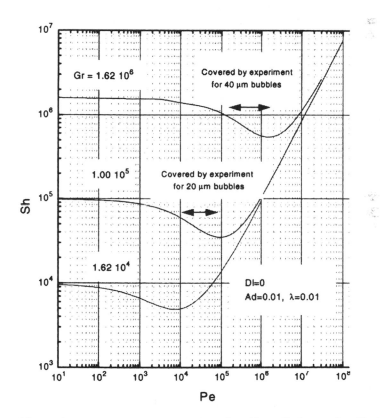

Figure 4   Dependence of Sherwood number, Sh, on Peclet number, Pe, for three values of the gravity number, Gr. The horizontal arrows show the range of Peclet numbers covered by the experimental conditions for 20- and 40-μm bubbles.

to note that as the Peclet number increases, which is usually associated with more intensive flow, the Sherwood number decreases. This phenomenon is related to conditions that occur when a steady state develops for small Peclet numbers. Even if the energy barrier is absent, local concentrations in this region increase in order to preserve the flux, whereas bubble mobility decreases in the vicinity of the wall due to hydrodynamic interactions. If the flow intensity increases such that Pe ≈ Gr, the bubbles are carried away by the flow and the local concentration of bubbles decreases; accordingly, the flux to the collector also declines. If Pe ≫ Gr, bubble transport is controlled by convection and the Sherwood number is proportional to Pe. The arrows in Figure 4 show the range of Peclet numbers covered by the experiments described later.

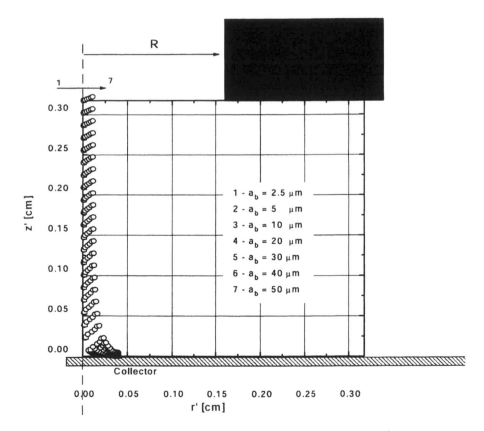

Figure 5 Limiting trajectories of bubbles of various sizes in the impinging jet region. The gravity acts downward. The radius of the area under observation is 400 μm and Re = 30. R is the radius of the tube.

C. Limiting Trajectory Method

For large Peclet numbers, the diffusion layer is compressed to a thin region in the vicinity of the surface where particle behavior is determined by specific forces. If repulsive forces are absent or are weaker than hydrodynamic forces acting on the particle or bubble, the collector intercepts the bubble. The interception rate can be determined by the method of limiting trajectory analysis. The limiting trajectory of a bubble can be found by following its deterministic path from the boundary of the region under observation back to the exit of the tube. The bubble velocity at a given position can be calculated by taking into account the velocity components due to the fluid flow and the external and colloidal forces acting on the bubble. All bubbles that enter the region enclosed by the limiting trajectory

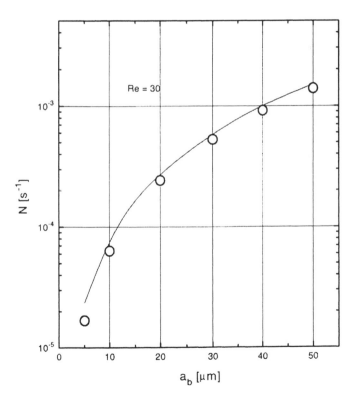

Figure 6 Calculated number of bubbles captured per second in a circle with radius 400 μm, normalized by the bulk concentration, as a function of bubble size, a_b. The solid line was calculated using the limiting trajectory method, and the open symbols represent solutions of the continuity equation.

end up on the collector surface in the region under observation. Assuming a uniform concentration of bubbles in the central part of the tube and an appropriate velocity profile, it is possible to calculate the flux of bubbles to the collector surface. The trajectories calculated for bubbles of various diameters are shown in Figure 5.

Figure 6 compares the results obtained by solving the continuity equation and by using the limiting trajectory method for Re = 30, where Re is the Reynolds number. As one can see, the agreement is good for all bubble sizes. The small systematic differences between the two methods can be related to the approximate nature of the analytical expressions used to calculate the fluid velocity [Eq. (1)] as well as possible systematic errors that occur during the calculation of the velocity in the vicinity of the wall. This comparison shows that the method based on the continuity equation provides correct results for very large values of the Peclet number, even if the diffusion-layer thickness is extremely small compared to the bubble radius.

III. EXPERIMENTS

A. Materials

1. Preparation of Glass Plates

Microscope slides (75 × 50 × 1 mm, Fisher Scientific) were used to prepare clean, methylated, and bitumen-film-coated glass plates.

> *Clean glass.* Glass plates were first cleaned using a detergent and then placed in chromicsulphuric acid. The plates were then washed intensively and stored in deionized water for later use.
>
> *Methylated glass.* Glass methylation was done by soaking cleaned glass plates in 20% v/v dimethyldichorosilane-in-toluene solution for 4 h. The plates were then rinsed with methanol and kept dry in air.
>
> *Bitumen-coated glass.* Methylated glass plates were immersed in a 20 wt% bitumen in naphtha solution and pulled out. After drying in air, a thin bitumen film remained on the glass, one side of which was cleaned using toluene. The bitumen film is stable and can remain on the methylated glass in pH 10 solutions for several days. Coker-feed bitumen and naphtha were supplied by Syncrude Canada Ltd.

All experiments were carried out in aqueous NaCl solutions. The solution pH was adjusted using NaOH and HCl. The chemicals were supplied by Fisher Scientific. Deionized water was used throughout.

B. Methods

1. Gas-Bubble Attachment

Figures 7 and 8 show the experimental setup. Its major component is a cell, shown in detail in Figure 7, placed under a microscope. The cell is composed of upper and lower parts with an inner glass tube connecting the two. The upper part is covered by a glass plate that serves as a collector of gas bubbles, which are generated in the lower part. The tube is 50 mm long with an internal diameter of 3.3 mm. This length-to-internal diameter ratio is large enough to ensure a steady, parabolic velocity profile inside the tube. The tube outlet is 3.3 mm away from the glass plate.

  Water flows into the chamber at the bottom, carrying the electrolysis-generated gas bubbles upward, and then leaves the upper cell. The jet from the inner tube is directed against the surface of the glass plate. Bubbles that are not captured by the glass are swept away by the flow. The flow is induced by the head due to the level difference between the upper container and the lower container, and it is controlled by a flowmeter (Fig. 8). A peristaltic pump recirculates the water.

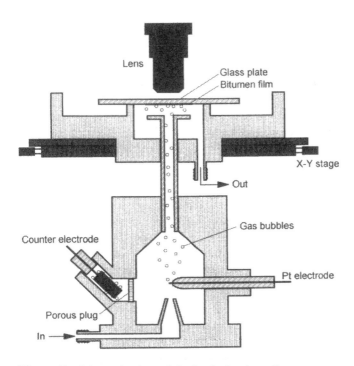

Figure 7   Schematic view of the impinging jet cell.

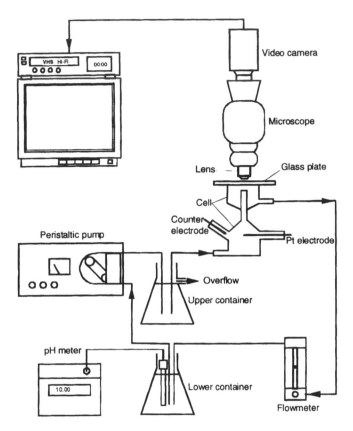

Figure 8   Schematic view of the experimental setup.

Under the flow conditions in our experiments, gas bubbles generated in aqueous NaCl solutions were typically 20–40 μm in diameter.

During an experiment, the microscope was first focused on gas bubbles attached to the glass surface and the scene was videotaped for around 10 min. Then, the focus was shifted to the outlet of the tube to record the bubbles passing through the tube exit. The videotape was subsequently played at a slow speed to determine the following:

1. The number of attached bubbles within a circle of radius 400 μm from the symmetry axis as a function of time. Depending on the experimental conditions, this number ranged from less than 10 (e.g., for a clean glass) to several hundreds within 10 min.

2. The bubble size distribution (based on a few hundred randomly selected bubbles rolling on or approaching the glass surface). Figure 9 shows an example of such size-distribution data.

3. The number of bubbles passing through the central circular section (100 μm in radius) of the tube exit per unit time. The number was in the range of 200–1000 per minute and is directly related to the bubble concentration at a certain flow rate, by which the attachment rate is normalized. The concentration has been calculated by dividing the number of bubbles counted in a given time by the volume of liquid flowing through this area in the same time, and applying a correction for buoyancy effects on the bubbles.

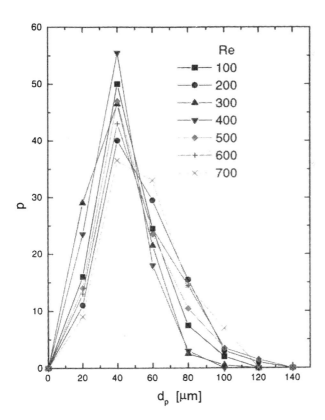

Figure 9 Examples of measured size distributions of gas bubbles at various Reynolds numbers. p is a number fraction of bubbles in a given size range, 20 μm wide.

The attachment experiment was repeated at two spots on each glass plate using two glass plates for each condition, which included the type of glass surface, pH, NaCl concentration, and flow rate.

IV. RESULTS AND DISCUSSION

A. Calculation of Theoretical Flux

The flux, j/c_0, is the number of gas bubbles attached to a collector surface per unit area, per unit time, normalized by the bulk bubble concentration, c_0. The theoretical flux, to which an experimental flux is compared, is calculated by solving the mass-transfer equation (1–3). However, the flux depends on bubble size, which ranges from 20 to 140 μm. Using the size distribution measured in a particular experiment, the average flux was calculated by adding contributions from various size fractions of bubbles according to the formula

$$<j> = \Sigma\, j(a_i)p(a_i)$$

where $p(a_i)$ is the normalized population of bubbles with diameters around a_i.

B. Siliconized Glass Collector

Let us define N as the number of bubbles attached per unit area. Figure 10 shows N, normalized by the bulk bubble concentration c_0, as a function of time, t, for methylated glass. The symbols on the graph represent the experimental values and the solid lines were fitted by least squares linear regression. As seen in Fig. 10, the number of bubbles attached to the methylated glass can be described by a linear function. This suggests that there is little blocking interfering with bubble attachment within the experimental time period (i.e., 10 min). Although most of the data follow a linear pattern, there is some scatter, especially at high Reynolds numbers. This could be due to the attachment of a few large bubbles early in the experiment, as it was sometimes observed, thereby reducing the area available for attachments. Also, large bubbles often result from the coalescence of smaller bubbles.

The slope of the N/c_0 versus time line is equal to the normalized flux, j/c_0. The experimental flux as a function of Re is shown for the methylated glass in Figure 11. Solid lines show theoretical flux calculated for a given bubble size. They exhibit minimum values for Reynolds numbers around 100, below which the flux is strongly influenced by gravity. For higher Reynolds numbers, the flux is controlled mainly by flow. Solid symbols represent the theoretical flux averaged to account for the actual bubble-size distribution. The scatter shown in the solid symbols is due to the variation in the bubble-size distribution measured for each experiment at a particular Re.

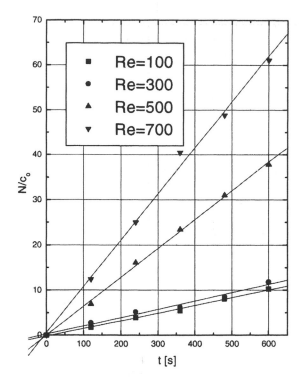

Figure 10 The coating density N (cm^{-2}), normalized by the bulk bubble concentration c_0 (cm^{-3}), as a function of time for various Reynolds numbers.

Because the assumed bubble–collector interaction is attractive, the theoretical flux can be considered to be close to the maximum flux. Compared to the theoretical flux (solid symbols), the experimental flux (open symbols) is the same order of magnitude. By contrast, the flux in the case of clean glass under the same condition is about an order of magnitude lower [11]. Because solid symbols (Fig. 11) represent the maximum flux expected in the system, the ratio between flux value represented by a solid symbol and the value represented by an open symbol defines the stability factor.

C. Bitumen Collector

Similar experiments were carried out using bitumen-coated glass as the bubble collector at various pH levels in the Re range of 100–700. Data were analyzed in the same way as discussed earlier. Results for bitumen at pH 6,and methylated

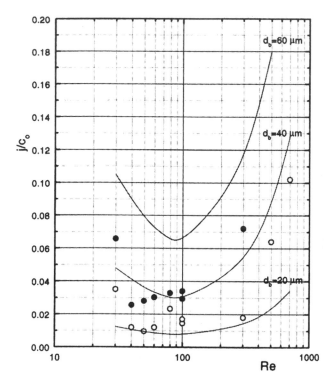

Figure 11 Bubble flux j (cm^{-2} s^{-1}) to a methylated glass surface, normalized by the bulk bubble concentration c_0 (cm^{-3}), as a function of Reynolds number. Open symbols are experimental data. Solid symbols are values calculated after accounting for the actual size distribution of bubbles. Solid lines are calculated values for a given bubble size, d_b.

glass for comparison, are shown in Figure 12. The normalized flux for pH 9 and 10 is shown in Figure 13. At all pH levels investigated, in general, the dependence of experimental flux on Re was similar to that determined theoretically; that is, the flux increased with Re for Re $>$ 100. In other words, increased flow intensity enhanced the bubble–bitumen attachment. As can be seen in Figure 12, at pH=6, the experimental flux values were similar to those obtained for methylated glass. Moreover, bubbles were captured by the bitumen surface in a similar way as with methylated glass surface: The attachment usually led to the rupture of the liquid film between the bubble and the collector surface. As the pH was increased to 9, the flux decreased substantially and the bubble behavior was visibly different. Many bubbles rolled over the bitumen surface and were swept away by the flow. In terms of its low affinity for gas bubbles, the bitumen surface at both

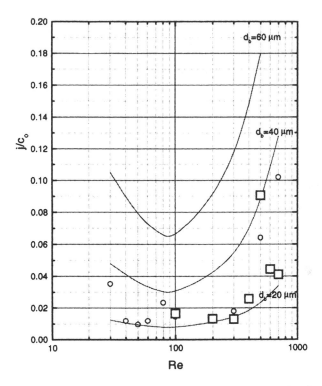

Figure 12   Open squares show the bubble flux j(cm^{-2} s^{-1}) to a bitumen-coated glass surface, normalized by the bulk bubble concentration c_0 (cm^{-3}), as a function of Reynolds number at pH 6. Open circles are the measured flux of bubbles for methylated glass. Solid lines are the calculated values for a given bubble size, d_b.

pH=6 and 9 very much resembled the surface of clean glass, for which the flux is lower than that of the methylated glass surface and the bitumen surface at pH 6 and 10 as well. The decrease in the flux with increasing pH was due to the fact that the bitumen surface is more negatively charged at higher pH levels, provided that the solution is not strongly alkaline [12]. The higher negative charge on the bitumen surface leads to a stronger repulsive electrical double-layer force between the bitumen surface and the bubble surface, which is also negatively charged. When the pH is increased to 10, the flux increases to the same level as the case of pH 6. This cannot be explained in the same way, as earlier, because the trend of bitumen surface charge changing with pH is not reversed until at least pH 10. Therefore, one would expect an even lower flux at this pH than at pH 9. In the previous study [11], it was found that, at pH 9, increasing the NaCl concentration from 0.01 to 0.1 mol/L increased the bubble flux, indicating a

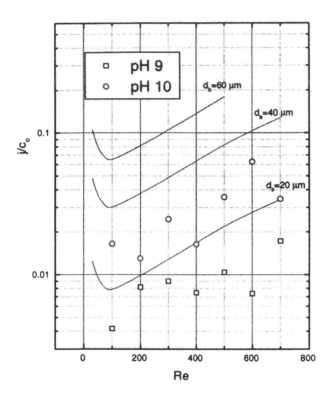

Figure 13 Open squares and circles show the bubble flux j (cm^{-2} s^{-1}) to a bitumen-coated glass surface, normalized by the bulk bubble concentration c_0 (cm^{-3}), as a function of Reynolds number at pH 9 and 10, respectively. Solid lines are calculated values for a given bubble size, d_b.

significant effect of the ionic strength on bubble–collector interactions. However, the change in the ionic strength due to adjusting the pH to 10 was negligible compared to the base ionic strength due to the presence of 0.01 M NaCl solution. Thus, the reason for the increase in flux at pH 10 is not clear at this stage.

V. CONCLUSIONS

1. The impinging jet technique has been successfully used to study the attachment of small gas bubbles to solid surfaces from flowing solutions.
2. Experimental results were compared to theoretical predictions based

on solutions of the continuity equation and calculated from the limiting trajectory approach.

3. The measured flux of bubbles to methylated glass is in reasonable agreement with theoretical predictions.

4. At 0.01 M NaCl and pH 6, the flux of bubbles to the bitumen surface was similar to that for methylated glass.

5. At pH 9, the flux is similar to that for methylated glass only when the salt concentration is higher.

6. The mode of bubble attachment is determined by the stability of the liquid film between the bubble and the collector surface. Therefore, the hydrophobic properties of the surface are significant in the attachment process.

REFERENCES

1. T Dabros, TGM van de Ven. Colloid Polym Sci 261:694–707, 1983.
2. C Yang, T Dabros, D Li, J Czarnecki, JH Masliyah. J Colloid Interf Sci 219:69–80, 1999.
3. C Yang, T Dabros, D Li, J Czarnecki, JH Masliyah. J Colloid Interf Sci 208:226–239, 1998.
4. AD Gosman, WM Pun, AK Runchal, DB Spalding, M Wolfshtein. Heat and Mass Transfer in Recirculating Flows. London: Academic Press, 1969.
5. VG Levich. Physicochemical Hydrodynamics. Englewood Cliffs, NJ: Prentice-Hall, 1962.
6. Z Adamczyk, T Dabros, J Czarnecki, TGM van de Ven. Adv Colloid Interf Sci 19:183–252, 1983.
7. Z Adamczyk, T Dabros, J Czarnecki, TGM van de Ven. J Colloid Interf Sci 97:91–104, 1984.
8. LG Leal. Laminar Flow and Convective Transport Processes. Boston: Butterworth–Heinemann, 1992.
9. E Ruckenstein, RK Jain. Faraday Trans 70:132–147, 1974.
10. A Sharma, E Ruckenstein. J Colloid Interf Sci 137:433–445, 1990.
11. T Dabros, Q Dai, HA Hamza, J Czarnecki in ''Processing of Hydrophobic Minerals and Fine Coal,'' pp 413–424. Proceedings of the 1st UBC–McGill Biannual International Symposium on Fundamentals of Mineral Processing Vancouver, 1995.
12. K Takamura, EE Isaacs. In: LG Hepler, C Hsi, eds. Technical Handbook on Oil Sands, Bitumen and Heavy Oils. AOSTRA Tech. Pub. Series 6, Edmonton, Alberta, Canada, 1989.

25
Rheological Instabilities in Waterborne Coatings

Syed Qutubuddin, Alexander M. Jamieson, Jingrong Xu, and Shi-Qing Wang
Case Western Reserve University, Cleveland, Ohio

ABSTRACT

A successful latex emulsion formulation for coatings requires that the resulting system be stable during the processing and application stages. In certain situations, an emulsion that is stable at the formulation stage may become unstable in the manufacturing process, which generally involves subjecting the emulsion to a wide range of shear and/or extensional deformation rates. To understand such phenomena in coatings and adhesives application of latex emulsions, dynamic processes must be considered in addition to the thermodynamic stability of such multiphase systems. Therefore, rheological characterization of emulsion stability under conditions appropriate to the processing and application of such materials is of interest. In this chapter, the rheological behavior of waterborne latex dispersions used in coatings and adhesives is reviewed relative to the most commonly used application conditions. Rheological characterization of several commercial emulsions is described. Two types of instability behaviors were identified: irreversible particle flocculation and reversible shear thickening. The influence of experimental parameters was studied with a view to improving the rheological performance of these emulsions for coatings applications.

I. INTRODUCTION

The rheology or flow behavior of a coating is of crucial importance to its processing (mixing, stirring, pumping), ease of application (brushing, roll or spray coating), and to phenomena such as splatter, leveling, and sagging, which influence the final appearance. An enormous amount of literature exists on the rheological behavior of coatings. It is well understood that the flow behavior over a wide range of shear and extensional deformation rates is pertinent to optimal performance of coatings [1] (see Table 1). Thus, an ideal coating should exhibit strong shear-thinning characteristics, such that it flows easily at higher shear rates relevant to application conditions, and a relatively high, although not too high, viscosity at low shear rates to facilitate leveling and avoid sagging [2]. An inordinately high extensional viscosity may result in excessive splattering of coatings in roll applications [3]. Also, waterborne coatings may exhibit time-dependent rheological behavior due to interparticle structure buildup (rheopexy–viscosity increase) or loss of interparticle structure (thixotropy–viscosity decrease) [4]. Under such circumstances, the flow characteristics may become dependent on the geometry of the rheometer (e.g., gap size, parallel plate versus cone and plate, etc.). This is generally an indication that macroscopic structures are being created and destroyed in the flow cell. Such an effect can lead to flow-instability behavior, which may adversely affect coating deposition.

Bearing in mind that the flow fields and flow histories encountered in coatings applications are complex, important insight into flow behavior necessary to optimize coatings performance can be gained from well-designed rheological characterization [4,5].

As noted, shear-thinning and thixotropy behaviors commonly observed in latex coatings and adhesives are due to the breakdown of preexisting structure

Table 1 Shear Rates Typical of Some Coating Processes

Coating process	Typical shear rates range (s^{-1})
Leveling and sagging	10^{-3}–10
Mixing and stirring	10^1–10^3
Dip coating	10^1–10^2
Spraying and brushing	10^3–10^4
Brushing	10^4–10^6
High-speed coating (paper)	10^5–10^6

Source: After Ref. 1.

in the system upon the application of shear, and they are generally desirable during manufacture and application. More problematic are the opposite effects (i.e., shear thickening and rheopexy) that may lead to flow instabilities during application. However, a modest rheopectic behavior may be beneficial in the final stages of film formation to prevent sagging [6].

II. MATERIALS AND PROCEDURES

All latex dispersions under investigation were commercial systems for application as coatings, adhesives, or sealants. They consisted of natural or synthetic polymers of high concentration (50–60% by volume) dispersed in water with electrostatic (anionic) stabilization. Unitex is a natural rubber latex, and SBR is a styrene–butadiene–rubber latex, both provided by Tesa Tape (Charlotte, NC). H5 (provided by BF Goodrich, Cleveland, OH) is an acrylic ester latex with the narrowest particle size distribution among all systems. More detailed description of H5 can be found in a previous publication [7].

Rheological characterization was carried out using a Carri-Med controlled-stress rheometer of cone and plate geometry. A water seal was used around the flow cell to prevent loss of water from the sample during experiments.

III. RESULTS AND DISCUSSION

Rheopectic and shear-thickening behaviors were encountered in several commercial waterborne emulsions. These effects fall into two distinct categories: The first is an irreversible time-dependent viscosity increase due to flow-induced particle aggregation which can occur at low shear rates; the second is a reversible shear-thickening phenomenon which results from particle jamming effects and occurs at relatively high shear rates.

Among the latex systems that show irreversible agglomeration under shear are the natural rubber (Unitex) and styrene–butadiene–rubber (SBR2) latices. In these materials, one observes an abrupt increase of viscosity after an induction period at a constant shear rate, which is always accompanied by the formation of visible aggregates. The results are shown in Figures 1 and 2. The experiments were usually terminated when the aggregates grew large enough to span the gap, at which time, large fluctuations appear in the viscosity. Particle coagulation in flow is a well-known phenomenon and occurs because particles in different streamlines in a simple shear field move at different velocities. According to the original analysis of Smoluchowski [8], the collision rate between particles is

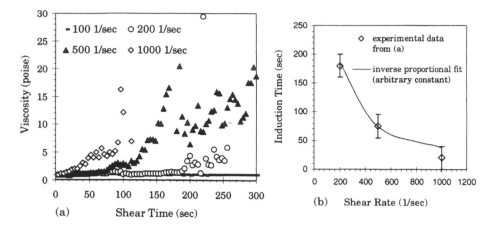

Figure 1 Comparison of (a) viscosity evolution with shear time and (b) induction period of Unitex latex (50% solids content) at different constant shear rates.

proportional to the particle concentration and to the shear rate. Based on this, it is possible to deduce that the induction period is inversely proportional to the shear rate [9].

$$t_{ind} \sim \frac{1}{J} \propto \frac{1}{\dot{\gamma}}$$

where t_{ind} is the induction time, J is the collision rate, and $\dot{\gamma}$ is the shear rate. The experimental results of shear-rate dependence of the agglomeration induction period for Unitex latex showed qualitative agreement with the above prediction, as illustrated in Figure 1b.

Besides shear rate, the probability that collisions will lead to coagulation will decrease with the increase in the strength of the interparticle repulsion forces; that is, the induction period depends on the same factors that control colloidal stability under quiescent conditions. The fact that coagulation occurs indicates that during close encounters in shear flow, the particles are able to cross the repulsion energy barrier and locate in the first primary potential energy minimum. Thus, if the interparticle repulsion is enhanced, the shear-induced instability should be delayed or eliminated. This is indeed confirmed by the experimental results depicted in Figure 2, where the anionic surfactant sodium dodecyl sulfate (SDS) was added to the emulsion. As evident in Figure 2, with addition of 22 mM SDS, the stability of SBR2 latex is greatly improved, as indicated by the increased induction period before the viscosity buildup. Upon addition of 50 mM

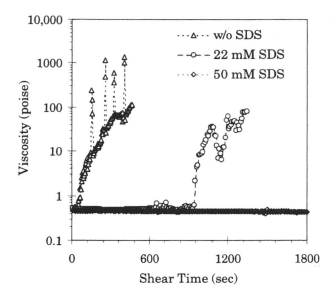

Figure 2 Prolonged shear behavior of SBR2 latex with and without the addition of SDS.

SDS to the emulsion, the irreversible increase of viscosity was eliminated at a shear rate of 200 s^{-1} within the experimental time limit (2 h). The SBR particles in this latex are stabilized using a synthetic anionic surfactant and are, therefore, negatively charged. The SDS increases the negative charge density if it adsorbs onto the particle surface. Therefore, enhancing the electrostatic repulsion between the particles by adding SDS proved to be an effective way to eliminate the irreversible instability of the SBR2 latex.

In a second study, the reversible shear-thickening rheology of a commercial aqueous polyacrylic ester latex (designated H5) was studied in detail. The polymer particles in this latex were also anionically stabilized and were nearly monodisperse with average particle diameter 270 nm, calculated from quasielastic light-scattering measurements. To induce the shear-thickening effect at shear rates in the range accessible to the Carri-Med rheometer, the solids content of the latex was increased to above 56.7% by volume.

The essential details of the reversible shear-thickening behavior of the H5 latex are described earlier [7]. The results showed many of the characteristics associated with a well-known phenomenon observed in concentrated charge-stabilized colloidal dispersions with narrow size polydispersity, and they are interpreted in terms of a mechanism involving a flow-induced order-disorder transition (ODT) process. At low shear rates, the particles form ordered layer assemblies in the flow velocity–vorticity plane, stabilized by the strong interparticle forces, so that the particles can flow with few collisions, giving a low viscosity. Above a critical shear rate, hydrodynamic forces overcome the interparticle repulsions which stabilize the layered state, and particles move out of the shear plane, causing particle jamming, and a sudden large increase in viscosity [10]. Some interesting features were observed as the prolonged shear was applied to H5 latex above the critical shear rate of shear thickening, $\dot{\gamma}_c$. The time-dependent behavior was explained by the formation of particle clusters due to high shear. The fact that the viscosity in such state is partially recoverable, given enough relaxation time, seems to suggest that the secondary minimum in the potential energy curve is encountered. The details of the experimental results and proposed mechanism can be found in an earlier article [7]. Additional evidence from cryogenic scattering electron microscopy can be found in the Appendix.

In the studies of shear-thickening behavior of the charge-stabilized latex H5, in the absence of added salt, it was demonstrated [7] that the critical shear rate, $\dot{\gamma}_c$, exhibited a strong dependence on particle concentration (solids content), pH, ionic strength, and particle polydispersity in a manner consistent with the OTD mechanism. Thus, as shown in Figure 3, $\dot{\gamma}_c$ varies inversely with particle

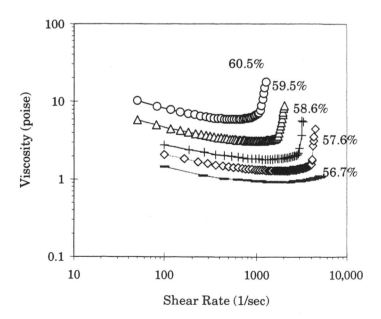

Figure 3 Shear-rate sweep of H5 latex at different solids content. (From Ref. 7, with permission.)

concentration (solids content), as predicted by a theoretical model which assumes that shear-thickening occurs when hydrodynamic forces overcome the charge repulsions which stabilize the layered state [11,12]. Upon the addition of salt, the attendant increase in ionic strength will diminish the strength of the electrostatic forces which stabilize the layered state. We anticipate that this will decrease $\dot{\gamma}_c$ for two reasons. First, it becomes easier for the hydrodynamic forces to disrupt the ordering [12], and, second, if the layered state disappears, the viscosity increases and the efficacy of the shear flow to create "hydroclusters" is enhanced [13]. Such behavior is indeed seen in the H5 latex when sodium chloride is added to the latex, as shown in Figure 4, where it is evident that the critical shear rate for shear thickening is significantly reduced when the NaCl concentration is increased. At a NaCl concentration of $0.5M$, the critical shear rate declines from 2000 s^{-1} to 200 s^{-1} for a salt-free system.

The weakening of interparticle repulsions at a higher salt concentration not only shifts the onset of shear thickening to a lower shear rate but also alters the nature of the flow-induced disordered structural state. This is demonstrated when H5 latex with added $0.5M$ NaCl is subjected to repeated shear-rate sweeps (Fig. 5). After six consecutive sweeps (not shown in the figure), the low shear viscosity starts to increase until eventually the total stress exceeds the machine limit, therefore terminating the experiment. Large particle agglomerates are visible after the test, demonstrating the irreversible nature of the shear-generated structure. This

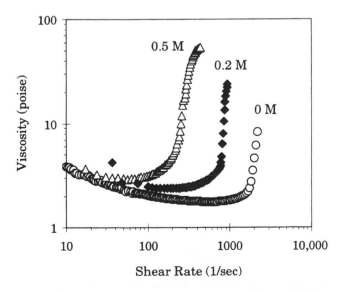

Figure 4 Shear-rate sweep of H5 latex with different NaCl concentrations.

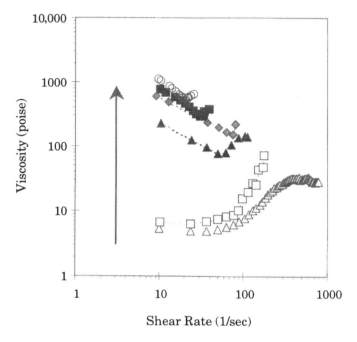

Figure 5   Repetitive shear sweep of H5 latex with 0.5*M* NaCl (arrow indicates the sweep sequence: △—7th; □—8th; ▲—9th; ◆—10th; ■—11th; ○—12th).

indicates that the particles are able to locate in the primary minimum of the potential energy curve. Increasing the ionic strength decreases the height of the repulsive energy barrier to coagulation. It appears that in charge-stabilized waterborne latices, upon an increase of ionic strength, there is a smooth transformation in the mechanism of shear thickening from hydrocluster formation via an ODT, to hydrocluster formation from a disordered colloidal fluid, to an irreversible coagulation process.

Two types of shear-induced instability effects have thus been identified in waterborne latices. One process is a reversible shear-induced "hydrocluster" formation, which may or may not involve a prior ODT; the other mechanism is an irreversible flow-induced colloidal agglomeration. In the latter, the primary particles evidently fuse together to form tightly bound aggregates, and the electrical double layer between the individual primary particles is presumably destroyed. In the former, "hydroclusters" are transient structures that consist of primary particles weakly bound together, and their reversible nature indicates that the double layer is preserved or can be easily restored by the Brownian motion. Irreversible clustering is facilitated by a low interparticle repulsion energy barrier so that particles can easily pass into the primary minimum of the potential energy curve. The reversible clustering may involve either coupling of particles during close encounters via lubrication forces or via the presence of a secondary minimum. The possibility of irreversible aggregation appears to exist in the H5 latex, even under conditions of low ionic strength (no added salt). As previously reported [7], on prolonged shear at very high shear rates (i.e., above $\dot{\gamma}_c$), hysteresis appears in the shear-thickening behavior, accompanied by an increase in the low-shear viscosity, suggesting that some degree of long-lived particle association occurs. Evidently, when the hydrodynamic forces are strong enough to produce "hydroclusters," prolonged shearing drives some particle clusters into the primary minimum.

IV. CONCLUSIONS

Two types of shear-instability phenomena were observed in aqueous latex dispersions: a reversible and an irreversible process. The reversible shear instability was observed in a dispersion of charge-stabilized particles in the absence of added salt and showed rheological behavior consistent with a flow induced ODT transition. Irreversible shear thickening was observed in poorly stabilized systems and was characterized by permanent viscosity buildup with time after an induction period, accompanied by formation of visible agglomerates. The addition of salt to the highly stable emulsion produces a smooth transition from the situation where the high-shear reversible process is dominant to one where the low-shear irreversible phenomenon controls the rheological response. To avoid the irreversible-shear-instability phenomenon, one can employ standard methodology to improve colloidal stability (i.e., by increasing interparticle repulsion). The critical shear rate for reversible shear thickening can be moved to higher values by decreasing particle concentration, increasing particle size polydispersity, or by increasing interparticle repulsion.

APPENDIX

Scanning electron microscopy was used in an effort to interpret the rheological behavior by characterizing the microstructure of the latex. Due to the moisture content of the latices, a cryogenic technique has to be applied, which requires the sample temperature to be around $-190°C$ during scanning electron microscopic (SEM) analysis. To prevent the crystallization of water when quenching the samples in liquid nitrogen, the jet-propane technique, which has been proven to be the fastest way to freeze samples, was applied. The SEM instrument was a JEOL JSM840 microscope equipped with a cryogenic unit and samples were frozen in a MF7200 "Gilkey-Staehelin" Propane Jet Ultrarapid Freezer. The latex samples frozen between small copper hats were fractured under high vacuum in the SEM chamber and the fractured surface was sputter-coated before observing under the microscope.

The postulated mechanism for time-dependent rheological behavior in the shear-thickening region involves a particle-clustering process at high shear rates. Figure 6 is a schematic of such mechanism, which shows that the clustering follows the well-known ODT shear-thickening transition and it will only occur when the shear rate is beyond $\dot{\gamma}_c$ and that it is a more gradual process compared to the former sudden transition. Recovery of structure—dissociation of clusters—is also possible with time (see Ref. 7 for details). SEM measurements were carried

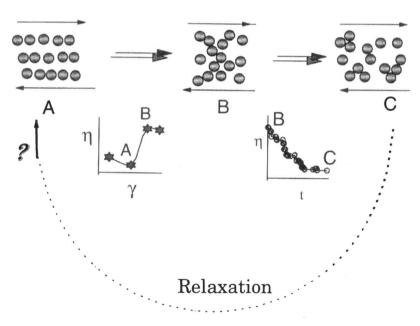

Figure 6 Schematic illustration of the proposed mechanism of shear thickening and cluster formation. (A) Low-viscosity ordered layered state of particles at low shear; (B) high-viscosity jammed state above critical shear of shear thickening; (C) low-viscosity clustered state after prolonged high shear. The transition from state A to B can be induced by a shear-rate sweep, and the transition from B to C is usually induced by a prolonged shear at high shear rates.

out to obtain support for such a mechanism. The SEM specimens were taken from samples after being subjected to a specified shear history, with the expectation to preserve any shear-induced structures by rapid freezing. The long relaxation time (>2.0 h) of the shear-induced clusters previously discussed made such an attempt realistic. Shown in Figure 7 is a sample frozen right after it had been sheared at 3000 s^{-1} for 18 min. At lower magnification, many particle clusters can be seen, shown by the brighter spots. For comparison, a sample sheared to just above the $\dot{\gamma}_c$ but without prolonged high shear was also analyzed. At lower magnification, no structure was observed. Higher magnification reveals that the particles remain individual (Fig. 8), which agrees with the reversible ODT mechanism. Spherical particles (half buried) of lighter shade are easily distinguished from the darker background—water. The results obtained from cryo-SEM agree well with the proposed mechanism. Although far from conclusive, this technique has proven to be helpful in characterization of shear-induced microstructures in emulsions.

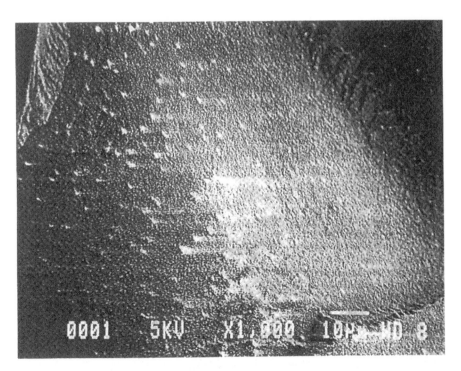

Figure 7 Cryo-SEM micrograph of H5 latex, 59.5 vol%, frozen immediately after prolonged shear at 3000 s^{-1} for 30 min at a magnification of 1000 (15 min etching of fractured surface).

444   **Qutubuddin et al.**

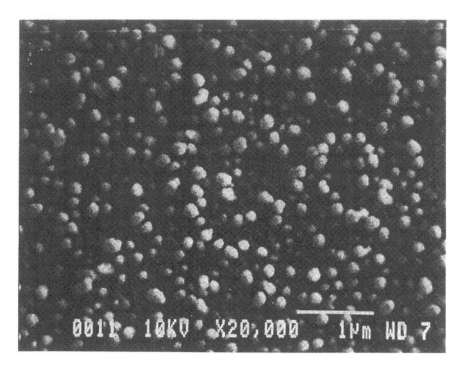

Figure 8 Cryo-SEM micrograph of H5 latex, 59.5 vol%, frozen immediately after shear ramp to 2000 s^{-1} at a magnification of 20,000 (without sublimation).

ACKNOWLEDGMENT

These studies were supported by the Edison Polymer Innovation Corporation, Akron, OH.

REFERENCES

1. HA Barnes, JF Hutton, K Walters. An Introduction to Rheology. Amsterdam: Elsevier, 1989.
2. RD Hester, DR Squire, Jr. J Coatings Technol 69(864):109, 1997.
3. JE Glass. J Coatings Technol 50(641):56, 1978.
4. DC Cheng. In: H Benkreira, ed. Thin Film Coating. London: Royal Society of Chemistry, 1993, p. 3.
5. M Power, S Smith. In H Benkreira, ed. Thin Film Coating. London: Royal Society of Chemistry, 1993, p. 34.

6. RL Powell. In: BG Higgins, ed. Coating Fundamentals: Suspension Rheology for Coating. Atlanta, GA: Tappi Press, 1988.
7. J Xu, AM Jamieson, SQ Wang, S Qutubuddin. J Colloid Interf Sci 182:172, 1996.
8. H Sonntag, K Strenge. Coagulation Kinetics and Structure Formation. New York: Plenum, 1985.
9. Y Hu. Ph.D. dissertation, Case Western Reserve University, Cleveland, OH, 1994.
10. HM Laun, R Bung, S Hess, W Loose, K Hahn, E Hadicke, R Hingmann, F Schmidt, P Lindner. J Rheol 36:743, 1992.
11. RL Hoffman. J Colloid Interf Sci 46:491, 1974.
12. WH Boersma, J Laven, HN Stein. AIChE J 36:321, 1990.
13. J Bender, NJ Wagner. J Rheol 40:899, 1996.

26

Preparation and Characterization of Thin Metal Oxide Layers Used for Producing Angle-Dependent Optical Effects

Martin Stech
Darmstadt University of Technology, Darmstadt, Germany

Peter Reynders
Merck KGaA, Darmstadt, Germany

ABSTRACT

Pearlescent pigments are the most important industrial products in the field of angle-dependent optical effects. Their preferred production route includes a calcination step of thin metal oxide films and consequently a densification. The influence of different process temperatures on the densification behavior and the grain growth of the nanocrystalline metal oxide layers has been studied. Finally a change in the optical properties was assigned to the microstructural development.

I. INTRODUCTION

Angle-dependent optical effects are applied on industrial products [1] primarily for two reasons: They can fulfill the special functional requirements and are used because of their decorative properties. Functional applications are, for instance, security patterns on banknotes, ID cards, or on tickets, which cannot be copied by conventional copiers or by photography. Forging of such products is substantially more difficult than without angle-dependent optical characteristics. A second functional application is optical filters and glass coatings which can be used as a means for antireflection or to change the light spectrum in greenhouses. Infrared reflection coatings are found in public buildings as well as in factory plants to reduce the energy influx from direct sunlight.

The application as a decorative medium makes use of the fact that these special optical effects are commonly perceived as appealing or even valuable. Therefore, the paint coating of about 40% of the cars in the United States contains so-called pearlescent pigments. These pigments imitate the luster of natural pearls and also find application in packaging and cosmetics.

Pearlescent pigments are, commercially, the most important products in the field of angle-dependent optical effects. They take advantage of thin-film interference [2]. Other optical systems that produce angle-dependent effects and are not the subject of this chapter are holograms, gratings, and liquid crystals.

Thin-film interference is the interference of light reflecting off the top surface of a film, with the light waves reflecting off the bottom surface (Fig. 1). If the light falls perpendicularly upon a planar thin film of a thickness in the range of a few hundred nanometers, the intensity R of the reflected light is a function of the refractive indices of the film n_1 and the surrounding medium n_0:

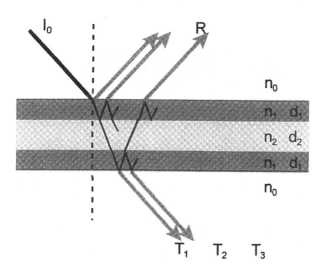

Figure 1 Thin-film interference: I_0 = incident light intensity; R = intensity of reflected light; T = intensity of transmitted light; $n_0, \ldots, n_2$ = refractive indices of surrounding medium, layer material, and substrate material; d_1 and d_2 = thicknesses of layer and substrate, respectively.

$$R = \left(\frac{n_1^2 - n_0^2}{n_1^2 + n_0^2}\right) \tag{1}$$

This equation shows that the square of the difference in the refractive indices determines the amount of reflected light and, consequently, the quality of the end product.

If the incident light is a broad spectrum—like white sunlight—there is always at least one wavelength that has maximum reflectance depending on the refractive index and the thickness of the layer. If the refractive index of the surrounding medium is set to a value lower than the refractive index of the film, a simple relationship for the wavelength of maximum reflectance λ_{max} can be found:

$$\lambda_{max} = \frac{4n_1 d}{2m + 1} \tag{2}$$

with n_1 the refractive index of the film material, d the thickness of the film, and m the order of interference.

In general, a pearlescent pigment consists of a disk-shaped substrate coated on both sides with an optical layer [3,4] to generate the interference color (Fig. 2). For a maximum effect according to Eq. (1), the substrate should have a low refractive index. It should also be transparent and, ideally, colorless. Small platelets of the substrate material serve as a template for the coating process and also provide a mechanical support to the thin, often brittle, optical layers. The most

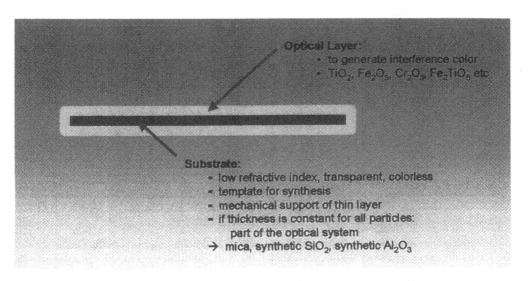

Figure 2 General scheme of a pearlescent pigment.

used substrate is muscovite mica, which can be cleaved and ground into very thin platelets. Although some other substrates show an interference color themselves, muscovite mica does not. The reason is that the muscovite platelets have a broad thickness distribution from particle to particle. Therefore, the single-substrate particles show the defined interference colors of single-layer systems and the ensemble of particles does not. Because of the low refractive index of muscovite and its usual thickness of 200–600 nm, the wavelength of maximum reflection from Eq. (2) is normally outside the visible range or the resulting maximum is weak compared to the reflection of the optical layer. The thickness of synthetic substrates like SiO_2 flakes [5] or Al_2O_3 flakes [6] can be controlled during their synthesis to yield a narrow thickness distribution at a desired value. This results in a defined interference color.

As a coating material, titania or iron oxides are typically chosen, both showing a high refractive index. Basically every coated substrate particle is then a three-layer system in which the substrate also contributes to the interference color. For the muscovite mica substrate, however, the overall effect is that of a single layer on a thick mica substrate due to the broad substrate particle thickness distribution mentioned earlier. If the pearlescent pigments consist of coated synthetic substrates with a narrow particle thickness distribution like SiO_2 or Al_2O_3, the ensemble of pigment particles acts as a true optical three-layer system. These products, therefore, show a stronger and clearer interference color [2].

II. MATERIALS

There are different alternatives for the coating process. One possibility is the evaporation of a precursor that is oxidized in a gas flow and, consequently, simultaneously deposited on a substrate. This chemical vapor deposition (CVD) process can be successfully performed with $Fe(CO)_5$ on mica at 400°C [7], but it does not work well for titania. In general, this technique is rather costly for the mass production of pigments.

Another route is the well-known sol–gel technique using organometallic precursors such as $Ti(OBu)_4$. This process is widely used for glass coatings and is well described in the scientific literature [8]. However, it is also too expensive for the production of pigments and does not offer significant advantages over the aqueous process described next.

The method that is used for the synthesis of more than 95% of the pearlescent pigments is based on aqueous chemistry. The unbeatable advantages of this production route are inexpensive, relatively harmless precursors like $TiOCl_2$ or $FeCl_3$ and environmentally uncomplicated side products like NaCl. A typical coating process for an anatase–mica pigment is as follows: (1) Mica flakes are coated in an aqueous suspension by titration with a $TiOCl_2$ solution and simultaneous adjustment of the pH to about 2 by NaOH. (2) Then, a drying step followed by calcination at 600–950°C is performed. The calcination reduces the chemical activity of the TiO_2 layers. An optional last step might be a special surface treatment that improves the photostability, weather resistance, and/or the wettability in the desired application. This chemical route leads to the anatase modification of titania. When the mica flakes are precoated with a thin SnO_2 layer, the subsequent TiO_2 precipitation results in rutile, which has a higher density and a higher refractive index than anatase.

There have been many efforts by Merck KGaA in the last few years to obtain a better understanding of the coating process [9]. The aqueous chemical route takes advantage of the fact that $TiOCl_2$ is not soluble at a pH above 1.5. Therefore, such a pH is adjusted in the mica suspension via the addition of NaOH. X-ray diffraction of room temperature dried but not calcined pigments shows crystalline titania with crystallite sizes in the range of 3–8 nm in the case of anatase and 10–17 nm for rutile. The scanning electron microscopic (SEM) and

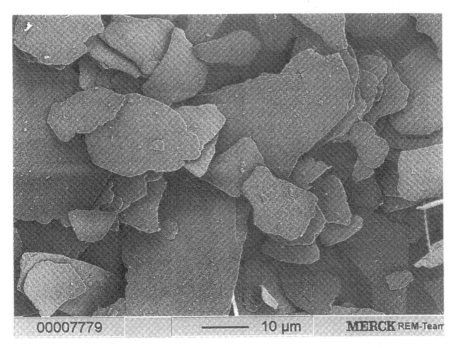

Figure 3   SEM image of a titania–muscovite pigment with a structure according to Figure 2, having a diameter distribution of 10–50 μm, a muscovite thickness distribution of 100–1000 nm with a mean value of 400 nm, and titania-layer thickness on each side of about 140 nm. (Courtesy Merck REM-Team.)

transmission electron microscopic (TEM) images of dried samples were taken with the coating process interrupted at early stages to trap the initial states. The micrographs show small seeds of crystalline TiO_2 spread all over the planes and edges of the mica platelets. Brunauer, Emmett, Teller (BET) surface-area measurements of the dried pigments result in very high surface-area values in relation to the relatively coarse powder characteristics (100 m^2/g for a 140-nm anatase layer thickness on mica platelets with a diameter in the range of 10–50 µm). Assuming a spherical shape, the diameter of the particle was derived from the BET surface-area values and the crystallite size determination via X-ray was confirmed this way.

Based on this and other experimental evidence, we believe that in microscopic dimensions, the coating process is described as follows. A low concentration of nanocrystalline TiO_2 is generated at the inlet for the $TiOCl_2$ solution. These small seeds may collide with a mica platelet to form the desired optical layer or with themselves to form undesired agglomerates (side precipitation). There is no strong energetic preference for directly hitting a mica particle or

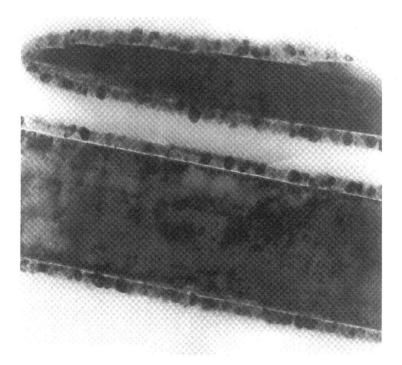

Figure 4 TEM cross-sectional image of a pearlescent pigment consisting of muscovite coated with an anatase layer of 65 nm thickness.

another seed. During the coating process, the seeds build up the titania layer on the mica substrate through fractal growth. On bare mica, only anatase is formed. If the mica is precoated with a thin layer of SnO_2, the relatively flexible anatase nanocrystals transform into rutile during attachment to the substrate.

Figures 3 and 4 show typical SEM and TEM images of a titania–mica pigment (here, anatase on muscovite). In Figure 3, one can see not only the regular and flat nature of the titania coating but also the irregular shapes of the platelet edges. This is a consequence of the mica being a natural product. Figure 4 shows a fracture through a single pigment particle. The anatase layers on both sides of the mica consisting of individual grains are visible. The very flat surface of the mica particle can also be seen.

III. METHODS AND RESULTS

As mentioned earlier, the dried titania layers are chemically as well as photo-chemically active. Therefore, a calcination step follows after the coating process to make the coatings more stable. The heat treatment also leads to densification of the TiO_2, which results in a higher refractive index. As pointed out earlier, a higher refractive index difference between the layer material and the substrate material causes better coloristics (e.g., a stronger reflection).

The densification process can be described in terms of sintering of nano-crystalline material. The substrate acts like a geometrical constraint to the sin-tering [10]. Therefore, we face the case of constrained film sintering of nanocrys-talline titania. Different questions arise from this complicated system. How does the constraint hinder the densification? How do grain sizes and pore sizes develop in relation to density? How does that influence the coloristics of the pigment? Does the constraint influence the phase transformation of anatase to rutile? To get answers to these questions, a comparison to the microstructural development of a freely sintering TiO_2 sample was made by titrating $TiOCl_2$ without mica but under otherwise identical conditions.

Different experimental techniques were used: nitrogen adsorption measure-ments to obtain the BET surface-area values, total pore volumes, and pore size distributions according to the method proposed by Barrett et al. [11]; SEM images for grain size determination with the linear intercept method [12]; and x-ray dif-fraction patterns to detect phase modifications.

Figure 5 shows the BET surface area as a function of temperature for an anatase–mica pigment with a layer thickness of 140 nm. The annealing time was 30 min in each case. Starting from a value of almost 100 m^2/g for the dried material, the surface area drops linearly to 30 m^2/g at 600°C. At higher tempera-tures, the decrease is weaker, and at 900°C, a minimum of about 7 m^2/g is reached. This points to a considerable coarsening during calcination.

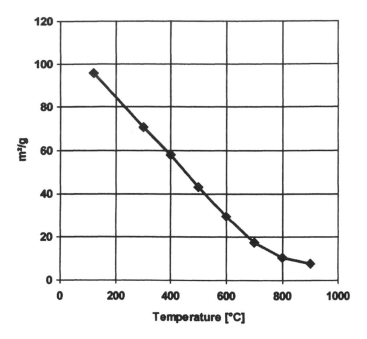

Figure 5   The BET surface area as a function of annealing temperature for an anatase–muscovite pigment, with a diameter distribution of 10–50 μm and an anatase layer thickness of 140 nm (see Fig. 3) after annealing at 850°C for 30 min.

In Figure 6, the nitrogen adsorption isotherms of pigments calcined at different temperatures are represented. The dried pigment shows the isotherm of a highly microporous material with almost no mesopores (pores with diameters in the range 2–50 nm). Already the 600°C sample shows a pronounced hysteresis loop typical for mesoporous materials. At 700°C, the absorbed nitrogen volume is reduced further in the whole pressure range, which means that the total porosity decreases. The hysteresis loop shifts to higher relative pressures, corresponding to larger mesopore radii, and changes to a shape typical of a porous material consisting of agglomerates or compacts of uniform spheres in a regular array [13]. At 900°C, the hysteresis starts to vanish and the porosity reaches a very low level.

Figure 7 depicts the development of the pore size distribution at 700°C with increasing sintering time. The diagram should be read from bottom to top. For the distribution of the only-dried sample, the maximum is cut off at the lower limit of the validity range of the Kelvin equation at 2 nm. With increasing sintering time, the maximum moves to higher values and loses in height. Already

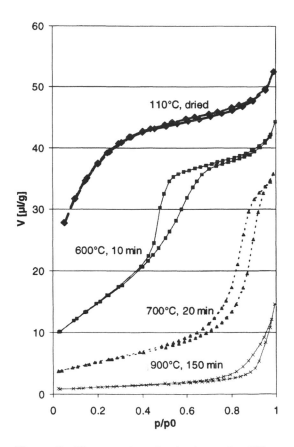

Figure 6 Nitrogen adsorption isotherms for different calcination temperatures, for an anatase–muscovite pigment, with a diameter distribution of 10–50 μm and an anatase layer thickness of 140 nm after annealing at 850°C for 30 min.

at 30 min, the maximum has reached a stationary value of about 17 nm. For freely sintering TiO_2, a different development was observed: The maximum moves further to higher pore sizes with increasing sintering time and exceeds the upper limit of the Kelvin region at 50 nm at sintering temperatures above 800°C.

One useful graphical representation for judging the development of the densification process is to plot the grain size versus the relative density, as a so-called sintering path. The main parameters of coarsening and densification can be observed simultaneously this way. Grain sizes were measured by applying the linear intercept method on SEM images. (The density was calculated from the

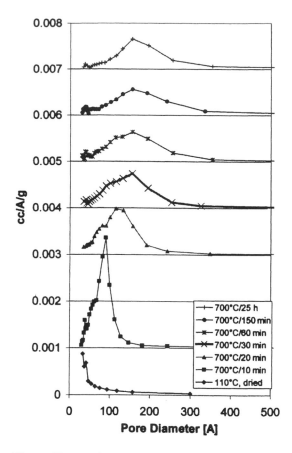

Figure 7 Development of pore size distribution with increasing calcination time at 700°C for an anatase–muscovite pigment with a diameter distribution of 10–50 μm and an anatase layer thickness of 140 nm after annealing at 850°C for 30 min.

total pore volume of the nitrogen adsorption at the highest pressure point of the isotherm $p/p_0 = 0.995$). In Figure 8, a comparison of the sintering paths of the layer material and freely sintering TiO_2 is represented. The layer material starts with a green density of 62% and a grain size of 7 nm. The sintering path increases almost linearly up to a grain size of about 140 nm at a relative density of 98%. At that point, the grain size is of the magnitude of the layer thickness. The green body of the freely sintering sample has a grain size of 8 nm and a density of 50%. In the beginning, the sintering path rises linearly up to a relative density of 85%, at which an accelerated grain growth starts. The grain growth

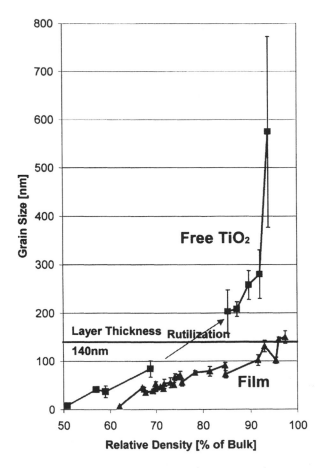

Figure 8 Comparison of sintering path for layered material of Figures 5–7 and the freely sintered TiO₂ prepared according to the same procedure without muscovite.

of this material is not limited by the layer thickness and, consequently, grain sizes of up to 600 nm at 92% relative density are reached. Furthermore, x-ray diffraction measurements reveal that in the freely sintering sample, a phase transformation from anatase to rutile takes place at 700°C, which also contributes significantly to the grain growth and densification process. Therefore, a break in the data between 70% and 85% relative density is observed. This phase transformation does not happen for the constrained film material below 1000°C.

To evaluate the influence of the microstructural development on the coloristics of the pigment, draw-down cards of every sample were produced. These were

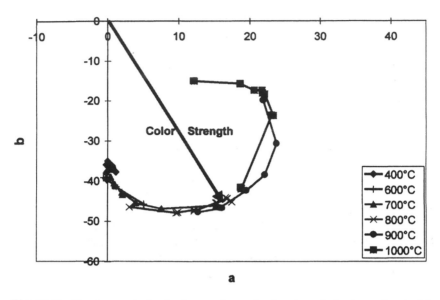

Figure 9 Development of color impression and color strength for layered material of Figures 5–7, for different annealing temperatures and times, measured on a draw-down card in nitrocellulose lacquer. The data points on each curve correspond to annealing times of 10, 20, 30, 60, 150, and 3600 min, starting at the left end of the curve. The CIELAB nomenclature is used; see text. The distance to the origin corresponds to the color strength.

taken to perform color measurements according to the CIELAB standard [14]. The results of such measurements are L-a-b values that describe the color impression and the brightness in a three-dimensional coordinate system. In a plot of the data in the a-b plane (Fig. 9), the a-axis stands for a color impression between green (negative values) and red (positive values). The b-axis covers the range from blue (negative) to yellow (positive). The distance of a data point from the origin represents the color strength of this sample. From Figure 9, it can be seen that with increasing calcination temperature and time, the color strength first increases while the color impression changes from bluish to violet. The color strength reaches a maximum at 900°C and sintering times between 10 and 20 min. Longer calcination or higher temperatures result in a dramatic decrease in color strength.

IV. DISCUSSION

It has been shown that the calcination process leads to a considerable densification both for the layers constrained by a rigid substrate and for freely sintering samples. In both cases, the densification is accompanied by a pronounced coarsening of the microstructure. In the freely sintering case, the stronger grain growth in the later sintering stages results in a reduction of the densification rate [15], which leads to a limited maximum density of about 95% of the theoretical density. In the constrained film, the densification is slowed down by the stress that builds up in the plane of the substrate during sintering [16]. These stresses occur because the rigid substrate does not allow lateral shrinkage of the layer material. They counteract the sintering stress, which is the driving force for the densification process. Therefore, higher temperatures are needed to achieve densification similar to that in free sintering.

The maximum grain size of the layer material is limited by the layer thickness. When the grain size reaches the dimensions of the layer thickness, the grain growth rate decreases. For the freely sintering material, the grains can grow without limits. The grain growth also influences the development of the pore size distribution. Both cases start with a highly microporous state where the maximum in the pore size distribution lies beneath the limits of the Kelvin equation, which is the theoretical basis of a pore size calculation. Again, in both cases, the maximum moves to higher values and loses in height. However, in the constrained case, the maximum reaches a stationary value, whereas the pores of the freely sintering sample can grow further. This points to a limiting effect of the geometrical constraint on pore growth.

Another—negative—effect of the substrate is that it impedes the phase transition. Although in the free case, rutilization was observed at 700°C, there was no transformation of anatase to rutile in the TiO_2 layers in the constraint case up to a processing temperature of 1000°C (at which muscovite–mica decomposes). The stresses that build up in the layer due to the geometrical constraint hinder the volume contraction by 10%, which accompanies the phase transformation.

The optical properties of pearlescent pigments are critical for their economic success. Therefore, attention was focused on the L-a-b measurements. Although higher densities corresponding to higher refractive indices can be obtained at 1000°C, the optical characteristics worsen rapidly at 900°C and sintering times above 20 min. This is probably caused by grain sizes that grow into a range where diffuse scattering takes effect or due to a decomposition of the mica substrate. Hence, too high sintering temperatures reduce the desired luster of pearlescent pigments.

V. SUMMARY

Three parameters primarily determine the microstructure development of a dried pigment: the calcination temperature, the layer thickness, and the rigidity of the substrate. We have seen that higher temperatures lead not only to better densification but also to accelerated grain growth. Larger grains have a negative influence on the coloristics of pearlescent pigments because they cause diffuse scattering instead of the desired directed reflection. Therefore, one goal would be to achieve the same densification with less grain growth (i.e., a "flat" sintering path). This has been reported by using dopants that retard the movement of the grain boundaries. Also, the calcination atmosphere can affect the grain growth rates.

The layer thickness has a limiting effect on the grain size development (which is desired for commercially more attractive pearlescent pigments). When the grains grow into the thickness range, the material transport is limited perpendicular to the substrate surface so that only two dimensions contribute to diffusion processes instead of three. Hence, the final grain size after a fixed calcination time at a set temperature should be a function of the layer thickness. This has to be confirmed in further experiments.

The critical parameter in the densification process is the stiffness and the thermal expansion of the substrate. The geometrical constraints of the substrate platelets do not permit a lateral shrinkage of the layer material. This can lead to crack formation already in the dried pigments. The constraint also prevents the rutilization of the TiO_2, which would have a positive effect on the refractive index of the coating. Therefore, the optimal substrate that had no retarding influence on densification would shrink by the same amount as the layer material during the calcination process. For synthetic substrates, the microstructural properties can be partially "designed" to yield superior optical properties at somewhat higher production costs.

REFERENCES

1. G Pfaff, K-D Franz, R Emmert, K Nitta. In: Ullmann's Encyclopedia of Industrial Chemistry. 6th ed. Weinheim: Section 4.1.3, VCH, 1998, Sect. 4.1.3 (electronic release); R Maisch, M Weigand. Pearl Lustre Pigments. Landsberg/Lech, Germany: Verlag Moderne Industrie, 1991; R Glausch, M Kieser, R Maisch, G Pfaff, J Weitzel. In: U Zorll, ed. Special Effect Pigments. Hannover: Vincentz Verlag, 1998.
2. C Schmidt, M Friz. Kontakte (Darmstadt) (2):15, 1992; M Born, E Wolf. Principles of Optics: Electromagnetic Theory of Propagation, Interference and Diffraction of Light. 6th ed. Cambridge: Cambridge University Press, 1998.
3. LM Greenstein. In: PR Lewis, ed. Pigment Handbook Volume 1: Properties and Economics. 2nd ed. New York: John Wiley & Sons, 1988, p. 829.
4. FC Atwood. U.S. Patent 2,278,970 (1942) (to Atlantic Research Associates, Inc.); HR Linton. U.S. Patent 3,087,828 (1963) (to DuPont); HR Linton. U.S. Patent 3,087,829 (1963) (to DuPont); CA Quinn, CJ Rieger, RA Bolomey. U.S. Patent

3,437,515 (1969) (to Mearl); HW Kohlschütter, H Rössler, H Getrost, W Hörl, W Reich. U.S. Patent 3,553,001 (1971) (to Merck KGaA).

5. H Schröder. U.S. Patent 3,138,475 (1959) (to Jenaer Glaswerke Schott); CN Schmidt, G Bauer, K Osterried, M Uhlig, N Schül, R Brenner, R Vogt. World Patent 93/08237 (1993) (to Merck KGaA).

6. K Nitta, TM Shau, J Sugahara. European Patent Application 763573 (1997) (to Merck KGaA).

7. W Ostertag, K Bittler, G Bock. European Patent 33457 (1981)(to BASF AG); W Ostertag, N Mronga, P Hauser. Farbe Lack 93:973, 1987.

8. H Schroeder. In: G Hass, RE Thun, eds. Physics of Thin Films, Vol. 5. New York: Academic Press, 1969, p. 87.

9. VD Hildenbrand. Doctoral thesis, Darmstadt University of Technology, 1995; VD Hildenbrand, S Doyle, H Fuess, G Pfaff, P Reynders. Thin Solid Films 304:204, 1997.

10. RK Bordia, R Raj. J Am Ceram Soc 68:287, 1985; RK Bordia, GW Scherer. Acta Metall 36:2393, 1988.

11. EP Barrett, LG Joyner, PP Halenda. J Am Chem Soc 73:373, 1951.

12. ASTM E-112-88, Standard Test Methods for Determining Average Grain Size.

13. KSW Sing, DH Everett, RAW Haul, L Moscou, RA Pierotti, J Rouquérol, T Siemieniewska. Pure Appl Chem 57:603, 1985.

14. F Hofmeister, H Pieper. Farbe Lack 95:557, 1989; F Hofmeister. Farbe Lack 93: 799, 1987; R Emmert. Cosmet Toiletr 104(7):57, 1989; F Hofmeister. Eur Coatings J (3):80, 1990.

15. U Eisele, RJ Brook. In: RW Cahn, P Haasen, EJ Kramer, eds. Materials Science and Technology. Weinheim: VCH, 1996, p. 83.

16. TJ Garino, HK Bowen. J Am Ceram Soc 73:251, 1990.

27

Polymeric Surfactants Based on Oleic Acid: I. Polymerization in Bulk Phase

Zhiqiang "George" Zhang,* Fang Li, and Stig E. Friberg
Clarkson University, Potsdam, New York

Patricia A. Aikens
Uniqema, Wilmington, Delaware

ABSTRACT

Oleic acid was polymerized by reaction with 3-chloroperoxybenzoic acid to form epoxide, followed by a ring-opening to form polyether, or by an acid-catalyzed condensation to form polyestolide. The structure of the polymer products was confirmed by infrared and ^{1}H-NMR spectroscopies, and gel permeation chromatography.

Surface tension of the aqueous solutions of the polymer soaps was measured. Lamellar liquid crystal formation and the interlayer spacing in the three-component system consisting of the polymer, triethanolamine, and water were investigated through small-angle x-ray diffraction.

The results showed that the polymer soaps could bring down the surface tension of water to lower than that of the precursor, i.e., sodium oleate, while the polyether soap has higher critical micellization concentration than sodium oleate.

---

* *Current affiliation*: The Valvoline Company, Lexington, Kentucky.

I. INTRODUCTION

Fatty acids from a variety of sources are important not only to the food industry but to such nonfood applications as drying oils in paint, soap, cosmetics, pharmaceuticals, synthetic rubber, and emulsifiers [1]. The degree of unsaturation of the fatty acids has a pronounced influence on their physical properties, such as melting point, viscosity, and chemical properties [2]. Among them, oleic acid is one of the main components of human sebum [3]; it bears one double bond in the hydrocarbon chain and, thus, can serve as a model compound for the study of properties and reactions of unsaturated fatty acids [4–11].

The polymerization of a double bond located in the middle of a long hydrocarbon chain, such as the case of oleic acid, is notoriously difficult. Recent attempt by Isbell et al. [11] using mineral acids as catalysts resulted in the reaction of the carboxylic group of one oleic acid monomer with the double bond of another monomer to form an estolide, and a polyestolide with the continuation of the reaction. The difficulties of separating the estolides from the reaction mixture due to the formation of emulsions during the quenching of excess mineral acid led us to search for other methods of reaction on the middle double bond. The ring-opening polymerization has been well documented lately [12–16], which employs the epoxide as the intermediate, and initiates the opening of the ring to form polymers.

In this chapter, the epoxide polymerization of oleic acid was performed using 3-chloroperoxybenzoic acid as the initiator and trifluoroacetic acid as the ring-opening agent, and the reaction product was analyzed by infrared (IR) and ^{1}H-NMR spectroscopies and gel permeation chromatography (GPC) and then compared with the product prepared by acid-catalyzed condensation [11]. The surfactant properties of the polymer were evaluated through surface tension measurements and phase diagram determination in the system containing the polymer, triethanolamine, and water.

II. EXPERIMENTAL

A. Materials

Oleic acid (95%) and NaOH (99%) were purchased from Fisher Scientific (Pittsburgh, PA). 3-Chloroperoxybenzoic acid (57–86%), chloroform (99.8%), $NaHSO_3$ (99%), $NaHCO_3$ (99%), trifluoroacetic acid (CF_3COOH, 99+%), and perchloric acid ($HClO_4$, 70%) were obtained from Aldrich (Milwaukee, WI). NaCl (99.5%) and anhydrous $MgSO_4$ was purchased from J.T. Baker Chemical Co. (Phillipsburg, NJ). Water was deionized and doubly distilled.

B. Epoxide Polymerization

Fifteen grams of oleic acid was placed in a 500-mL three-neck flask and 24 g of 3-chloroperoxybenzoic acid in 300 mL chloroform was added into the flask over the course of 2 h. The reaction mixture was heated to reflux for 6 h, and CF_3COOH (10 wt% of oleic acid) was added and the mixture was kept at room temperature overnight. The flask was cooled in an ice-water bath and the precipitated 3-chloroperoxybenzoic acid was removed by filtration. The filtrate was washed with an aqueous solution containing 20% $NaHSO_3$, 10% $NaHCO_3$, and saturated NaCl, and dried over anhydrous $MgSO_4$. The final product, light yellow and viscous, was obtained through rotary evaporation.

C. Acid-Catalyzed Condensation

The polyestolide from oleic acid was prepared by using perchloric acid as the catalyst, according to the procedure by Isbell et al. [11].

D. Infrared Spectra

The IR spectra were recorded on a Mattson Galaxy 202 Fourier-transform infrared spectrometer after casting the chloroform solutions of oleic acid and the reaction product onto CaF_2 disks and letting the solvent evaporate.

E. ^{1}H-NMR

^{1}H-NMR spectra were obtained through a Bruker AC250 Fourier-transform nuclear magnetic resonance spectrometer (250 MHz). Samples were prepared in deuterated chloroform ($CDCl_3$) using tetramethylsilane (TMS) as an added reference.

F. Gel Permeation Chromatography

Molecular weight was determined using a Waters 590 pump module system. Samples and standards were dissolved in tetrahydrofuran (THF). Eight polyethylene glycol standards (Polymer Laboratories, Amherst, MA) were used from an average molecular-weight range of 106–23600. The GPC separation was obtained with five PL gel columns (300 mm long, 7.5 cm in diameter) in series at 50°C and a refractive index detector (Waters 410 refractometer).

G. Surface Tension

The aqueous solution of the polymer was adjusted to pH 10 using NaOH, and solution surface tension was measured by means of a Fisher Model 21 Surface Tensiomat.

H. Phase Diagram Determination

The partial phase diagram of the three-component system containing the polymer, triethanolamine, and water was determined by optical observation of the sample both visually and with the aid of a microscope equipped with crossed polarizers. The boundary of the liquid crystalline phase was confirmed by the small-angle x-ray diffraction.

I. Small-Angle X-Ray Diffraction

A small amount of the sample was drawn into a glass capillary of 0.7 mm in diameter and placed in a brass sample holder. The instrument is from Philips Electronic using a Kiessig low-angle camera (Richard Seifert) and an ORDELA detection system. Lead stearate with an interlayer spacing of 4.82 nm was used as the calibration standard.

J. Optical Microscopy

Liquid-crystal samples were observed by an Olympus microscope (Model BHA-P) equipped with crossed polarizers, and microphotos were obtained with a Polaroid microcamera.

K. Molecular Modeling

Geometry optimization of the basified polyether molecule was performed on an IBM workstation by the molecular mechanics method using the software SPARTAN IBM version 5.0.3.

III. RESULTS AND DISCUSSION

It is well known and patented that when the double bond is treated with peroxide, the three-membered ring containing two $C-O$ bonds and one $C-C$ bond, namely epoxide, forms [16–18]. This three-membered ring is unstable, and it might open under certain conditions [18]. Ring-opening polymerization is one of the consequences. In the present contribution, 3-chloroperoxybenzoic acid was used as the initiator to form the epoxide, and the epoxide was expected to open the ring and form a $C-O-C$ connection with other molecules using trifluoroacetic acid [15].

The IR spectra of oleic acid and the polymers obtained through the epoxide procedure and acid-catalyzed one are shown in Figure 1. The main features of the IR spectrum of oleic acid (Fig. 1a) consist of the C—H stretching band in the double bond (3004 cm^{-1}), the CH$_3$ antisymmetric stretching vibration (2960 cm^{-1} as a shoulder), the CH$_2$ antisymmetric and symmetric stretching bands (2924 and 2854 cm^{-1}), the C=O stretching mode of the acid group (1709 cm^{-1}), the CH$_2$ scissoring band (1464 cm^{-1}), and the C—O stretching in the carboxylic group (1285 cm^{-1}) [19,20].

After the epoxide polymerization, the 3004-cm^{-1} peak disappears completely (Fig. 1b), indicating the absence of the C=C double bond in the product. Other noticeable changes in the IR spectra are (1) the appearance of three bands in the region 1160–1050 cm^{-1} and (2) the reduction of the free acid C=O stretching absorption and the appearance of a broad peak centered around 1570 cm^{-1}. The

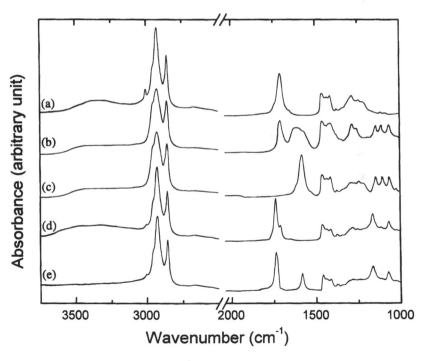

Figure 1   Infrared spectra of oleic acid (a), the product by epoxide polymerization (b) and acid-catalyzed condensation (d), and corresponding soaps from the polymers [(c) for epoxide reaction and (e) for acid-catalyzed one] obtained by adjusting their pH to 10 through NaOH.

former is known to be caused by the C—O—C stretching vibrations, and the latter indicates that some of the carboxylic groups have undergone reaction during the process. To further test whether the carboxylic groups are involved in the polymerization or not, the pH value of the final product was adjusted to around 10 by sodium hydroxide, and the IR spectrum recorded (Fig. 1c). The acid C=O band disappears and the strong peak of antisymmetric COO$^-$ stretching shows up at 1560 cm^{-1}, which is of nearly the same intensity as the C=O stretching band in oleic acid (Fig. 1a). The above results imply that the carboxylic groups are unlikely to be involved in the epoxide polymerization, and the change of the C=O absorption after polymerization (Fig. 1b) is resulted by washing the reaction mixture with the alkaline NaHCO$_3$ and NaHSO$_3$ solution.

Figure 1d is the IR spectrum of the product obtained by acid-catalyzed reaction of oleic acid. The 3004-cm^{-1} peak strongly decreases, indicating that most, but not all, of the double bonds have undergone reaction. The C=O stretching band splits into two: a strong peak at 1735 cm^{-1}, which is the typical absorption of the C=O group in an ester, and a weak one at 1709 cm^{-1}. The absorbance of the C—O stretching in the carboxylic group, 1285 cm^{-1}, decreases compared with that of oleic acid (Fig. 1a), whereas two new peaks appear in the region below 1200 cm^{-1}: 1170 cm^{-1}, corresponding to the C—O—C (ester) antisymmetric stretching band, and 1030 cm^{-1}, corresponding to the C—O—C (ester) symmetric band [19,20]. After adjusting the pH value of the product to around 10, the absorption of the C=O band of acid (1709 cm^{-1}) disappears, whereas that of ester group (1735 cm^{-1}) stays intact, and the COO$^-$ band shows up at 1560 cm^{-1} (Fig. 1e). Therefore, in the acid-catalyzed polymerization, both the C=C double bond and the carboxylic group participate in the reaction.

Figure 2 shows the ^{1}H-NMR spectra of oleic acid and the polymer products. Detailed assignment of the proton absorption in the spectrum of oleic acid (Fig. 2a) can be given as follows: 0.88 (t, 3H, CH$_3$); 1.27 and 1.31 (m, 20H, C$_{(4-7,12-17)}$ H$_2$); 1.63 (m, 2H, C$_{(3)}$H$_2$); 2.00 and 2.02 (m, 4H, C$_{(8,11)}$H$_2$); 2.35 (t, 2H, C$_{(2)}$H$_2$); 5.35 (t, 2H, =CH). Distinct changes in the nuclear magnetic resonance (NMR) spectrum after the epoxide polymerization, (Fig. 2b) are the absence of the =CH signal and the appearance of a signal at around 3.56 ppm, which is the absorption of the proton connected to the ether carbon [21]. For the spectrum of the product from an acid-catalyzed reaction (Fig. 2c), the absorption of the proton on the vinyl carbon does not disappear but is strongly reduced compared with Figure 2a. A new signal turns up at around 4.2 ppm, which is attributed to the hydrogen absorption in the ester group:

(O=C—O—CH—)
  |

Combining the information obtained from IR and ^{1}H-NMR spectra, the reactions of oleic acid undergone in the case of epoxide polymerization can be deduced as (1) epoxide formation initiated by 3-chloroperoxybenzoic acid,

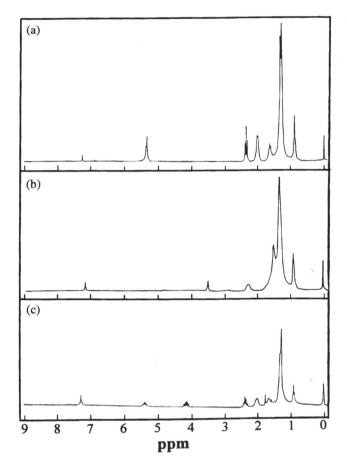

Figure 2   ¹H-NMR spectra of oleic acid (a) and the polymer from epoxide polymerization (b) and that from the acid-catalyzed reaction (c).

(2) ring opening induced by superacid, and (3) coupling of the hydrocarbon
chains by ether bonds, as depicted in Scheme 1 (the product is referred to as
polyether thereafter).

The GPC analysis shows that the molecular weight of the product obtained
is around 1400 (weight average) with the polydispersity index as 1.26, which
means that the average structure contains about five oleic acid monomers. These
monomers are linked together through the C—O—C bond with their hydrocarbon chains "parallel" to each other, leaving the carboxylic groups intact.

The mechanism of the acid-catalyzed condensation of oleic acid has already

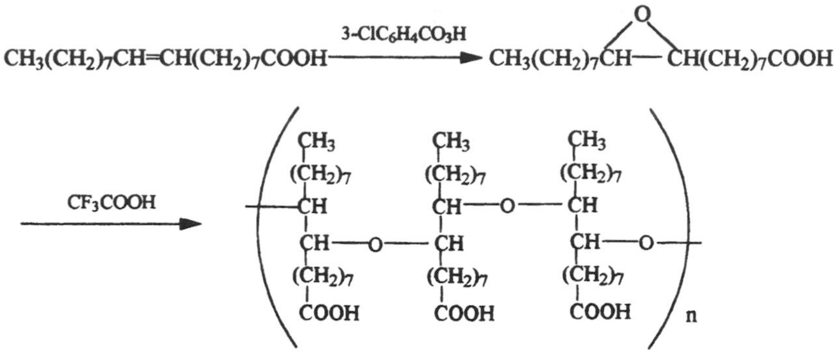

$$CH_3(CH_2)_7CH=CH(CH_2)_7COOH \xrightarrow{\text{3-ClC}_6\text{H}_4\text{CO}_3\text{H}} CH_3(CH_2)_7CH \overset{\displaystyle O}{\overset{\diagup \diagdown}{\text{———}}} CH(CH_2)_7COOH$$

$$\xrightarrow{\text{CF}_3\text{COOH}} \left(\begin{array}{ccc} CH_3 & CH_3 & CH_3 \\ (CH_2)_7 & (CH_2)_7 & (CH_2)_7 \\ CH & CH\text{—O—}CH \\ CH\text{—O—}CH & CH\text{—O—} \\ (CH_2)_7 & (CH_2)_7 & (CH_2)_7 \\ COOH & COOH & COOH \end{array} \right)_n$$

Scheme 1

been reported [11], which is summarized in Scheme 2, in which $n = 0$–10 [11]. The formation of polyestolide is strongly supported by the IR and NMR results presented earlier. Compared with oleic acid, the unsaturation of the polyestolide decreases and the transformation of part of the carboxylic groups into esters increases the hydrophobicity of the molecule.

As shown earlier, a lamellar liquid-crystal (LLC) phase will form when oleic acid is combined with triethanolamine, containing three components: triethanolammonium oleate, oleic acid, and triethanolamine [22]. With the addition of a third component, a polar solvent, into the binary acid–amine system, the liquid crystal extends into a large single-phase region in the three-component phase diagram [23]. The interaction of the carboxylic group in oleic acid with the amine is of importance for the formation of such a liquid-crystal structure. In the polymers obtained earlier, the carboxylic groups are either kept intact or left part unreacted; therefore, they might be able to undergo the liquid-crystal formation when combined with triethanolamine and a polar solvent (e.g., water).

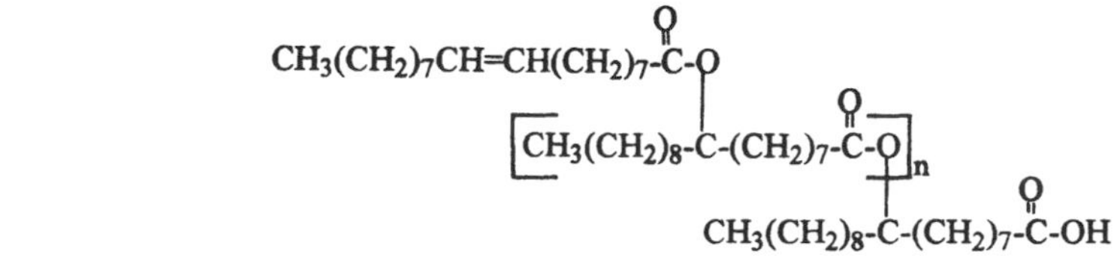

$$CH_3(CH_2)_7CH=CH(CH_2)_7COOH \xrightarrow{\text{HClO}_4}$$

$$CH_3(CH_2)_7CH=CH(CH_2)_7\text{-}\overset{\displaystyle O}{\overset{\|}{C}}\text{-}O$$

$$\left[CH_3(CH_2)_8\text{-}\overset{\displaystyle O}{\overset{\|}{C}}\text{-}(CH_2)_7\text{-}\overset{\displaystyle O}{\overset{\|}{C}}\text{-}O \right]_n$$

$$CH_3(CH_2)_8\text{-}\overset{\displaystyle O}{\overset{\|}{C}}\text{-}(CH_2)_7\text{-}\overset{\displaystyle O}{\overset{\|}{C}}\text{-OH}$$

Scheme 2

The LLC regions of the polymers, triethanolamine, and water are shown in Figure 3. For polyether (solid line in Fig. 3), along the polymer–triethanolamine axis, the compositions between 60 wt% and 77 wt% polymer form a lamellar liquid crystal, and by adding water, this phase extends to around 50 wt% water. When the polymer is substituted for oleic acid, no practical changes in the LLC phase boundary are observed (phase diagram not shown). In the case of polyestolide (dashed line in Fig. 3), the whole LLC region shifts upward to the polymer corner, which is reasonable because there are less free carboxylic groups in the product. Figure 4 is the microphotographs of the sample marked by × in Figure 3 viewed under crossed polarizers. Apparent features of the photos are the "mosaic" texture and the Maltese crosses in Figure 4a and the "oily streaks" in Figure 4b, which are typical for lamellar liquid crystals [24].

The interlayer spacings, d, from the LLC phase in Figure 3 plotted against the volume ratio of water, R, are given in Figure 5. Two important parameters of LLC phase are obtained through linear fit of the data: the interlayer spacing at zero water content, d_0, and the fraction of water penetrating from the water zone to the hydrophobic region, α, according to Eq. (1) [25]:

$$d = d_0[1 + (1 - \alpha)R] \tag{1}$$

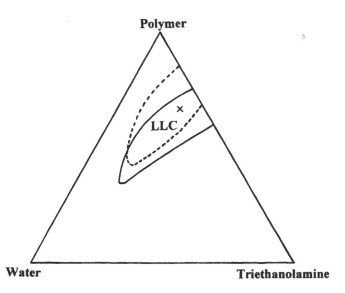

Figure 3 The partial phase diagram for the system of water, triethanolamine, and the polymer (solid line for polyether and dashed line for polyestolide). The sample marked as × was prepared for microscopic observation (Fig. 4).

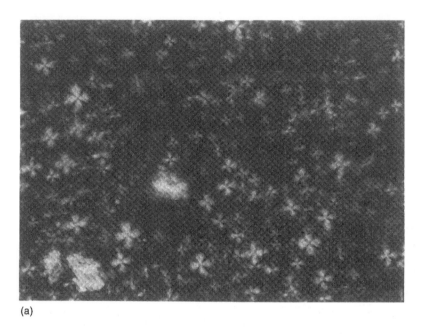

(a)

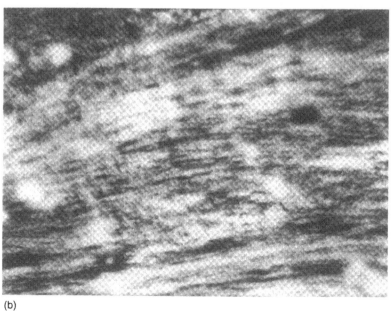

(b)

Figure 4   Microphotos of the sample marked as ✕ in Figure 3: (a) polyether; (b) polyestolide. Crossed polarizers, magnification: ✕200.

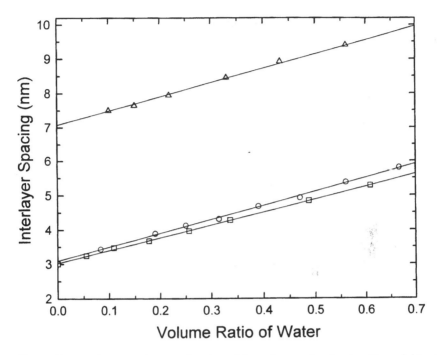

Figure 5 Interlayer spacing as a function of the volume ratio of water. The weight ratio of the acid (oleic acid, polyether, and polyestolide) to triethanolamine is 73:27. □: oleic acid; ○: polyether; △: polyestolide. The linear fit results are listed in Table 1.

The d_0 and α values are listed in Table 1. Compared with the LLC of oleic acid, the LLC of polyether has a similar interlayer spacing (3.10 versus 3.03 nm), but the penetration parameter α changes from -0.23 of oleic acid to -0.30 of polyether. It implies that the liquid crystal from the polyether–triethanolamine–water system has a similar thickness to that of the oleic acid–triethanolamine–water

Table 1 Interlayer Spacing and Water-Penetration Values for Lamellar Liquid Crystals

System	d_0 (nm)	α
Oleic acid–triethanolamine–H$_2$O	3.00 ± 0.01	-0.23 ± 0.02
Polyether–triethanolamine–H$_2$O	3.10 ± 0.02	-0.30 ± 0.04
Polyestolide–triethanolamine–H$_2$O	7.08 ± 0.03	0.42 ± 0.07

system, but better organized due to the local lateral connection of the amphiphilic molecules. Thus, increasing the water amount does not lead to penetration of water into the hydrophobic layer; instead, something, probably triethanolamine, comes out of the hydrophobic layer. For the LLC of polyestolide, both d_0 and α increase drastically (Table 1). The formation of polyestolide, Scheme 2, involves the reaction of the carboxylic group at the end of one oleic acid molecule with the middle double bond of another molecule to form esters, which extends the molecular length. As a simple and direct estimation, the polyestolide contains on average five to six monomers, because it is known that the product obtained by $HClO_4$-catalyzed condensation of oleic acid is the mixture of oligomers from 2–12 molecules [11]. The molecular length of stearic acid is about 2.35 nm [26,27] and the polyestolide molecule is actually a substituted stearic acid with the substitution position very near the center. Therefore, the total length of an extended polyestolide molecule is about 7.4 nm. From Table 1, the interlayer spacing of the polyestolide LLC is 7.08 nm, which implies that the hydrocarbon chains completely interdigitate in the bilayer in the LLC. The cross-sectional area occupied by each of the polyestolide molecule in the liquid crystal can be estimated according to

$$A_{LLC} = \frac{1}{\rho} \frac{M}{N_A d_0} \times 10^{21} \quad \text{(molecule nm}^2\text{)} \tag{2}$$

In Eq. (2), ρ and M are the density and molecular weight of the amphiphile, N_A is the Avogadro number, and d_0 is the interlayer spacing of LLC at zero water content. The density of oleic acid is 0.89 g/cm^3, and the molecular weight is taken as 1400 (pentamer). Substituting these values into Eq. (2), A_{LLC} is calculated as 0.37 nm^2/molecule. Both the d_0 and A_{LLC} values indicate that the polyestolide molecules are organized in the liquid crystal, as shown schematically in Figure 6a. Such an arrangement is very unusual in a liquid crystal. It has been found in gels with an α-crystalline structure [28] and in cast films from surfactant aqueous solutions [29–31], but a literature search has failed to find liquid-crystalline structures of this kind. The introduction of the ester group ($O{=}C{-}O{-}$) enables the

water molecules to penetrate deeper into the hydrophobic zone as compared to the case of oleic acid, thus leading to a much larger α value (0.42 compared with −0.23).

The corresponding polymer soaps are obtained by using sodium hydroxide to neutralize the polymers. Polymer soaps, or polysoaps, have been known for a long time [32], and their association behavior has been investigated extensively [33–36]. Plots of surface tension versus log(concentration of the aqueous soap solutions) are shown in Figure 7. The concentration unit used here is gram per liter, instead of the more common molar concentration, because the polymers

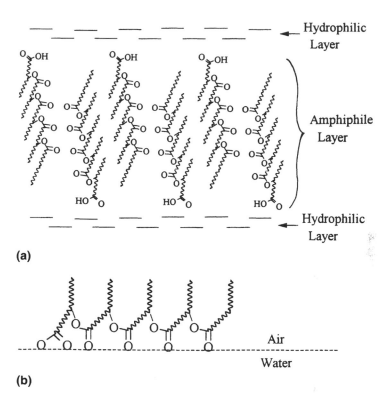

Figure 6 Schematic drawing of the polyestolide molecules in the lamellar liquid crystal (the triethanolamine molecules are not shown here) (a) and the corresponding soap at the air–water interface (b).

synthesized are mixtures of oligomers. Both the polymer soaps bring the surface tension of water down to lower than 30 mN/m, showing superior surface activity. The critical micelle concentrations (cmc's) are 1.3 g/L for polyestolide and 2.5 g/L for polyether. For comparison, the surface tension of the sodium oleate solution is also included in Figure 1 and the cmc is obtained as 0.33 g/L (i.e., $1.1 \times 10^{-3}\,M$), which is consistent with earlier research by Klevens [37]. The maximum surface excess (Γ_{max}) can be calculated from the slope of the curve near the cmc, and, therefore, the minimum area per molecule (A_{min}) is obtainable from the following equations [38]:

$$\Gamma = -\frac{1}{2.303RT}\left(\frac{\partial \gamma}{\partial \log C}\right)_T \qquad (3)$$

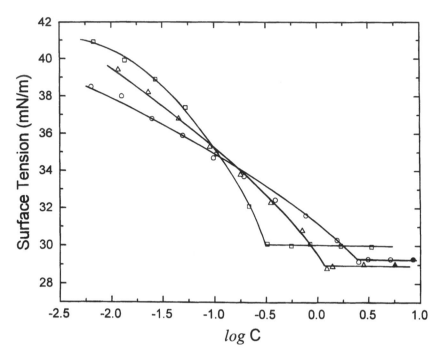

Figure 7 Plots of surface tension versus log(concentration) (in g/L). □: sodium oleate; ○: polyether soap; △: polyestolide soap. Precipitates were observed for the points with solid symbols. Ionic strength was kept constant by adjusting the concentration of Na⁺ as 0.03 M.

$$A_{min} = \frac{1}{\Gamma_{max}} \tag{4}$$

The A_{min} values calculated from Figure 7 are 0.52, 1.92, and 2.56 nm²/molecule for sodium oleate, polyestolide, and polyether, respectively.

The A_{min} for the polyestolide soap is much larger than the cross-sectional area of the molecule in lamellar liquid crystal (A_{LLC}), 1.92 versus 0.37 nm²/molecule. As there are additional hydrophilic parts along the hydrocarbon chain of the amphiphile, the possible conformation of the polyestolide soap molecule at the air–water interface is a "fold-up" one (Fig. 6b) rather than the "stretch-out" one in the LLC (Fig. 6a).

As an estimation, the average molecular weight of polyestolide and polyether is considered to be 1400; hence the corresponding cmc's are $9 \times 10^{-4}\ M$ and $1.8 \times 10^{-3}\ M$. The cmc value gives one a measure of the competition between

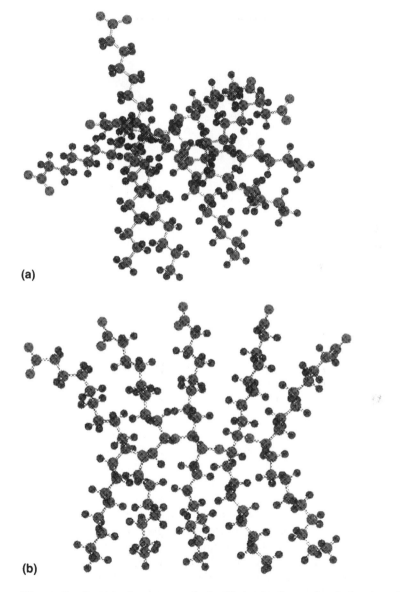

(a)

(b)

Figure 8 Optimized geometry of a basified polyether molecule (pentamer) calculated by the molecular mechanics method using SPARTAN 5.0.3. Darkest circles: oxygen atoms; lightest circles: hydrogen; circles with intermediate darkness: carbon. (a) gaseous phase (free conformation); (b) distance restraint of 0.65 nm on the carboxylate carbon atoms; (c) top view of (b).

(c)

Figure 8 Continued.

two tendencies of a surfactant molecule in an aqueous medium: the association with other surfactant molecules or with water molecules. Surprisingly, the cmc of the polyether is higher than that of its precursor (i.e., sodium oleate), although there are more hydrophilic headgroups in the former, which is contrary to the results of dimeric surfactants [39–41]. In order to clarify this fact, geometry optimization of a model molecule, a fully basified pentamer, was performed using the molecular mechanics method. Figure 8a shows the geometry of the model polymer in a gaseous phase: The carboxylate headgroups are not restricted to one end of the molecule, as in the case of single-chain or dimeric surfactants; instead, they have geometrical freedom and are distant from each other, which would help the polysoap molecules dissolve, rather than aggregate, when they are put into water. Increasing the concentration of the polysoap will cause them to associate in an aqueous solution to form micelles. Under such circumstances, the headgroups are located at, or at least near, an interface and, therefore, restricted within a certain distance from each other. To mimic this environment at an interface, the carboxylate carbon atoms are restrained within 0.65 nm from each other and the geometry optimization is performed; the final structure shown in Figure 8b. Now, the carboxylate groups are preferrentially in one direction. Figure 8c gives the top view of the surfactant in Figure 8b, showing that the headgroups are aligned in a zigzag fashion to minimize the electrostatic repulsion between the negatively charged headgroups. The area covered by the molecule from the top view (Fig. 8c) can be given by the SPARTAN software as ~2.5 × 1.0 = 2.50 nm^2, which is similar to the value 2.56 nm^2 obtained from the surface tension versus concentration plot (Fig. 7). The energy difference between the two conformations, from "free" conformation (Fig. 8a) to the "restricted" one (Fig. 8b), is given as 102.9 kcal/mol = 430.1 kJ/mol. Such a huge energy gap makes the conformation change from the "free" to the "restricted" one impossible if there are no other driving forces. Early calculations using CNDO/2 methods by

Friberg and Flaim [42] showed that hydration of the carboxylate group could lower the free energy of the system, up to 84 kJ/mol per carboxylate group. In the case of five carboxylate groups per molecule, the system free energy is reduced by 420 kJ/mol by the hydration of the water molecules—at the same level as the energy difference in the conformation change mentioned earlier.

IV. SUMMARY

Oleic acid was polymerized through epoxide polymerization to form polyether, or through acid-catalyzed condensation to obtain polyestolide.

ACKNOWLEDGMENTS

This research was financed in part by ICI Surfactants (now Uniqema), Wilmington, DE, through a grant from the National Soybean Board. The authors would like to thank Professor James Peploski and Professor Yuzhuo Li of Clarkson University for helpful discussions.

REFERENCES

1. T Applewhite (ed.). Bailey's Industrial Oil and Fat and Oil Products, Vol. 3, New York: John Wiley & Sons, 1985.
2. TWG Solomons. Organic Chemistry. 3rd ed. New York: John Wiley & Sons, 1984, pp. 976–984.
3. L Stryer. Biochemistry. 3rd ed. New York: W.H. Freeman & Co., 1988.
4. RA Korus, TL Mousetis. J Am Oil Chem Soc 61:537, 1984.
5. MA Grompone, P Moyna. J Am Oil Chem Soc 63:550, 1986.
6. T Sato, S Kawano, M Iwamoto. J Am Oil Chem Soc 68:827, 1993.
7. WE Neff, TL Mounts, WM Rinsch, H Konishi. J Am Oil Chem Soc 70:163, 1993.
8. J Berthelot, P-L Desbène, A Desbène-Monvernay. J Am Oil Chem Soc 70:277, 1993.
9. XJ Liu, MR Myers, WE Artz. J Am Oil Chem Soc 70:355, 1993.
10. G Knothe, D Weisleder, MO Bagby, RE Peterson. J Am Oil Chem Soc 70:401, 1993.
11. TA Isbell, R Kleiman, BA Plattner. J Am Oil Chem Soc 71:169, 1994.
12. JV Crivello, M Fan. J Polym Sci A: Polym Chem 30:1, 1992.
13. A Duda. J Polym Sci A: Polym Chem 30:21, 1992.
14. LM Gan, KS Ool, SH Goh, KK Chee. J Appl Polym Sci 46:329, 1992.
15. Y-L Liu, G-H Hsiue, Y-S Chiu. J Polym Sci A: Polym Chem 32:2543, 1994.
16. P Daute, G Stoll. German Patent No DE 92-4223581 920717 (1992).
17. RE Parker, NS Isaacs. Chem Rev 59:737, 1959.

18. RC Larock. Comprehensive Organic Transformations. New York: VCH, 1989, pp. 456–461, 505–512.
19. LJ Bellamy. The Infrared Spectra of Complex Molecules. 3rd ed. London: Chapman & Hall, 1975.
20. K Nakanishi, PH Solomon. Infrared Absorption Spectroscopy. 2nd ed. San Francisco: Holden-Day, 1977.
21. DA Skoog. Principles of Instrumental Analysis. 3rd ed. Philadelphia: Saunders College Publishing, 1985.
22. SE Friberg, P Liang, FE Lockwood, M Tadros. J Phys Chem 88:1045, 1984.
23. SE Friberg, CS Wohn, FE Lockwood. J Pharm Sci 74:771, 1985.
24. F Rosevear. J Am Oil Chem Soc 31:628, 1954.
25. SE Friberg, B Yu, GA Campbell. J Polym Sci A: Polym Chem 28:3575, 1990.
26. GL Clark, PW Leppla. J Am Chem Soc 58:2199, 1936.
27. J Umemura, S Takeda, T Hasegawa, T Takenaka. J Mol Struct 297:57, 1993.
28. JM Vincent, A Skoulios. Acta Crystallogr 20:432, 441, 447, 1966.
29. T Kunitake, M Shimomura, T Kajiyama, A Harada, K Okuyama, M Takayanagi. Thin Solid Films 121:L89, 1984.
30. X Lu, Z Zhang, Y Liang. J Chem Soc Chem Commun 2731, 1994.
31. M Shimomura, S Aiba, N Tajima, N Inoue, K Okuyama. Langmuir 11:969, 1995.
32. R Varoqui, UP Strauss. J Phys Chem 72:2507, 1968.
33. W Binana-Limbelé, R Zana. Macromolecules 23:2731, 1990.
34. HE Rios, JS Rojas, IC Gamboa, RG Barraza. J Colloid Interf Sci 156:388, 1993.
35. SE Friberg, T Moaddel, ED Sprague. J Phys Chem 98:13414, 1994.
36. A Laschewsky. Adv Polym Sci 124:1, 1995.
37. HB Klevens. J Am Oil Chem Soc 30:74, 1953.
38. MJ Rosen. Surfactants and Interfacial Phenomena. 2nd ed. New York: John Wiley & Sons, 1989.
39. R Zana, M Benrraou, R Rueff. Langmuir 7:1072, 1991.
40. M Frindi, B Michels, H Lévy, R Zana. Langmuir 10:1140, 1994.
41. R Zana, H Lévy. Colloids Surfaces A 127:229, 1997.
42. SE Friberg, T Flaim. In: SL Holt, ed. Inorganic Reactions in Organic Media. ACS Symposium Series 177. Washington, DC: American Chemical Society, 1982, pp. 1–17.

Index